Bioremediation of Petroleum Contaminated Sites

Eve Riser-Roberts

C. K. SMOLEY

Library of Congress Cataloging-in-Publication Data

Riser-Roberts, Eve.
    Bioremediation of petroleum contaminated sites / Eve Riser-Roberts
      p. cm.
    Includes bibliographical references and index.
    1. Oil pollution of soils. 2. Oil pollution of water.
  3. Petroleum--Biodegradation. 4. Bioremediation. I. Title.
TD879.P4R57 1992
628.5'36--dc20   92-10166
ISBN 0-87371-832-1 (alk. paper)

Direct all inquiries to CRC Press, Inc.
2000 Corporate Blvd. N.W.
Boca Raton, FL 33431

PRINTED IN THE UNITED STATES OF AMERICA
1 2 3 4 5 6 7 8 9 0

Printed on acid-free paper

Preface

This review will provide background information on the major aspects of technologies and related research dealing with the use of biodegradation for treatment of environmental contamination by toxic organic substances. It is intended to serve as a broad reference base for development of a program for in situ biorestoration of fuel-contaminated soil and groundwater.

The book is organized as follows: Sections 1-7 contain the main text, which is an overview of the many different subject areas involved in bioremediation. This summary is intended for managers and others who wish to learn about the technology without having to confront the technical details.

The Appendix is a Supplementary Text, which contains the technical information for readers who need indepth coverage. The section headings in Sections 1-7 correspond to those in the Appendix to allow convenient referral to the supplementary text when more details are required.

Acknowledgment

This technology review was conducted under the auspices of the Naval Civil Engineering Laboratory (NCEL) in connection with the installation restoration effort at Patuxent River Naval Air Station (NAS Patuxent River), MD. The author wishes to acknowledge the helpful input of Mr. R. Hoeppel, NCEL Project Engineer, and Dr. M. Swindoll (previously of NCEL), as well as that of Dr. D.L. Sorensen and Dr. R.R. Dupont, of the Utah Water Research Laboratory, who reviewed the manuscript.

Information for this report was obtained, in part, through API, NTIS, DTIC, and Dialogue searches, and by extensive use of the library facilities and the University of California at Santa Barbara.

Notice

Table of Contents

List of Tables

List of Illustrations

Glossary

Term/Acronym	Description
Aerobic	In the presence of oxygen
Anaerobic	In the absence of oxygen
Anthropogenic	Of man-made origin
Autochthonous	Indigenous or native bacteria found in soil in relatively constant numbers that do not change rapidly in response to the addition of specific nutrients
Autotrophic	The ability to fix carbon for growth from carbon dioxide and obtain energy from light or oxidation of inorganic compounds
Bioaugmentation	See Enhanced biodegradation
Biodegradation	Breakdown of organic substances by microorganisms
Bioreclamation	Use of biodegradation to restore soil and groundwater to uncontaminated condition
BDAT	Best demonstrated available technology
BOD	Biochemical oxygen demand
BTX	Benzene, toluene, and xylenes
CEQ	Council on Environmental Quality
CERCLA	Comprehensive Environmental Response, Compensation, and Liability Act (Superfund)
CGA	Colloidal gas aprons
COD	Chemical oxygen demand
Commensalism	Sequential degradation of a compound by two or more microorganisms in a relationship that may benefit only one partner
Cometabolism or Cooxidation	The indirect metabolism of a recalcitrant substance; the process by which microorganisms, in the obligate presence of a growth substrate, transform a nongrowth substrate
Cyanobacteria	Blue-green algae, which are actually bacteria
Denitrifying bacteria	Bacteria that reduce nitrate using the oxygen of nitrate as a hydrogen acceptor; requires source of hydrogen andlimitation on supply of free oxygen
DOC	Dissolved organic carbon
DOD	Department of Defense
E^o	Standard reduction potential
EPA	Environmental Protection Agency

Enhanced biodegradation	Stimulation of microbial degradation of organic contaminants by addition of microorganisms, nutrients, or optimization of environmental factors on-site or *in situ*
Enrichment culturing	Addition of a specific hydrocarbon to a minimal medium to select for degraders of that compound
Eukaryotic	Nucleus is surrounded by a membrane, as in fungi and higher organisms
Facultative	The ability to adapt to the conditions specified with this term
Facultative anaerobes	Microorganisms that are metabolically active under aerobic/anaerobic conditions
Heterotrophic	The ability to derive energy and carbon for survival and growth from decomposition of organic materials
HSWA	Hazardous and Solid Waste Amendments
Hydrocarbonoclastic	Ability to degrade and utilize hydrocarbons
In situ bioreclamation	Biodegradation operations taking place in the contaminated soil or groundwater without excavation or withdrawal
In vitro	In a test tube
In vivo	In life
Lithotrophic	The ability to obtain energy from oxidation of inorganic compounds
Mesophilic	The ability to grow at temperatures from 10° to 45°C, with optimum growth around 20° to 40°C; most human pathogens grow best at 37°C
Methanogenic consortia	Groups of microorganisms that function under highly reducing conditions and produce methane from degradation of small or low molecular weight organic compounds
Microaerophilic	The ability to survive on very low levels of oxygen
Mineralization	Complete biodegradation of organic molecules to mineral products
MLSS	Mixed liquor suspended solids
Nitrogen demand	Amount of nitrogen required for degradation of a given amount of contaminant
Normal flora	Mixed population of microorganisms occurring in nature
NRV	Nitrogen requirement value; the amount of nitrogen required by organisms to decompose or degrade a particular organic chemical
Obligate	Strict dependence upon the conditions specified with this term
Oligotrophic	The ability to survive on very low concentrations of nutrients
On-site bioreclamation	Biodegradation operations that occur above ground at the site of contamination
Operon	A DNA region that codes for several enzymes in a reaction pathway; it enables or prevents repression of structural gene function by controlling synthesis of mRNA by RNA polymerase enzyme
OTA	Congressional Office of Technology
PAHs	Polycyclic aromatic hydrocarbons, also called polyaromatic hydrocarbons and PNAs

PCBs	Polychlorinated biphenyls
Photosynthetic	The ability to obtain energy from light
Pleomorphs	Bacteria having multiple shapes
PNAs	Polynuclear aromatic hydrocarbons, also called PAHs
Procaryotic	"Nucleus" is a single chromosome without a membrane, as in bacteria
Psychrophilic	The ability to grow best at temperatures from -5° to 30°C, with optimum growth between 10° and 20°C
RAS	Return activated sludge
RBC	Rotating biological contactor
RCRA	Resource Conservation and Recovery Act
Recalcitrant	Resistant to microbial degradation
SARA	Superfund Amendments and Reauthorization Act
SITE	Superfund Innovative Technology Evaluation Program
Sulfate-reducing bacteria	Bacteria that use sulfate as a hydrogen acceptor, conferting sulfate to sulfide
Superfund	See CERCLA
Tetren	Tetraethylenepentamine, a chelator
TSDFs	Treatment Storage and Disposal Facilities
Thermophilic	The ability to grow at temperatures from 25° to 80°C, with optimum growth at 50° to 60°C
Turnover time	The amount of time required to remove the concentration of substrate present
Vadose zone	Unsaturated soil above water table
Xenobiotic	Compounds that are man-made or are unique in nature
Zymogenous	Soil bacteria that increase rapidly when furnished with certain nutrients and then diminish in numbers when the material is exhausted

Bioremediation of Petroleum Contaminated Sites

SECTION 1

Introduction

1.1 BACKGROUND

1.1.1 Environmental Contamination

Environmental pollutants are defined as chemicals of natural or synthetic origin that are released by man's activity into the environment where they have an undesirable effect on the environment or on man via the environment (Hutzinger and Veerkamp, 1981). There is a growing concern in the United States as a wide variety of synthetic organic chemicals are being introduced inadvertently or deliberately into waters and soils (Alexander, 1980). More than five million chemical compounds were described in Chemical Abstracts by the year 1980 (Ghisalba, 1983). This does not necessarily encompass all naturally occurring substances, nor does it include those resulting from the degradation of synthetic chemicals released to the environment (Richards and Shieh, 1986). Some 70,000 substances are in U.S. commerce (Ghisalba, 1983). Around 1000 new chemicals are brought on the market annually, while 150 chemicals are produced in excess of 50,000 tons/year. The total world production of synthetic organic chemicals is estimated at 300 million tons/year. As a by-product of industrial production, some 265 metric tons of hazardous waste are generated every year by 14,000 U.S. industrial plants (Nicholas, 1987). The U.S. Environmental Protection Agency (EPA) and the Congressional Office of Technology Assessment (OTA) estimate that 80% of this waste is

1

disposed of in landfills. Many of these wastes are toxic and persistent in the environment. Ecotoxicological data are available for less than 1000 compounds (Ghisalba, 1983). The dangers arising from the environmental exposure to such chemicals and their decomposition are not known for most of them (Schmidt-Bleek and Wagenknecht, 1979).

A nationwide survey conducted for the EPA indicated that organic compounds are the most common environmental contaminants, being found at 358 of the 395 hazardous waste sites included in the survey (Steelman and Ecker, 1984). Another study was made of the contaminants at 114 high priority Superfund sites (Ellis, Payne, and McNabb, 1985). The classes of chemical wastes found at the greatest number of sites, in order of decreasing prevalence, were: slightly water-soluble organics (e.g., aromatic and halogenated hydrocarbon solvents, chlorophenols), heavy metal compounds, and hydrophobic organics (e.g., polychlorinated biphenyls, aliphatic hydrocarbons).

Organic solvents were reported at nearly 28% of all hazardous waste sites (JRB and Associates, Inc., 1984b). These solvents appear to degrade only slowly, if at all, once they enter the groundwater. Hydrocarbon compounds, such as fuels, lubricating oils, and creosote, were found at nearly 15% of hazardous waste sites. Some of these contaminants are toxic (Alexander, 1980). Naphthalene and its methyl-substituted derivatives are some of the most acutely toxic, water-soluble components of crude oils (Anderson, Neff, Cox, Tatem, and Hightower, 1974). As the molecular size of polyaromatic hydrocarbons (PAHs) increases up to four or five fused benzene rings, their lipophilicity, environmental persistence, and mutagenicity also increase (Jacob, Karcher, Belliardo, and Wagstaffe, 1986; Miller and Miller, 1981). Some PAHs may be converted to toxic substances in the environment (Alexander, 1980). Many represent classes of molecules not previously investigated, and some have no close structural analogs in nature. Many have been intentionally developed to be resistant to microbial attack, and, not surprisingly, tend to persist in nature.

Chemicals may enter the environment directly as a result of accidents, spills during transportation, leakage from waste disposal or storage sites, or industrial facilities (Ghisalba, 1983). Oils and fuels generally enter the environment as a result of leaking storage tanks and associated piping (both above and below ground) or accidental spills (JRB and Associates, Inc., 1984b). A sudden and increasing frequency of tank failures has been experienced from the leaking of gasoline from underground gasoline storage tanks, primarily at automobile service stations. Steel tanks have an average life expectancy of 25 years and, after their first 20 years, have demonstrated an increase in the rate of test failure with age (Robison, 1987). It is estimated that of the approximately 1.4 million buried gasoline and fuel tanks, 75,000 to 100,000 are currently leaking, and the final count could be as high as 350,000 (Brown, Norris, and Brubaker, 1985). The EPA estimates that service station gasoline tanks alone could be losing 11 million gal/year of product.

Actual tank leaks are only a small part of the problem. Freezing and rain affect the surrounding soils and create stresses that cause buoyancy and the shifting of pipes and other installations. Because of these stresses a fueling system can actually start leaking from the day it is installed. In pipeline accidents, 157,760 barrels of gasoline and crude oil were lost to the environment in 233 reported incidents in 1971 (Meyer, 1973). Added to this are the unreported or undetected leaks, e.g., well casings and tanks.

For decades, the Department of Energy (DOE) and its predecessor agencies have engaged in a wide range of operations that generate organic chemical wastes (Steelman and Ecker, 1984). The vast majority of Department of Defense (DOD) hazardous waste sites are contaminated with organic chemicals, the most common being fuels and chlorinated solvents. Past disposal practices have led to migration of organics from the disposal sites via the groundwater pathway. DOD sites are most often contaminated by underground tank and line leaks, sludge burial pits, chemical evaporation pits, fire training areas, and landfills. The DOD is developing in situ treatment technologies as part of a general effort to determine the most effective and least costly methods to clean up its hazardous waste sites.

Subsurface contamination is complex and difficult to treat (FMC Aquifer Remediation Systems, 1986). The soil is particularly difficult to treat because of the tendency for contaminants to tightly bind or sorb to the soil particles. Soil contamination is also generally more severe than groundwater contamination, ranging from a few thousand ppm to greater than 10 to 15% by weight in the soil, as compared with a maximum of a few hundred ppm in groundwater.

It is very difficult to estimate the quantity and scope of hydrocarbon contamination from spills that have been reported (Raymond, Jamison, and Hudson, 1976). In spills involving light and more soluble hydrocarbons, e.g., gasoline, the adsorbed product could slowly leach into a water supply for many years, if the removal processes were not enhanced. Ultimately, the result of subsurface soil contamination is groundwater contamination (FMC Aquifer Remediation Systems, 1986).

1.1.2 Groundwater Contamination

Groundwater is water present in the soil within a zone of saturation; it is generally derived from precipitation or stream infiltration (Bitton and Gerba, 1985). Water-bearing formations that will transmit and yield significant quantities of water to wells and springs are called aquifers. Within the zone of saturation, groundwater occurs under either water table or artesian conditions. Under water table conditions, the groundwater is not confined. The upper surface of the saturated zone is called the water table and may be near the land surface or hundreds of feet below it. The water table rises and falls with changes in volume. Under artesian conditions, the groundwater is confined between an upper and lower confining bed. Water in a well will then rise above the bottom of the upper confining bed. Some background information about groundwater is given in Table 1.

In recent years, there have been many cases of groundwater contamination (Wilson, McNabb, Balkwill, and Ghoirse, 1983). Contamination of groundwater with hydrocarbons is a very common problem in the United States (Brown, Loper, and McGarvey, 1986). A study by the Council on Environmental Quality (CEQ) showed that major groundwater contamination problems in many states were caused by synthetic organic chemicals resulting from industrial and manufacturing operations (Ellis, Payne, and McNabb, 1985). Contamination of groundwater by these compounds was a major problem in at least 34, and possibly as many as 40, states (Steelman and Ecker, 1984). The problems exist in all states east of the Mississippi River and in many less industrialized states, such as Arizona and Idaho. Once an aquifer used as a water supply is contaminated, 100% of its available groundwater must be considered contaminated. Environmental "watchdog" organizations are demanding that if a groundwater supply is contaminated, it must be restored (Brown, Loper, and McGarvey, 1986).

Groundwater is an important natural resource (Brown, Loper, and McGarvey, 1986), constituting approximately two-thirds of the world's freshwater. Its use has increased 250% from 1950 to 1980, with over 50% of the population depending upon it as a source of drinking water. A portion of the U.S. population drinks untreated groundwater. In Florida, for instance, almost 92% of the population depends upon groundwater for drinking, and 20% of those receive untreated groundwater (Florida Environ. Service Center, 1982).

Table 1. Facts About Groundwater (Bitton and Gerba, 1985)

Some Parameters:	
Age	Few hours to thousands of years
Temperature	Increases with depth (c.3°C/100 m)
Conductivity	Presence of dissolved ions (e.g., Na^+, $Ca^=$, Mg^{2+}, Cl^-, SO_4^{2-}, NO_3^-...)
	Concentration: hundreds of mg/l
Hydrostatic pressure	Increases with depth (1 atm/10 m)
Major Sources of Groundwater Contamination (U.S.)	Domestic Water
	3 billion M^3/yr
	More than 0.6 million wells receive 850 billion gal/yr
	Industrial Sources
	100,000 abandoned industrial waste sites
	75,000 active industrial landfills
	173,000 waste lagoons
	Agricultural Practices
	Pesticides and fertilizers
Groundwater as a Resource (U.S.)	25% of all water used
Groundwater Uses (U.S.)	Irrigation 69%
	Industry 14%
	Urban and rural drinking 17%
Groundwater for Drinking (U.S.)	Supplies the needs of 50% of U.S. population
	Supplies 95% of drinking needs in rural areas
	U.S. public water systems: 75% rely on groundwater
Major Public Health Concerns	Waterborne Diseases: 54% of diseases occurred in noncommunity water systems relying solely on groundwater
	Inorganic and Organic Chemicals: Concern about toxic, mutagenic, carcinogenic, and teratogenic compounds. Not well documented.
Groundwater Legislation	Groundwater quality addressed in numerous pieces of legislation. Acts must be coordinated.

Groundwater is also a vulnerable resource. Its protection becomes critical as more and more of the surface waters are contaminated and greater reliance is placed on groundwater for domestic, agricultural, and industrial use (Florida Environ. Service Center, 1982). It is estimated that about 1% of the nation's groundwater supply is currently contaminated to some degree (Josephson, 1983a). One-third of all large public water systems have man-made contamination and over 7000 private, public, and industrial wells have been closed or seriously affected by contamination from 1980 to 1986.

Groundwater is a fragile resource; once contaminated, it is difficult to remediate (Steelman and Ecker, 1984). Contaminants can remain in concentrated areas of the groundwater for long periods of time (Ahlert and Kosson, 1983). The average residence time of a pollutant in groundwater is 200 years (Meyer, 1973). Once aquifers are contaminated, it can be very difficult and costly to restore them to their original quality. The treatment will probably be time consuming and expensive, with costs ranging from tens of thousands of dollars for simple treatment programs, up to tens of millions of dollars for complex, large sites (Lee and Ward, 1986). Groundwater also supplies 40% of the nation's irrigation water (Plumb, 1985). Yet, despite the importance of groundwater and the potential effect that contamination could have on its use, there has been no concerted effort to define the impact of the estimated 20,000 to 50,000 waste disposal sites in existence on this resource.

Groundwater may contain a wide range of organic substances, some of which are of natural (e.g., humic and fulvic acids) and some of anthropogenic (man-made) origin (Bitton and Gerba, 1985). Many chemicals now found in groundwater have not yet been tested for health effects, and no standards exist for their maximum concentration in water (Pye and Patrick, 1983). Some man-made chemicals are resistant to microbial degradation in the soil and reach the groundwater (Bitton and Gerba, 1985). The occurrence of these compounds in groundwater supplies has raised concern over their toxicity, mutagenicity, teratogenicity, and carcinogenicity.

Groundwater contamination is a complex phenomenon (FMC Aquifer Remediation Systems, 1986). Material, such as gasoline, will enter the subsurface, travel through the vadose zone (the unsaturated soil), and leave behind soil containing residual contamination. When the material reaches the water table or an impermeable medium (bedrock, clay), it spreads out on the surface (Texas Reserch Institute, Inc., 1982). If the material is less dense than water, such as petroleum hydrocarbons, it will form a "pool" of mobile free-phase liquid that travels along the groundwater table following the hydraulic gradient (FMC Aquifer Remediation Systems, 1986). The water-soluble components of the gasoline begin to dissolve in the water, contaminating a water resource (Texas Research Institute, Inc., 1982). As the water table rises and falls, the free-phase material will be smeared within the soil, resulting in a wide band of adsorbed hydrocarbon (FMC Aquifer Remediation Systems, 1986). Water in contact with the free phase and adsorbed contaminant will leach hydrocarbons creating a soluble plume of contamination that travels downgradient. Rainwater continues to contribute to the leaching of waste components into the aquifer, as it percolates through the gasoline-contaminated area. This will be a problem as long as any residual product remains in the soil (Texas Research Institute, Inc., 1982).

In gasoline spills, the monoaromatic hydrocarbons have relatively high pollution potential based upon their concentration in gasoline, their aqueous solubility, and their estimated toxicity or carcinogenicity (Barker and Patrick, 1985). The monoaromatics of particular concern are the water-soluble and mobile gasoline components, referred to as BTEX; i.e., benzene, toluene, ethylbenzene, and the xylene isomers, m-, p-, and o-xylene. It is likely that gasoline spills will have the most serious and immediate impact upon highly permeable, shallow, water table aquifers. Such aquifers are also favored for private and municipal drinking water supplies.

The impacts of losing hydrocarbons to the subsurface will vary (FMC Aquifer Remediation Systems, 1986). Fumes from free product and contaminated soils will migrate up through the vadose zone and can collect in basements, creating explosion hazards. Leaching of hydrocarbons will impact the quality and use of the groundwater. With extreme losses of material, the free product layer can actually seep to the surface in groundwater discharge areas.

Most site investigations have been confined to the property, or near vicinity of the suspected source of contamination. However, an extensive analysis of water supply wells by the Council on Environmental Quality (CEQ) in 1981 (Council on Environmental Quality, 1981) indicated that the migration of contaminant plumes is more widespread than this. An example involved a contaminant plume in Nashua, NH, that was 1500 ft long, 110 ft deep, and covered an area of about 1.29×10^6 ft^2 (Josephson, 1983b). The estimated movement of the plume was about 1.7 ft/day.

Groundwater in the vicinity of hazardous waste disposal facilities may become contaminated with a larger number of chemicals than previously thought (Plumb, 1985). Monitoring data from 358 hazardous waste disposal sites identified 720 substances, including 558 organic compounds, that have been reported in the groundwater at one or more sites. The organic compounds established by the EPA as priority pollutants are listed in Table A.1-1. Most classes of priority pollutants (volatiles, base/neutrals, acid extractables, pesticides, and inorganics) have been detected in the groundwater in the vicinity of disposal sites. However, since this information has never been centralized, contamination from hazardous waste disposal sites has been treated as a series of isolated examples rather than symptoms of a more extensive problem. By comparison, 1259 contaminants have been reported in surface water (Shackelford and Keith, 1977). The organic priority pollutants have been detected in the groundwater of all 10 EPA regions (Plumb, 1985). The most commonly encountered are listed in Table A.1-2. Volatile compounds have been identified most frequently and at the greatest concentrations.

There are more than 200,000 various types of surface impoundments that are potential future sources of pollution (FMC Aquifer Remediation Systems, 1986). In 1978, the EPA estimated there were 25,000 active industrial impoundments and landfills in the U.S. and that many of these could impact usable aquifers (Pye, Patrick, and Quarles, 1983). Many of these sites may require remedial action to restore the groundwater to usable condition. An estimated 50 to 60% of the landfills with interim status under the Resource Conservation and Recovery Act (RCRA) may be contaminating groundwater, including several RCRA facilities that have received materials from Superfund cleanups (Hileman, 1984).

Groundwater has been assumed to be pristine due to the protective role of the overlying soil mantle (Thornton-Manning, Jones, and Federle, 1987). In reality, it is as contaminated as surface waters in some areas (Page, 1981).

Unlike streams and rivers, groundwater moves very slowly (Brown, Loper, and McGarvey, 1986). Its rate and direction of flow are influenced by factors, such as substrate composition, recharge rates, and gravity. It has been widely held that once groundwater is contaminated, it is permanently lost. However, groundwater systems do, in fact, have mechanisms to regenerate themselves. Contaminant transport is a dynamic process. Natural biodegradation is an active phenomenon; however, biodegradation mechanisms are limited by high levels of contamination. Nevertheless, biodegradation offers an alternative strategy for dealing with groundwater contamination by accelerating attenuation through in situ treatment.

1.1.3 Pollution Legislation

The June 1976 Consent Decree of the National Resources Defense Council resulted in the U.S. Environmental Protection Agency (EPA) identifying 65 classes of toxic chemicals (Richards and Shieh, 1986). This toxic pollutants list presented the core pollutants for which pretreatment standards had to be established under the Clean Water Act of 1977. The list comprised thousands of individual chemicals and proved too voluminous as a guideline for chemical analyses to be performed on environmental samples and industrial effluents (Leisinger, 1983). Therefore, the list was reduced to 129 compounds and elements (114 organic compounds plus cyanide, asbestos, and 13 metals) from the 65 pollutant classes (Keith and Telliard, 1979). These compounds and elements are referred to as "priority pollutants." In 1979, the U.S. EPA published the original two-volume report Water-Related Environmental Fate of 129 Priority Pollutants (1979) (Environmental Protection Agency, 1979), which identified the nine pollutant groups of concern. These included aliphatics, halogenated aliphatics, nitrosamines, aromatics, chloroaromatics, polychlorinated biphenyls (PCBs), nitroaromatics, polynuclear aromatic hydrocarbons, and pesticides (see Table A.1-1).

Proper management of hazardous waste has become a top priority of government officials, industrial managers, and private citizens (Scholze, Wu, Smith, Bandy, and Basilico, 1986). Since 1976, considerable effort has been expended toward this goal by identifying the origin of hazardous wastes, developing appropriate technologies for management of these wastes, and implementing regulatory schemes that provide greater protection of public health and the environment. However, for many large users, the threat of liability is more powerful than any regulations that can be written (Robison, 1987).

There are numerous institutional limitations that can affect if, when, what, and how remediation will be selected and carried out (Wilson, Leach, Henson, and Jones, 1986). Groundwater remediation may require compliance with a multitude of local, state, and Federal pollution control laws and regulations. If the response involves handling hazardous wastes, discharging substances into the air or surface waters, or the underground injection of wastes, federal pollution laws apply. These laws do not exempt the activities of federal, state, or local officials or other parties attempting to remediate contamination events. They apply to generators and responding parties alike, and it is not unusual for these pollution control laws to conflict. For example, a hazardous waste remediation project may be slowed, altered, or abandoned by the imposition, upon the party undertaking the effort, of elaborate RCRA permit requirements governing the transport and disposal of hazardous wastes.

Federal Regulations

Environmental laws pertinent to in situ bioreclamation of fuel oil are the Resource Conservation and Recovery Act (RCRA), Comprehensive Environmental Response, Compensation, and Liability Act (CERCLA), Superfund Amendments and Reauthorization Act (SARA), Clean Water Act, and Safe Drinking Water Act (Public Law 99-499, 1986). RCRA was passed by Congress in 1976 to begin to cope with unsafe disposal of hazardous waste (Nicholas, 1987). RCRA mandated the establishment of a "cradle to grave" program for safe disposal of hazardous waste and directed EPA to adopt criteria for identifying hazardous waste and to establish standards for handling and disposal of hazardous waste by generators, transporters, and disposers. Operators of hazardous waste generators and Treatment Storage and Disposal Facilities (TSDFs) were required to obtain permits from EPA. Congress attempted to force alternative solutions to landfills by making the use of landfills expensive. RCRA also provided more direct incentives for resource conservation and recovery to reduce the volume of waste and to facilitate development of alternative technologies.

Under the Safe Drinking Water Act of 1974, EPA was directed to establish national drinking water standards to protect public health (Miller and Miller, 1981). This act contains an underground injection control program to prevent groundwater contamination by waste injection wells and to protect sole-source aquifers for drinking water (Pye and Patrick, 1983).

The Clean Water Act of 1977 requires the EPA to establish, equip, and maintain a water quality surveillance system for groundwater, as well as surface water (Pye and Patrick, 1983). The Toxic Substances Control Act of 1976 and the Surface Mining Control and Reclamation Act both have provisions that could offer a measure of protection for groundwater.

The Hazardous and Solid Waste Amendments (HSWA) were passed to reduce the dependence upon land disposal of hazardous wastes (Scholze, Wu, Smith, Bandy, and Basilico, 1986). As a result, the EPA determined which wastes should be restricted from land disposal unless treated. The EPA is considering a proposal to establish performance standards for treatment of all listed hazardous wastes and would have to identify the Best Demonstrated Available Technology (BDAT) for a particular listed waste category. The highest achievable performance level for that BDAT process would become the standard for that waste. Waivers for specific wastes would require use of the next BDAT for that waste. Two BDAT performance standards may have to be established: one for concentrated and one for dilute wastes. The former would continue to be incineration for most organic constituents and some type of solidification for inorganics. The latter may be based on biological treatment, which would provide many opportunities for commercial development of a wide range of microbial applications.

In 1980, Congress passed CERCLA, commonly referred to as Superfund. CERCLA authorizes the federal government to clean up contamination caused by inactive waste disposal sites or spills, many of which pose immediate threats to groundwater quality (Pye and Patrick, 1983). Superfund put the burden on the generator, transporter, and disposer to clean up existing, nonpermitted disposal sites. It established a mechanism to identify sites that present significant threats to health or to the environment; it required EPA to develop a list of high-priority sites for cleanup; it empowered EPA to clean up the priority sites, if the responsible party was unidentifiable; it created an emergency response fund, with a small Federal contribution and a tax on petrochemical feedstocks and certain other chemicals, for EPA cleanups; and it authorized EPA to recover cleanup costs from the responsible parties. The original Superfund program was established for five years at a cost of $1.1 billion. RCRA and Superfund were to provide a strong national mandate for the development of alternative ways to eliminate or lessen the toxicity and volume of hazardous waste.

In 1980, EPA proposed a national groundwater strategy (Environmental Protection Agency, 1980). Its goal was to prevent groundwater contamination rather than provide remedial action. The suggested approach included the development of state management and protection strategies; the development of a groundwater classification system; and EPA coordination of existing federal programs for groundwater protection. The proposed strategy was aimed at protecting groundwater quality according to its value and use, and the technical approach adopted included the use of siting and design criteria, best management practices, effluent standards, innovative and alternative technologies to achieve performance standards, and, to a lesser extent, numerical groundwater quality standards and economic incentives.

In 1981, because of its lack of action on the proposed national groundwater strategy, EPA was compelled to create two separate groundwater task forces, one for policy and one for technical purposes, to develop a consistent agencywide strategy. This objective has still not been realized and the EPA is revising its groundwater policy statement.

The EPA has also established a formal program to enhance the development and use of new or innovative technologies for mitigating the problems caused by releases of hazardous substances at uncontrolled hazardous waste sites (Sanning and Olfenbuttel, 1987). In the U.S., the program is called the Superfund Innovative Technology Evaluation, or SITE, program. In 1986, the NATO-CCMS formally adopted a U.S. proposal for a new pilot study entitled, "Demonstration of Remedial Action Technologies for Contaminated Land and Groundwater." The following NATO countries are participating: Canada, Denmark, Federal Republic of Germany, Greece, Italy, The Netherlands, Norway, Spain, and the United States. Two non-NATO countries, Australia and Japan, have also expressed an interest in participating. This study is a logical international extension of the U.S. EPA SITE program and will field demonstrate and evaluate new technology or existing systems for remedial action at uncontrolled hazardous waste sites.

SARA is the Superfund Amendments and Reauthorization Act of 1986, which extends and amends CERCLA. On a federal level, not later than 18 months after the enactment of SARA, the administrator must take steps to assure that a preliminary assessment is conducted for each facility on the Federal Agency Waste Compliance Docket (Public Law 99-499, 1986). The facilities must be evaluated in accordance with the criteria established, in accordance with Section 104 under the National Contingency Plan for determining priorities among releases, and these facilities must be included on the National Priorities List.

On a state level, not later than six months after the inclusion of any facility on this list, the department, agency, or instrumentality responsible for such facility must begin a remedial investigation and feasibility study for the facility (Public Law 99-499, 1986). Within 180 days after the results of an investigation and study are reviewed, an agreement must be reached to complete all remedial action necessary. Substantial continuous physical on-site remedial action must begin not later than 15 months after completion of the investigation and study.

The 1984 amendments to RCRA and the 1986 reauthorization of Superfund reaffirm the policy of limiting land disposal of hazardous waste and of supporting the development of alternative technologies (Nicholas, 1987). According to the 1984 RCRA amendments, "certain classes of land disposal facilities are not capable of assuring long-term containment of certain hazardous wastes, and to avoid substantial risk to human health and the environment, reliance on land disposal should be minimized or eliminated, and land disposal, particularly landfill and surface impoundment, should be the least favored method for managing hazardous wastes." The act declares as a national policy "minimizing the generation of hazardous waste and the land disposal of hazardous waste."

Amendments to Superfund call for establishment of major new programs to stimulate waste reduction and development and demonstration of new technologies (Nicholas, 1987). The amendments authorize three new programs at a cost of approximately $250 million over the next five years. They establish a fund, administered by the National Institute of Environmental Health Sciences, for basic research and training grants. They also create a new office at EPA, called the Office of Technology Demonstration, to direct the "research, evaluation, testing, development, and demonstration of alternative or innovative technologies to be used at Superfund sites." And they call for the establishment of at least five university-based centers to conduct research and training on various aspects of hazardous waste disposal.

In situ remediation procedures may be subject to permitting or other requirements of federal or state underground injection control programs (Wilson, Leach, Henson, and Jones, 1986). Withdrawal and treatment approaches may be subject to regulation under federal or state air pollution control programs or to pretreatment requirements, if contaminated groundwater will be discharged to a municipal wastewater treatment system. Pumping from an aquifer may involve a state's groundwater regulations on well construction standards and well spacing requirements, as well as interfere with various competing legal rights to pump groundwater. Other factors influencing remediation decisions are the availability of alternate sources of water supply, political and legal pressures, and the availability of funds.

By executive order, all federal agencies must comply with the letter and intent of all federal, state, and local environmental regulations (Heyse, James, and Wetzel, 1986). The Department of Defense is complying with the requirements of the Superfund legislation with the Installation Restoration Program (IRP). The EPA has a backlog of 15,000 sites that may present a groundwater problem, and those are estimated to be less than half the sites in the country that may be contaminated (Brooks and McGinty, 1987). For FY 87, about $1.1 billion was allocated for the EPA to spend on Superfund projects, many of which will entail treating groundwater. The appropriation is part of a \$9 billion, five-year program authorized by Congress under SARA, which establishes a preference for remedial actions that permanently and significantly reduce the volume, toxicity, and mobility of hazardous wastes. Estimates indicate that over the next 50 years, approximately \$100 billion will be spent on groundwater reclamation programs (Fredrickson and Hicks, 1987). The National Priorities List is a compilation of hazardous waste sites for priority cleanup maintained by EPA (Steelman and Ecker, 1984). After spending more than \$1 billion, the Superfund had completed remediation of only six sites on this list (Brown, Loper, and McGarvey, 1986).

The EPA's Risk Reduction Engineering Laboratory has initiated a Chemical Countermeasures Program to define technical criteria for using chemicals and other additives at contamination sites, such that the combination of the contaminant plus the chemical or other additive, including any resulting reaction products, results in the least overall harm to human health and to the environment (Ellis, Payne, and McNabb, 1985). Under this program, the efficacy of in situ treatment of large volumes of subsurface soils, such as found around uncontrolled hazardous waste sites, and treatment of large, relatively quiescent waterbodies contaminated with spills of water-soluble hazardous substances, will be evaluated.

There are other federal statutes that could be used to protect groundwater, but the above mentioned are the most important (Pye and Patrick, 1983).

State Regulations

States have long been involved with groundwater allocation law and water rights, but it is only in the last decade that they have made large efforts to prevent, abate, and monitor groundwater pollution (Pye and Patrick, 1983). State regulations that may affect groundwater quality fall into three main categories: 1) those dealing with particular sources of pollution, such as septic tank systems and waste disposal sites, 2) those establishing and implementing water quality standards for aquifers, and 3) those regulating the use of land in areas overlying critical aquifer recharge zones.

A waste service industry has emerged in response to the national concern for management of hazardous waste under RCRA (Scholze, Wu, Smith, Bandy, and Basilico, 1986). Biological processes have been employed for treatment of hazardous wastes for many years. One goal of the Superfund remedial program is to promote on-site cleanup of contaminated sites, and the biological techniques are receiving increasing attention as a cost-effective remedial action.

Biotechnology can provide a useful tool to develop new and safer approaches for hazardous waste disposal and cleanup (Nicholas, 1987). However, such developments have been slowed by regulatory uncertainties and the lack of appropriate incentives to promote alternatives to land disposal of hazardous waste. There have been few reports of biotechnology companies entering the bioreclamation field. Most companies will not fund biotechnology research on waste treatment. Much of the research interest and funding have centered on laboratory studies of the use of indigenous microorganisms for waste treatment, in preference to development of a process technology.

1.2 BIODEGRADATION AS A TREATMENT ALTERNATIVE

During the past decade, government, industry, and the public have recognized the need to greatly reduce the volume and toxicity of waste and to develop safe, effective, and economic alternatives for its disposal (Nicholas, 1987). Problems associated with the cleanup of disposal sites and spills of toxic substances further demonstrate the need for the development of new technologies.

Protection of aquifers from contamination is probably the best solution to the problem, since cleanup of contaminated groundwater is likely to be difficult and expensive (Josephson, 1980). However, once groundwater pollution has occurred, there are a number of techniques that can be used to clean up the contamination (Lee and Ward, 1985; Lee and Ward, 1986; Lee, Wilson, and Ward, 1987). Techniques currently in use include controlling the flow of the polluted groundwater or physical containment of the material; removal of contaminated soil to a secure site; in situ treatment by physical, chemical, or biological techniques; or withdrawal of groundwater, followed by surface treatment by chemical, physical, and biological processes. Often, the volatile or soluble contaminants are sorbed onto activated carbon or resin, which can be very costly.

Traditional methods of treating groundwater contamination have relied upon removal or containment; however, these methods have had limited effectiveness (Brown, Loper, and McGarvey, 1986). Traditional remediation efforts at hazardous waste sites have been partially effective 54% of the time and completely successful only 16% of the time (Lee and Ward, 1985; Neely, Walsh, Gillespie, and Schauf, 1981). Most treatment schemes currently in use are not completely effective and do not offer permanent solutions for containment or remediation. Some methods may create additional uncontrolled hazardous waste sites.

Removal and/or containment were found to be the most common techniques in a survey of 169 remedial actions (Neely, Walsh, Gillespie, and Schauf, 1981). Removal involves excavation and transport of the contaminated soil to a secure site, such as a landfill. Excavation is usually accomplished by a dragline, which can reach a maximum depth of 60 ft (18.3 m) or a backhoe that can go to 70 ft (21.3 m) (Ehrenfeld and Bass, 1984). Excavation costs between $1.75 to 4.50/yd^3 ($2.30 to 5.90/m^3; plus additional costs for transportation of the contaminated material and disposal in an approved facility. Total costs are between $200 and 500/yd^3. It may be difficult to excavate all the contaminated soil in instances where contamination is widespread or when the contaminated soil is deep below the land surface. Physical plume containment measures include slurry trench walls, grout curtains, sheet piling, block displacement, infiltration controls, and passive interceptor systems (Lee, Wilson, and Ward, 1987; Canter and Knox, 1984).

There are some circumstances where excavation of contaminated soils is a practical cleanup method (FMC Aquifer Remediation Systems, 1986). For instance, in repairing or replacing underground tanks or pipe lines, contaminated soil must be removed to gain access to the faulty equipment. Excavation can also be practical in immediately dealing with high concentrations that pose a health or environmental hazard. Excavation can also be cost-effective for preventing groundwater contamination from substances limited to the surface. However, excavation raises the question of what to do with the contaminated soil. It essentially transfers the contamination from one site to another. In addition, soil excavation affects only some of the contaminated soil; it will not treat groundwater and may accentuate groundwater contamination through soil disturbance. Extractions of free contaminant product or contaminated groundwater will not address the material adsorbed on the soil and, therefore, are very long-term or incomplete cleanup methods.

Groundwater contamination by organic chemicals is a multiphase problem (Brown, Loper, and McGarvey, 1986). Organic contaminants in the subsurface can be present as three phases: mobile free product, residually saturated soil (adsorbed phase), and contaminated groundwater (dissolved phase). All three must be addressed to control groundwater quality. Free product must be removed, or it will continue to contaminate the soil through sorption and groundwater movement (vertical and horizontal) and continue to dissolve into the groundwater. If the residually saturated soil is untreated, it can also be a continuing source of contaminated groundwater. The gross contamination must be removed to the point that continued leaching of contamination does not exceed the attenuation capacity of the aquifer (FMC Aquifer Remediation Systems, 1986). Then, the impact of groundwater contamination will have been contained and reversed.

Biological processes have been used both in situ, without removal of contaminated groundwater, and on-site (above ground), following removal, to restore contaminated aquifers (Lee and Ward, 1985). In situ and on-site treatment processes avoid the economic and technical disadvantages, as well as environmental risks, incurred by transport of this type of hazardous material to alternate treatment facilities (Ahlert and Kosson, 1983). In situ treatment through chemical or biological transformation of the contaminant has the advantage of dealing with all aspects of the contamination problem and, by destroying the contaminant, of providing a permanent solution (FMC Aquifer Remediation Systems, 1986). Pump-and-treat methods (above ground) treat only the ground-water, while the contaminated soil continues to recontaminate the groundwater. In contrast, in situ treatment can reach organics trapped or sorbed by the soil matrix (Lee and Ward, 1985). In situ treatment is often less expensive than surface treatment or disposal methods and has been especially effective in remediating petroleum and volatile organic spills (FMC Aquifer Remediation Systems, 1986). In many cases, a combination of in situ and on-site treatment will achieve the most cost-effective results at an uncontrolled waste site (Environmental Protection Agency, 1985b). Therefore, both of these approaches will be dealt with in this report. The decision to use any individual procedure or combination of techniques will depend upon the contamination problems at a given site.

Natural attenuation in the subsurface environment is accomplished by biochemical degradation, evaporation, adsorption, metabolism, and transformation by microorganisms (Brown, Loper, and McGarvey, 1986). It is well known that microorganisms are capable of degrading a wide range of organic compounds (Pierce, 1982a).

Figure 1 summarizes the possible fate of xenobiotic compounds in the environment.

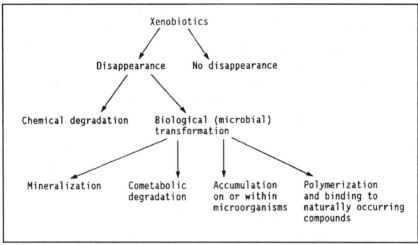

Figure 1. Possible Fate of Xenobiotic Compounds in the Environment (Leisinger, 1983)

An organic chemical may be subjected to nonenzymatic or enzymatic reactions brought about by microorganisms in the subsurface environment (Alexander, 1980); however, it is the enzymatic reactions that bring about the major changes in the chemical structure of these compounds. Extensive removal of organic materials is accomplished primarily through enzymatically mediated biological reactions, i.e., biodegradation (Thornton-Manning, Jones, and Federle, 1987). Few abiotic processes completely mineralize complex organic compounds in soil, and complete degradation depends upon microbial activity (Alexander, 1980). However, physical/chemical transformation processes may act synergistically with biochemical decomposition in this process.

The term "biodegradation" is often used to describe a variety of quite different microbial processes that occur in natural ecosystems, such as mineralization, detoxication, cometabolism, or activation (Alexander, 1980). It can be defined as the breakdown of organic compounds in nature by the action of microorganisms, such as bacteria, actinomycetes, and fungi (Sims and Bass, 1984). The microorganisms derive energy and may increase in biomass from most of the processes (Lee and Ward, 1985).

Mineralization occurs when there is complete biodegradation of an organic molecule to inorganic compounds; i.e., carbon dioxide, water, and mineral ions such as nitrogen, phosphorus, and sulfur (Sims and

Bass, 1984; JRB Associates, Inc., 1984b). Under anaerobic conditions, methane may be produced, and nitrate nitrogen may be lost as N_2 or N_2O gas through denitrification. Microorganisms can also transform hazardous organic compounds into innocuous or less toxic organic metabolic products. Chemical alteration can also be the result of cometabolism (cooxidation); i.e., growth on another substrate while the organic molecule is degraded coincidentally (Alexander, 1980). This process may be promoted by enzymes that catalyze reactions of chemically related substrates.

Contaminants in solution in groundwater, as well as vapors in the unsaturated zone, can be completely biodegraded or transformed to new compounds (Wilson, Leach, Henson, and Jones, 1986). Biodegradation converts petroleum components to compounds of lower molecular weight, while biotransforation can convert them to more polar compounds of a carbon number equal to the parent compound (Atlas, 1977). Although the compositions of refined oils, crude oils, and oily wastes are quite different, approaches toward their stimulated biodegradation have been similar.

Organisms that occur naturally in almost every soil system (the indigenous microbial populations) appear to be the chief agents involved in the metabolism of chemicals in waters and soils. Heterotrophic bacteria and fungi are responsible for most of the chemical transformations. Aerobic bacteria, including the actinomycetes and cyanobacteria, anaerobic bacteria, fungi, and some true algae have all been shown capable of degrading many classes of organic chemicals (Amdurer, Fellman, and Abdelhamid, 1985).

Many aerobic bacteria found in soil and water utilize biologically produced substances, such as the sugars, proteins, fats, and hydrocarbons of plant and animal wastes, and can also metabolize petroleum hydrocarbons, converting them to carbon dioxide and water (Brown, Norris, and Brubaker, 1985). Anaerobic bacteria are important for the biodegradation of many chlorinated pesticides and heavily halogenated organics (e.g., trichloroethylene, pentachlorophenol). During biodegradation, certain anaerobic bacteria (fermenters) commonly produce short-chain organic acids, while other microorganisms further break down these by-products to methane, carbon dioxide, and inorganic substances (Pettyjohn and Hounslow, 1983).

The ability of microorganisms to degrade or transform toxic substances has been applied to the treatment of environmental contamination by hazardous organic compounds. Biodegradation has been used, with reasonable success, to treat aquifers contaminated with petroleum hydrocarbons (Lee and Ward, 1985). Biorestoration is useful for hydrocarbons, especially water-soluble compounds and low levels of other compounds that would be difficult to remove by other means. Biodegradation is environmentally sound, since it destroys organic contaminants and, in most cases, does not generate problem waste products.

Natural soil bacteria may be present in a dormant or slow-growing state, but when stimulated by a specific set of environmental conditions, may multiply rapidly and subsequently adapt to the new environment (Buckingham, 1981). Some of the more common genera of bacteria involved in biodegradation of oil products include Nocardia, Pseudomonas, Acinetobacter, Flavobacterium, Micrococcus, Arthrobacter, and Corynebacterium. Tests have revealed that cultures containing more than one genus appear to have greater hydrocarbon-utilizing capabilities than many of the individual culture isolates.

Degradation in contaminated soil and aquifers may be affected by environmental constraints, such as dissolved oxygen; pH; temperature; toxicants; oxidation-reduction potential; availability of inorganic nutrients, including nitrogen, phosphorus, and others; salinity; and the concentration and nature of the organics. The number and type of organisms present in the environment will also play an important role in this degradation process.

Treatment, therefore, generally consists of optimizing conditions of pH, temperature, soil moisture content, soil oxygen content, and nutrient concentration to stimulate growth of the organisms that will metabolize the particular contaminants present (Sims and Bass, 1984). Optimum environmental conditions and nutrient application rates generally have to be established in laboratory/bench-scale studies and small field pilot tests. Some hazardous compounds may be degraded more readily under aerobic conditions; and some, under anaerobic conditions. Anaerobic conditions are always present at microsites in soil. Treatment might, therefore, consist of a combination of both aerobic and anaerobic conditions.

Biodegradation techniques are versatile and can be used at several different stages of treatment (Nicholas, 1987). Some applications include removal of contaminants from raw materials before processing, treatment of pipeline wastes before discharge from the factory, decontamination of soils and surface and groundwater, and cleanup of dump sites. These operations can be conducted in a number of ways: microorganisms or their active products (e.g., enzymes) can be released directly into the contaminated environment; microorganisms already present in the environment can be enhanced by the addition of suitable nutrients; or microorganisms can be used in contained or semicontained reactors.

If the locally occurring organisms are not effective for the given set of contaminants, the soil can be inoculated with microbial isolates that are specific for those compounds (Buckingham, 1981). Microorganisms can also be selectively adapted by growth on media containing the target chemicals. They can even be genetically engineered to enhance their ability to degrade these compounds. The use of genetically altered microorganisms may, in the future, expand the range of compounds that can be degraded and/or accelerate the rates at which degradation occurs (Walton and Dobbs, 1980).

Dissolved oxygen may promote biodegradation of wastes, such as aromatics, while some compounds, such as highly chlorinated compounds, can be degraded more readily without oxygen (Lee, Thomas, and Ward, 1985). Studies indicate that oxygen and nutrients are particularly essential to bacterial stimulation in contaminated sites and groundwater (Buckingham, 1981). Therefore, a successful biodegradation scheme requires careful site monitoring of these parameters, through proper sampling and testing procedures. Subsequent oxygen and nutrient management of soil or groundwater, or both, ensures that proper levels are maintained.

Biodegradation is a viable option for treatment of hazardous chemical spills (Buckingham, 1981). Its feasibility depends upon the availability of necessary equipment and manpower, as well as upon decontamination time restrictions. Biodegradation of massive spills may prove ineffective in terms of percent recovery per unit time, whereas it may be the only treatment alternative for low level quantities or concentrations of contaminants.

According to the Office of Technology Assessment (OTA), once developed and proved, biodegradation is potentially less expensive than any other approach to neutralizing toxic wastes (Nicholas, 1987). Such systems involve a low capital investment, have a low energy consumption, and are often self-sustaining operations. Biological means of decomposition require less energy than physicochemical processes and can be a competitive option under certain circumstances (Ghisalba, 1983).

Biological degradation of constituents in hazardous industrial wastewaters is now being practiced commercially (Scholze, Wu, Smith, Bandy, and Basilico, 1986). For commercial applications of biological technologies to achieve more widespread use, the biotechnologies must meet specific criteria:

1. Degradation

The end result of biological treatment should be to completely destroy the hazardous constituent of concern. This will allow the residues to be disposed of in nonhazardous landfills at a much lower cost.

2. Concentration

Dilute hazardous wastes can be difficult to manage cost effectively, since many chemical and thermal treatments are only cost-effective on concentrated waste constituents. Dilute hazardous wastes might be most cost-effectively treated by using biological treatment to concentrate the organic constituents, followed by thermal treatment of the residue.

3. Diverse Target Constituents

Treatment often has to include management of recalcitrant compounds or combinations of many hazardous constituents. Ideally, organisms in commercial applications should be able to degrade mixtures of organics, with an equal efficiency for all components. It would also be advantageous for the organisms to accomplish multiple tasks, such as degrading organics while they concentrated inorganics for further treatment or recovery.

4. Consistency

Variability in degradation efficiency among batches would be costly to a waste management firm; consistency in the composition of the residue is essential. The end product of the biological process must be predictable to keep monitoring costs down. Consistency in the residue also allows the biological process to be used in sequence with many solidification treatment processes, which may be sensitive to changes in constituent concentrations.

Also, many solidification treatment processes that could be used in sequence with biological concentration of hazardous constituents, often are sensitive to changes in constituent concentrations.

5. Relatively Low Cost

It has been difficult to reduce the use of landfills because of the high costs associated with alternative options. Biological treatment must be able to compete with other processes (e.g., chemical and thermal applications) in effectiveness and cost.

1.2.1 On-Site Biological Treatment Techniques

Conventional biological wastewater treatment techniques that employ microorganisms can be used to treat polluted groundwater (Lee and Ward, 1986). Both aerobic and anaerobic biological treatment systems are available (Roberts, Koff, and Karr, 1988). The systems that have used acclimated bacteria to restore contaminated aquifers typically have relied upon biological wastewater treatment techniques, such as activated sludge, rotating biological contactors, aerated and anaerobic lagoons, composting, waste stabilization ponds, trickling filters, and fluidized-bed reactors for the treatment of relatively large flow volumes with consistent loading characteristics (Lee and Ward, 1985; Lee and Ward, 1986). These processes are described in Appendix A. However, these designs are limited in their scope of application when dealing with groundwater treatment.   Contaminated groundwater is characterized by low-flow conditions, varying influent concentrations, decreasing system loading with time, and relatively short-term installation and operating periods.

Many of these systems recharge the effluent from biological treatment to the aquifer to create a closed loop of recovery, treatment, and recharge, which flushes the contaminants out of the soil rapidly and establishes hydrodynamic control to separate the contaminated zone from the rest of the aquifer (Lee and Ward, 1985; Lee and Ward 1986). Acclimated bacteria can also be added to the aquifer and can act in situ to degrade the contaminants. The recharge water can be adjusted to provide optimal conditions for the growth of the acclimated bacteria and of the indigenous populations, which may also act on the contaminants.

1.2.2 In Situ Bioreclamation

In situ treatment of contaminated soil is conducted without excavation and encompasses four methods: Treatment, involving in-ground chemical or biological transformation of contaminants to harmless products; Stabilization, involving the in-ground chemical immobilization of contaminants; Solidification, involving the in-ground physical immobilization of contaminants; and Extraction, involving a subsurface displacement of soil, followed (usually) by chemical treatment on the surface (Amdurer, Fellman, and Abdelhamid, 1985).

In situ biodegradation, commonly referred to as bioreclamation, is based upon the concept of using microflora to decompose the organic compounds contaminating a site to within acceptable levels by treating the soil and groundwater in place (Environmental Protection Agency, 1985ba). Enhanced bioreclamation involves stimulating the indigenous microbial population at the site to improve degradation of the contaminants (Lee and Ward, 1985). The contaminants must not be toxic to the microorganisms that would be available for degradation, and microbes specific for certain chemicals may be encouraged by manipulating environmental conditions.

The bioreclamation method that has received the most developmental effort, and that is the most feasible for in situ treatment, is one that relies on aerobic microbial processes (Environmental Protection Agency, 1985b). This technique provides essential nutrients and oxygen, which may be limiting factors, to stimulate microbial growth (Lee and Ward, 1985). Unpolluted water typically contains very low concentrations of organic and inorganic nutrients and oxygen (Brown, Loper, and McGarvey, 1986). (While unpolluted water would contain oxygen, the 8 to 10 ppm present at saturation is very low.) An organic contaminant introduced into the subsurface provides a source of carbon for the biosynthesis of new cellular material, as long as other constituents of cellular biomass and molecular oxygen are also available.

Fuel contaminants are hydrocarbons and are primarily carbon and hydrogen; cell material is composed of carbon, hydrogen, nitrogen, phosphorus, and oxygen (and minute amounts of other elements). Since most organic contaminants do not contain all the necessary elements for bacterial growth, adequate amounts of oxygen, nitrogen, sulphur, phosphorus, and certain trace minerals must be supplied to enhance the biodegradation rate.

The key to in situ bioreclamation is supplying the needed nutrients and oxygen to the contaminated area for aerobic degradation. Since normal groundwater flow ranges from 10 to 200 ft/year, it could take years for nutrients to traverse even a small site. By increasing the gradient via pumping (draw-down) and injection (mounding), transport times can be reduced from years to days or even hours. If nutrient availability is limiting the process at a particular site, the more quickly the nutrients can be added, the faster the remediation can begin. In situ treatment generally involves the installation of a bank of injection wells at the head of, or within the plume of, contaminated groundwater (Ward and Lee, 1984).

Anaerobic microorganisms are also capable of degrading certain organic contaminants (Environmental Protection Agency, 1985b). Methanogenic consortia, groups of anaerobes that function under very reducing conditions and produce methane as an end product, are able to degrade heavily halogenated aliphatics (e.g., perchloroethylene (PCE) or trichloroethylene (TCE)), while aerobic organisms have limited capability. Using anaerobic degradation as an in situ reclamation approach is, theoretically, feasible. However, the process is limited in substrate range and length of time required for degradation.

The following treatment variables or site conditions need to be determined to assess whether degradation could serve as an effective in-place treatment (Lee and Ward, 1986):

1. biodegradability of the waste constituents in the contaminated soil of interest or similar soil
2. soil pH
3. microorganisms present (type, population, genus, species)
4. soil oxygen content
5. oxidation-reduction potential
6. soil moisture
7. soil nutrient content (C:N:P ratio)
8. soil temperature
9. rates of biodegradation
10. potential for leaching and possible rates of movement of waste constituents through the soil by leaching
11. soil permeability/porosity

The problem must first be identified and defined. Before any treatment program can be implemented, a thorough investigation of the hydrogeology and contamination problems of the site must be made (Lee, Wilson, and Ward, 1987). Extensive monitoring from a number of monitoring wells is usually required to determine the aquifer characteristics and the extent of contamination. Geophysical techniques, such as ground-penetrating radar, electric resistivity soundings, borehole geophysics, seismic refraction profiling, remote sensing, and magnetometry, may fill the gaps in hydrogeologic information without the costs of additional well installation (Josephson, 1983b). For successful application of in situ technologies, aquifer flow patterns must be determined and channeled to the treatment area (Bove, Lambert, Lin, Sullivan, and Marks, 1984). This may require construction of hydraulic barriers (e.g., slurry walls) to ensure that all groundwater flows to or through the treatment area. The next step is to define the nature and magnitude of potential impacts (FMC Aquifer Remediation Systems, 1986). And the final step is to manage the problem. If there is no impact, this can be simple monitoring and plume management. If there is an impact, some degree of remediation must be undertaken.

Logical steps in the in situ bioreclamation process are 1) site investigation, 2) free product recovery, 3) microbial degradation optimization study, 4) system design, 5) operation, and 6) monitoring (Lee and Ward, 1985). These are described in detail in Appendix A, Section 5.5, Biodegradation Implementation Plan.

SECTION 2

Organic Compounds in Refined Fuels and Fuel Oils

2.1 CHEMICAL COMPOSITION OF FUEL OILS

Most petroleum products can be fractionated into a saturate or aliphatic fraction, an aromatic fraction, and an asphaltic or polar fraction (Brown, Ramos, Fiedman, and Macleod, 1969). Hydrocarbons within the saturated fraction include straight chain alkanes (n-alkanes), branched alkanes, and cycloalkanes (naphthenes) (Atlas, 1981). Since oils and fuels are seldom the same age or from the same source, rarely do they have the same chemical composition. Fuel oils of similar viscosity and volatility may vary widely in composition, dependent upon the source of the raw materials or method of manufacture (Naval Civil Engineering Laboratory, 1986). Due to chromatographic column inefficiency, a gas chromatogram of an oil may actually reveal fewer than one-tenth of 1% of all the components present in a sample (Blumer, 1975). Thus, there may be many more chemical compounds present in a sample than can be detected, and it is impossible to assign a standard composition to the different oils and fuels. However, to the extent possible, knowing what compounds are present in a contaminated site will help with correlating the degradative capabilities of different microorganisms for greater efficiency in biodegradation.

27

2.1.1 Naptha

Petroleum naphtha is a name sometimes given to a portion of the more volatile hydrocarbons distilled from petroleum. Naphtha is a generic term for a group of refined, partly refined, or unrefined petroleum products (primarily gasoline) in which not less than 10% of the product distills below 347°F (175°C) and not less than 95% distills below 464°F (240°C) (Holmes and Thompson, 1982). It forms highly complex mixtures of hydrocarbons that range from paraffin-type hydrocarbons to aromatics, which can be difficult to classify based upon chemical properties (Bland and Davidson, 1967).

2.1.2 Kerosene

Kerosene applies to that portion of petroleum boiling between the approximate limits of 180° and 320°C. It can be used as jet fuel. The chemical composition of kerosene depends upon its source and usually consists of a mixture of about 10 hydrocarbons containing 10 to 16 carbon atoms/molecule. The constituents include n-dodecane, alkyl derivatives of benzene, naphthalene, and derivatives of naphthalene.

2.1.3 Fuel Oil and Diesel #2

Diesel fuel is the fraction of petroleum that distills after kerosene (396° to 583°F or 202° to 306°C) (Holmes and Thompson, 1982). Other than for nitrogen content, the shale- and petroleum-derived diesel fuels are comparable (Ghassemi and Panahloo, 1984). The organic components of diesel fuel #2 can be found in Appendix A, Section 2.1.3.

2.1.4 Gasoline

Gasoline is composed of several hundred different hydrocarbon compounds in various proportions, primarily C_5 to C_9 hydrocarbon chain fractions with a boiling range of 80° to 400°F (23° to 204°C) (PEDCO Environmental, Inc., 1978). The basic types of hydrocarbons in gasoline are paraffins, olefins, cycloparaffins, and aromatics. There will also be some sulfur, nitrogen, and oxygen present and trace amounts of iron (0.07 to 13 μg/ml), nickel (0.001 to 2 μg/ml), and zinc (0.001 to 2 μg/ml) (Ghassemi and Panahloo, 1984). Appendix A provides a breakdown of the composition of gasolines.

2.1.5 JP-5

JP-5, "jet fuel," is made by blending naphtha, gasoline, and kerosene to meet military and commercial product standards (Ghassemi and Panahloo, 1984).   The content of its highly toxic components varies widely among different jet fuels.   One petroleum-derived JP-5 sample had a concentration of 0.04 $\mu g/g$ of benzo(a)pyrene, while shale oil-derived JP-5 contained none. JP-5 can contain high molecular weight normal paraffins or disulfides, and if it is prepared by hydrocracking, it will have a considerable degree of branching (Varga, Lieberman, and Avella, 1985).   See Appendix A, Section 2.1.5, for the organic compounds and trace elements present in JP-5.

2.1.6 JP-4

JP-4, "jet fuel," is made by blending various proportions of naphtha, gasoline, and kerosene to meet military and commercial product specifications (Gary and Hardwork, 1975). The content of highly toxic components in this fuel also varies. It has the highest hydrogen content, the highest volatility, and the lowest viscosity of any fuel widely used by the Air Force (General Electric Co. Aircraft Engine Business Gp., 1982). The organic compounds and trace elements present in JP-4 are listed in Appendix A.

2.2 FACTORS AFFECTING CONTAMINATION AND BIODEGRADATION

Crude petroleum and many of the products refined from petroleum contain thousands of hydrocarbons and related compounds (Cooney, Silver, and Beck, 1985). When such petrochemicals enter an aquatic system, they are subject to a number of physical, chemical, and biological factors that contribute to loss or alteration of some of the components. Volatile compounds can be selectively lost by evaporation; photochemical reactions may contribute to change; the oil may adsorb to detritus; wind, wave, or other mixing actions may cause formation of emulsions; and microorganisms can metabolize and cometabolize most of the hydrocarbons under proper environmental conditions.

The fate of a pollutant in the environment is determined by characteristics both of the chemical and the environment (Pfaender,

Shimp, Palumbo, and Bartholomew, 1985). Significant waste factors that will affect its biodegradation include (Parr, Sikora, and Burge, 1983):

1. Chemical composition of the waste
2. Its physical state (i.e., liquid, slurry, sludge)
3. Its carbon:nitrogen ratio
4. Water content and solubility in water
5. Chemical reactivity with and dissolution effects on soil organic matter
6. Volatility
7. pH
8. Biochemical oxygen demand (BOD)
9. Chemical oxygen demand (COD)

The behavior of toxic pollutants in the environment also depends upon a variety of processes and properties (Ghisalba, 1983; Josephson, 1983; Bitton and Gerba, 1985; Pettyjohn and Hounslow, 1983).

1. Chemical processes, e.g.:
- hydrolysis
- photolysis
- oxidation
- reduction
- hydration

2. Physical or transport processes and properties, e.g.:
- advection
- dispersion and diffusion
- sorption
- volatilization
- solubilization
- viscosity
- density
- dilution

3. Biological processes, e.g.:
- bioaccumulation
- biotransformation
- biodegradation
- toxicity

4. Combined environmental factors

There are major compositional variations between different crude and refined oils (Atlas, 1978c). Some oils contain toxic hydrocarbons, which may prevent or delay microbial attack; some refined oils have additives, such as lead, which can inhibit microbial degradation of polluting hydrocarbons. Oils contain varying proportions of paraffinic, aromatic, and asphaltic hydrocarbons. Both the rate and extent of biodegradation are dependent upon the relative proportions of these classes of hydrocarbons.

Under favorable conditions, microorganisms will degrade 30 to 50% of a crude oil residue, but degradation of such a residue is never complete and does not affect the different main hydrocarbon groups in the same manner (Solanas and Pares, 1984). Many synthetic organic compounds (Pettyjohn and Hounslow, 1983), and many complex hydrocarbon structures in petroleum (Atlas, 1978c), are not easily broken down by microbial action. However, with favorable conditions and the proper organisms, virtually all kinds of hydrocarbons--straight chain, branched chain, cyclic, simple aromatic, polynuclear aromatic, asphaltic--have been found to eventually undergo partial or complete oxidation (Texas Research Institute, Inc., 1982).

Each organic compound has unique characteristics that dictate which of the above mechanisms or combination of mechanisms controls its movement and degradability (Josephson, 1983b). Appendix A, Section 2.2, discusses in depth the role of these factors and processes in biodegradation.

2.2.1 Physical, Chemical, Biological, and Environmental Factors

Biodegradation can be affected by physical, chemical, biological, and environmental factors, such as solubility, advection, dispersion and diffusion, sorption, volatility, viscosity, density, chemical structure, environmental factors, toxicity, concentration, naturally occurring organic materials, hydrolysis and oxidation, and biological factors. These are discussed in depth in Appendix A, Section 2.2.1.

2.2.2 Rate of Biodegradation

It is important to have information on the rates of pollutant biodegradation under environmental conditions to be able to assess the potential fate of the compounds (Pfaender and Klump, 1981). The rate eralization of a number of organic compounds is directly proportional to their concentration over a wide range of concentrations (Alexander, 1985).

The parameters influencing the rate of biodegradation are of two types (Buckingham, 1981):

1. Those that determine the availability and concentration of the compound to be degraded or that affect the microbial population site and activity
2. Those that control the reaction rate.

The kinetics of biodegradation are zero-order, if the concentration is high relative to microbes that can degrade it, or first-order, if the concentration is not high enough to saturate the ability of the microbes (Kaufman, 1983). First-order kinetics apply where the concentration of the chemical being degraded is low relative to the biological activity in the soil; i.e., at any one time, the rate of chemical loss is proportional to its concentration in the soil (Hill, McGrahen, Baker, Finnerty, and Bingeman, 1955; Kearney, Nash, and Isensee, 1969). Michaelis-Menten kinetics seem to apply when the chemical concentration increases and the rate of decomposition changes from being proportional to being independent of concentration (Hamaker, 1966).

In most natural ecosystems, the numbers of hydrocarbon-utilizing microorganisms present will initially limit the rate of hydrocarbon degradation (Atlas, 1978c). But after a short period of exposure to petroleum pollutants, the numbers of hydrocarbon utilizers increase and will no longer be the principal rate-limiting factor.

In natural ecosystems, a variety of factors probably alter the shapes of substrate disappearance curves (Alexander, 1986). These factors may include predation by protozoa, the time for induction of the active organisms, the accumulation of toxins produced by other microorganisms, depletion of inorganic nutrients or growth factors, the presence of other substrates that may repress utilization of the compound of interest, and binding of the compound to colloidal matter. The impacts or interactions of such potentially important factors may make it difficult to predict the kinetics of mineralization or disappearance of a particular substrate.

The rates of degradation of long chain alkanes will depend upon the availability of the hydrocarbon to microorganisms (Atlas, 1978c). Availability will be greatly restricted by very low solubility and low surface area of contaminant to microbe for long chain alkanes, which are solid at normal environmental temperatures. Extensive branching also tends to reduce rates of hydrocarbon degradation.

Other factors that appear to have a major influence on rates of biodegradation include soil temperature and salinity (Palumbo, Pfaender, Paerl, Bland, Boyd, and Cooper, 1983).

There is considerable variation in the rate of biodegradation of specific contaminants that enter the subsurface environment (Wilson, Leach, Henson, and Jones, 1986). Rates can vary two to three orders of magnitude between aquifers or over a vertical separation of only a few feet in the same aquifer. Different biotransformation rates can be found for the same organic chemical tested by varying some parameter in the biotransformation feasibility testing (Leslie, Page, Reinert, Moses, Rodgers, Dickson, and Hinman, 1983). The following factors have to be considered when conducting biotransformation experiments: sediment concentration, sediment organic content, and amount of oxygen available to microorganisms in the environment.

Sections 2.2.2 and 2.2.3 in Appendix A further discuss the kinetics of biodegradation and provide the degradation rates for a number of organic chemicals.

2.2.3 Effect of Hydrocarbon Concentrations on Degradation

The concentration of a hydrocarbon can affect its biodegradability and toxicity to the degrading organisms (Environmental Protection Agency, 1985b). High concentrations of a hydrocarbon can be inhibitory to microorganisms, and the concentration at which inhibition occurs will vary with the compound (Alexander, 1985). Concentrations of hydrocarbons in the range of 1 to 100 $\mu g/ml$ of water or 1 to 100 $\mu g/g$ of soil or sediment (on a dry weight basis) are not generally considered to be toxic to common heterotrophic bacteria and fungi.

In some cases, relatively high concentrations of a pollutant (> 100 $\mu g/l$) can stimulate multiplication of the microbes that metabolize the organic contaminant (Wilson and McNabb, 1983). However, concentrations of less than 10 $\mu g/l$ of the material usually do not have this effect. It has been reported that if groundwater is taken from an anaerobic gasoline- or fuel oil-contaminated area with no significant degradation activity, and oxygen and nutrients are then added, rapid degradation of the hydrocarbons down to 1 $\mu g/l$ or less starts with little lag time (Jensen, Arvin, and Gundersen, 1985). This requires an initial hydrocarbon concentration below about 10 mg/l. At higher concentrations, a considerable adaptation time may be expected. If the concentrations are greater than 1 to 10 mg/l, metabolism of the compound can entirely deplete the oxygen or other metabolic requirements available in the groundwater (Wilson and McNabb, 1983).

Appendix A, Section 2.2.3, provides examples of concentrations of specific organic compounds that microorganisms have been able to tolerate and degrade. Both initial and final concentrations are given for target chemicals at contaminated sites treated with biodegradation. The effect of low concentrations of chemicals on biodegradation is also discussed in Appendix A.

Indigenous Microorganisms in Biodegradation

Microorganisms are the principal agents responsible for the recycling of carbon in nature. In many ecosystems there is already an adequate indigenous hydrocarbonoclastic microbial community capable of extensive oil biodegradation, provided that environmental conditions are favorable for oil-degrading metabolic activity (Atlas, 1977). This has been shown for many soil and marine and fresh-water environments (Atlas and Bartha, 1973c; Mironov, 1970; Cooney and Summers, 1976; Cooney, 1974; Mulkins-Phillips and Stewart, 1974a; Litchfield and Clark, 1973). It is suggested by some researchers (Atlas, 1977; McGill, 1977) that all soils, except those that are very acidic, contain the organisms capable of degrading oil products, that microbial seeding is not necessary, and that the problem is actually one of supplying the necessary nutrients at the site.

The ability to utilize hydrocarbons is widely distributed among diverse microbial populations (Atlas, 1981). Many species of bacteria, cyanobacteria, filamentous fungi, and yeasts coexist in natural ecosystems and may act independently or in combination to metabolize aromatic hydrocarbons (Cerniglia, 1984; Fedorak, Semple, and Westlake, 1984; Gibson, 1982). In general, population levels of hydrocarbon utilizers and their proportions within the microbial community appear to be a sensitive index of environmental exposure to hydrocarbons (Atlas, 1981). In unpolluted ecosystems, hydrocarbon utilizers generally constitute less than 0.1% of the microbial community; in oil-polluted ecosystems, they can constitute up to 100% of the viable microorganisms. This difference seems to reflect quantitatively the degree or extent of exposure of an ecosystem to hydrocarbon contaminants. Extensive degradation of petroleum pollutants generally is accomplished by mixed microbial populations, rather than single microbial species (Atlas, 1978c).

There are advantages to relying on indigenous microorganisms rather than adding microorganisms to degrade wastes (Environmental Protection Agency, 1985b). Through countless generations of evolution, natural populations have developed that are ideally suited for survival and proliferation in that environment. This is particularly true of uncontrolled hazardous waste sites where microorganisms have been exposed to the wastes for years or even decades.

3.1 MICROORGANISMS IN SOIL

Current evidence suggests that in aquatic and terrestrial environments, microorganisms are the chief agents of biodegradation of environmentally important molecules (Alexander, 1980). In 1946, ZoBell (Texas Reserach Institute, Inc., 1982) reported that nearly 100 species of bacteria, yeasts, and molds, representing 30 microbial genera, had been discovered to have hydrocarbon-oxidizing properties. Since that time, many other species and genera have been reported to have this ability (Texas Reserach Institute, Inc., 1982) and to be widely distributed in soils (Blakebrough, 1978; Atlas, 1981). Although many microorganisms appear limited to degradation of a specific group of chemicals, others have demonstrated a wide diversification of substrates that they are capable of metabolizing.

The metabolic diversity of microorganisms in the soil is an important factor in the biodegradation of xenobiotic or hazardous materials (JRB Associates, Inc., 1984b). Bacteria may be classified metabolically as heterotrophic, if they derive their energy and carbon for survival and growth from the decomposition of organic materials, or as autotrophic, if they fix the carbon they need for growth from carbon dioxide and obtain (usually) their energy from light (photosynthetic or phototrophic) or the oxidation of inorganic compounds (chemosynthetic or lithotrophic). Lithotrophic organisms may transform inorganic materials (especially nitrogen) into available nutrients for the heterotrophs (Focht and Verstraete, 1977; Painter, 1977). Lithotrophic or photosynthetic bacteria may also directly or indirectly transform toxic metals or metalloids.

In soil taken from the surface to 100 ft, facultative oligotrophic bacteria were dominant; these organisms can survive on low levels of nutrients, if necessary (Benoit, Novak, Goldsmith, and Chadduck, 1985). Obligate oligotrophic bacteria, which require low concentrations of nutrients, are the most frequent isolates from deep soil profiles (> 100 ft), and yeasts have been present in some samples. Obligate anaerobic bacteria and microaerophilic bacteria have been recovered from deep soil profiles; however, they are a small portion of the population.

The heterotrophic bacteria are the most important organisms in the transformation of organic hazardous compounds, and soil treatment schemes may be directed toward enhancing their activity (JRB Associates, Inc., 1984b). Heterotrophs can use the organic contaminants as sources of both carbon and energy (Knox, Canter, Kincannon, Stover, and Ward, 1968). Some organic material is oxidized for energy while the rest is used as building blocks for cellular synthesis. There are three methods by which heterotrophic microorganisms can obtain energy: fermentation, aerobic respiration, and anaerobic respiration. These processes are discussed in Section 4 and Appendix A.

Some compounds appear to be degraded only under aerobic conditions, others only under anaerobic conditions, and some under either condition, while others are not transformed at all. It has been concluded that hydrocarbons are subject to both aerobic and anaerobic oxidation (Dietz, 1980).

Stratification of microbes results from natural selection in response to chemical and physical gradients in the soil column (Ahlert and Kosson, 1983). Oxygen concentration is a principal gradient. It is greatest at the soil surface, where diffusion from the ambient air drives the gradient. The concentration of oxygen diminishes with depth and is depleted by aerobic respiration. This produces an aerobic community near the soil surface and anaerobic communities that dominate at greater depths. Other gradients are caused by available nutrients, temperature, and available organic carbon. At all levels in the soil, substrate is consumed and metabolites produced, which may then serve as substrate for other organisms at other locations in the column. As the environmental effects and constituents of waste vary, natural selection and adaptation occur. The biological community is continuously responding to all these dynamic variables.

Table A.3-1 lists a number of organisms and the compounds they are able to degrade.

Table A.3-2 shows the general groups of microorganisms that might be useful for treating contamination by specific types of organic compounds and the conditions that would be most favorable for their development (Kobayashi and Rittmann, 1982).

3.1.1 Bacteria

Numerous bacterial strains and species have been isolated that can degrade many of the petroleum-derived compounds of the greatest hazard to the environment and to health (Nicholas, 1987). The compounds that can be attacked include most of those of major environmental concern, including benzene, toluene, biphenyls, and naphthalene.

Bacteria are predominantly involved with degradation of those chemicals that have a higher degree of water solubility and are not strongly adsorbed. The binary fission-type reproductive methods of bacteria enable them to compete more successfully than fungi for readily available substrates. See Appendix A for a description of the small bacterial cells that are found in the subsurface.

3.1.1.1 Aerobes

Aerobic bacteria ultimately decompose most organic compounds into carbon dioxide, water, and mineral matter, such as sulfate, nitrate, and other inorganic compounds (Pettyjohn and Hounslow, 1983), and do not produce hydrogen sulfide or methane as reaction products (Amdurer, Fellman, and Abdelhamid, 1985).

The most commonly isolated organisms in areas of hydrocarbon contamination are heterotrophic bacteria of the genera Pseudomonas, Achromobacter, Arthrobacter, Micrococcus, Vibrio, Acinetobacter, Brevibacterium, Corynebacterium, Flavobacterium (Kobayashi and Rittmann, 1982), Mycobacterium (Gholson, Guire, and Friede, 1972; Soli, 1973), and Nocardia (Canter and Knox, 1984). Pseudomonas species appear to be the most ubiquitous and the most adaptable to the different pollutants, while Corynebacterium species may be major agents for decomposing heterocyclic compounds and hydrocarbons in contaminated aquatic environments (Kobayashi and Rittmann, 1982). Appendix A discusses hydrocarbon degradation by some of the more frequently encountered aerobic bacteria.

3.1.1.2 Anaerobes

Anaerobic decomposition of organic matter to carbon dioxide and methane involves interactions within consortia of obligate anaerobic bacteria (Kobayashi and Rittmann, 1982). At least four interacting trophic groups of bacteria are involved:

1. Hydrolytic bacteria that catabolize the major components of biomass, such as saccharides, proteins, and lipids
2. Hydrogen-producing, acetogenic bacteria that catabolize products from the activity of the first group, such as fatty acids and neutral end products
3. Homoacetogenic bacteria that catabolize multicarbon compounds to acetic acid
4. Methanogenic bacteria (Zeikus, 1980)

Obligate anaerobes require not only anoxic (oxygen-free) conditions, but also oxidation-reduction potentials of less than -0.2 V (Kobayashi and Rittmann, 1982). While many soil bacteria can grow under anaerobic conditions, most fungi and actinomycetes cannot grow at all (Parr, Sikora, and Burge, 1983). Anaerobic decomposition is performed mainly by bacteria utilizing either an anaerobic respiration or interactive fermentation/methanogenic type of metabolism. These processes are described in Appendix A, Section 4.1.2. The end products of anaerobic degradation are reduced compounds, some of which are toxic to microorganisms and plants. Table A.4-7 shows the end products formed from degradation of specific organic hydrocarbons by different microorganisms.

3.1.1.3 Oligotrophs

Organisms living at organic concentrations < 15 mg carbon/l are termed oligotrophs (Stetzenbach, Sinclair, and Kelley, 1983). They may be able to live under conditions of even lower carbon flux (< 1 mg/l/day) (Poindexter, 1981). They do not constitute a special taxonomic grouping of organisms, but come from almost any group of bacteria or chemotrophs. They are generally adapted to life under low nutrient conditions, but can readily be readapted to high nutrient conditions. Reverse adaptation to the low nutrient environment is, however, not readily achieved; therefore, oligotrophs are obtained only from low-nutrient environments.

Oligotrophs generally have a high surface/volume ratio and high affinity for substrate (Kobayashi and Rittmann, 1982). The minimum substrate concentration needed for measurable growth is lower than that required for eutrophic (high nutrient) organisms, but the maximum growth rate is also lower. Oligotrophs degrade xenobiotics more slowly than natural compounds (Alexander, 1985); however, their capability of surviving on low concentrations makes them potentially useful for removal of trace concentrations of organic contaminants from water, or effluent from wastewater treatment processes (Poindexter, 1981). Species of <u>Pseudomonas</u>, <u>Flavobacterium</u>, <u>Acinetobacter</u>, <u>Aeromonas</u>, <u>Moraxella</u>, <u>Alcaligenes</u>, and <u>Actinomyces</u> have been detected in water samples, with isolates surviving extended periods on low nutrient concentrations (Stetzenbach, Sinclair, and Kelley, 1983). These organisms are usually found living as biofilms (Poindexter, 1981). An important characteristic is that they often appear to have multiple inducible enzymes, are able to shift metabolic pathways, and can take up and use mixed substrates (e.g., a <u>Clostridium</u> sp.).

Some Actinomycetes (<u>Nocardia</u>) coryneforms, and mycobacteria have survived for 30 days under starvation conditions. Oligotrophic bacteria from surface water and those indigenous to the deeper subsurface fail to grow on complex media (Wilson, McNabb, Wilson, and Noonan, 1983). Apparently, many oligotrophic bacteria from surface water fail to use organic compounds that are used readily by eutrophic forms. There is an indication that the bacteria of the deeper subsurface will be active against a more limited range of organic compounds than are degraded in surface soil.

Microorganisms appear to have a threshold level below which some organic compounds cannot be converted to carbon dioxide (Alexander, 1985). This level is needed for growth, enzyme induction, and enzyme activity. At lower concentrations, substrate uptake by diffusion of the molecules will meet maintenance energy and survival requirements but will not support growth (Schmidt and Alexander, 1985).

It is not certain whether the trace amounts present in a contaminated site would be sufficient to allow microbial growth or propagation to numbers necessary for biodegradation of the material (Alexander, 1985). If not, this could explain the persistence of low levels of biodegradable organic substances, e.g., toluene, xylenes, naphthalene, and phthalate esters, in water or soil in nature (although this might also be due to oxygen limitation). It should be noted that if these organisms obtain energy and carbon for growth by using natural organic constituents of the environment, the threshold for a particular chemical contaminant may be below the level of detection possible using current analytical procedures.

Some organic compounds are mineralized even at trace levels below 1.0 pg/ml (Alexander, 1985). It is important to distinguish between compounds that can be transformed at low concentrations by large populations of nongrowing cells and substrates that must support growth for significant degradation to occur.

3.1.1.4 Counts in Uncontaminated Soil

Microorganisms are widely distributed in nature, but reports of the actual numbers present is confusing because of the methodological differences used to enumerate the microbes (Atlas, 1981). No place has been found in the U.S. or Canada--at depths to 400 ft--where sufficient organisms are not present to be brought up in 72 hr to a significant population (Rich, Bluestone, and Cannon, 1986). The bacteria are present; the problem is establishing the right conditions for their growth.

Early studies indicated that the number and activity of microbes decline dramatically with increasing depth (see Table A.3-3) (Markovetz, 1971). Recent reports indicate that active microbial populations can be found in deep, unsaturated and saturated subsoils (Pirnik, 1977). These counts show variation with depth when viable but not direct counts are used (Table A.3-4). Aseptic core samples from depths of 300 m (nearly 1000 ft) indicate an abundant and diverse population of microorganisms extending to well below surface soils and shallow aquifers (Fredrickson and Hicks, 1987).

It appears that different soil types vary in the distribution of biomass and enzymatic activity through their vertical profile (Federle, Dobbins, Thornton-Manning, and Jones, 1986). Biomass and activity are significantly correlated with each other and negatively correlated with depth. While biomass and activity decrease with increasing soil depth, the magnitude of decline differs for different soils. It is difficult to generalize on the level of biomass or activity to expect in a soil based on depth or horizon alone. Soil type is also important in determining the types of microbial populations present. Depth may be responsible for as much as 75% of the variation in biomass, but an additional 11% of the variation could be explained by pH and silt, clay, and organic contents. Depth also explained 78% of the variation in microbial activity; silt content explained another 4.5%.

Enumeration of microorganisms can be difficult, since most subsurface bacteria exist in an ecosystem low in organic carbon and do not grow well, if at all, in conventional growth media with high organic carbon concentrations (Wilson, Leach, Henson, and Jones, 1986). In addition, many organisms attach firmly to particles (Federle, Dobbins, Thornton-Manning, and Jones, 1986). Dilution plating techniques yield 1 to 10% of the number of cells determined by microscopic direct counting. An alternate approach is to determine biomass by analyzing the phospholipids extractable from soil. This method has been used to estimate microbial biomass in estuarine and marine environments (Gillan, 1983; White, 1983) and in subsurface soils.

It is important to be able to distinguish between viable and nonviable cells. However, it is believed that many organisms in the subsurface will be in a dormant state until stimulated by an appropriate concentration of a suitable substrate (Alexander, 1977). The deeper the soil, the more oligotrophic the organisms will become and, hence, the more fastidious their requirement for low nutrient concentrations. The proportion of hydrocarbon-degrading organisms to total heterotrophs is now considered to be a more significant indicator of the biological activity in the subsurface, with respect to hydrocarbon contaminants.

Direct counts tend to be fairly consistent with soil depth and sampling site and were found to be on the order of 10^6 to 10^7 bacteria/g dry wt of soil (see Appendix A). Plate counts, on the other hand, suggested highly variable and decreasing counts with depth, ranging from zero to 10^8.

3.1.1.5 Counts in Contaminated Soil

Microbial numbers and activity are initially depressed by even light hydrocarbon contamination (Odu, 1972). However, this is followed by a stimulation of activity. The number of hydrocarbon-utilizing organisms in a soil reflects the soil's past exposure to hydrocarbons (Atlas, 1981). These organisms are most abundant in places that have been chronically exposed to hydrocarbon pollution (Texas Research Institute, Inc., 1982). Few or none are found in unpolluted groundwater or petroleum directly from wells. Substantial adapted populations exist in contaminated zones, with the bacterial biomass increasing as the organic contaminants are metabolized (Environmental Protection Agency, 1985b).

Bacterial counts were 100 to 1,000 times higher inside than outside a zone of contamination of an aquifer containing JP-5 jet fuel (Ehrlich, Schroeder, and Martin, 1985). It appears that only a few species of specialized bacteria, presumably those able to assimilate the hydrocarbons, are preferentially selected for in a contaminated zone.

Appendix A discusses counts found in contaminated soil. Direct counts range from 10^3 to 10^8 organisms/g, while viable counts have been recorded from less than 100 to 10^5 CFU/g. Hydrocarbon degraders have been measured at naturally occurring levels of 10^2 to 10^5 organisms/g. These numbers increase after biostimulation of the contaminated sites (see Section 5.1.1.5).

3.1.2 Fungi

Fungi are eukaryotic microorganisms that lack photosynthetic structures and depend upon heterotrophic metabolism (Solanas and Pares, 1984). They may be filamentous or unicellular (yeast-like or amoeboid), or aggregate to form large structures, such as a plasmodium or fruit-bodies (mushrooms). Fungi are a large part of the microbial biomass in soil, especially in acidic conditions, and they contribute to most decomposition processes. When substrate or water availability is low, most of the fungal biomass is either dormant or dead. Fungal spores or other resistant structures can survive under adverse conditions for long periods of time and then quickly germinate and grow when environmental conditions become favorable. The species and diversity of fungi are affected by clay mineralogy, temperature, and other soil environmental conditions. Most filamentous fungi are aerobic, and yeasts are often facultatively anaerobic. The species of fungi that develop on plates are frequently those that produce spores in greatest abundance (Nannipieri, 1984). Additional colonies can, thus, grow from inactive spores or conidia and give an inaccurate estimate of population size.

Most soil fungi are mesophiles with temperature optima between 25° and 35°C, but with an ability to grow from about 15° to 45°C (Cooke and Rayner, 1984). Even thermophilic fungi do not grow above about 65°C (JRB Associates, Inc., 1984b). However, many fungi grow at temperatures below 10°C. The influence of soil acidity on fungal growth is difficult to assess because of the ability of fungi to radically alter the pH of their environment (Cooke and Rayner, 1984). This capability indicates that acidifying soil will not improve growth of the organisms, but does reduce competition from other organisms (Kirk, Schultz, Connors, Lorenz, and Zeikus, 1978).

Fungi play an important role in the hydrocarbon-oxidizing activities of the soil (Jones and Eddington, 1968). They seem to be at least as versatile as bacteria in metabolizing aromatics (Fewson, 1981). Their extracellular enzymes may help to provide substrates for bacteria, as well as for themselves, by hydrolyzing polymers. They are also important sources of secondary metabolites.

Some filamentous fungi, unlike other microorganisms that attack aromatic hydrocarbons, use hydroxylations as a prelude to detoxification rather than catabolism and assimilation (Dagley, 1981). These organisms do not degrade aromatic hydrocarbons as nutrients, but simply detoxify them. Several fungi (Penicillium and Cunninghamella) even exhibit greater hydrocarbon biodegradation than bacteria (Flavobacterium, Brevibacterium, and Arthrobacter). The ability to utilize hydrocarbons occurs mainly in two orders, the Mucorales and the Moniliales (Nyns, Auquiere, and Wiaux, 1968). Aspergillus and Penicillium are rich in hydrocarbon-assimilating strains. It has been concluded that the property of assimilating hydrocarbons is a property of individual strains and not necessarily a characteristic of particular species or related taxa.

The genera most frequently isolated from soils are those producing abundant small conidia; e.g., Penicillium and Verticillium spp. (Davies and Westlake, 1979). Oil-degrading strains of Beauveria bassiana, Mortieriella spp., Phoma spp., Scolecobasidium obovatum, and Tolypocladium inflatum have also been isolated. Fifty-six out of 500 yeasts studied were found to be able to degrade hydrocarbons; almost all of these were in the genus Candida (Komagata, Nakase, and Katsu, 1964). Hydrocarbonoclastic strains of Candida, Rhodosporidium, Rhodotorula, Saccharomyces, Sporobolomyces, and Trichosporon have been identified from soil (Ahearn, Meyers, and Standard, 1971; Cook, Massey, and Ahearn, 1973). Cladosporium resinae has been found in soil (Walker, Cofone, and Cooney, 1973) and has repeatedly been recovered as a contaminant of jet fuels (Atlas, 1981). This organism can grow on petroleum hydrocarbons and creates problems in the aircraft industry by clogging fuel lines. Specific fungi and some of the hydrocarbons they are able to degrade are discussed in Appendix A.

Bacteria and yeasts show decreasing ability to degrade alkanes with increasing chain length (Walker, Austin, and Colwell, 1975). Filamentous fungi do not exhibit preferential degradation for particular chain lengths and appear to be better able to degrade or transform hydrocarbons of complex structure or long chain length. Because they have nonspecific enzyme systems for aromatic structures, fungi (yeasts and filamentous) are believed to be capable of biodegrading PCBs better than bacteria can (Gibson, 1978). However, fungal metabolism often results in incomplete degradation that necessitates bacterial association for complete mineralization. Whereas bacteria oxidize aromatic hydrocarbons to cis-dihydrodiols, fungi convert them to trans-diols, with arene oxides (epoxides) as intermediates (Dagley, 1981). This suggests that fungi metabolize aromatic hydrocarbons in a manner similar to mammalian systems, i.e., via a monooxygenase-catalyzed reaction (Cerniglia, Hebert, Szaniszlo, and Gibson, 1978). It is probable that a cytochrome P-450 dependent reaction may be responsible for the initial oxygenation of naphthalene by these organisms. The products of fungal metabolism are often recognized carcinogens, a point that supports combining the fungi with bacteria for complete degradation.

Fungi appear to be predominantly involved in metabolizing those xenobiotics of lower water solubility and greater adsorptivity (Kaufman, 1983). The mycelial-type growth characteristics of fungi perhaps enable them to encapsulate and penetrate soil particles to which xenobiotics may be adsorbed. Soil fungi are generally believed to play a more important role in the formation, metabolism, and interactions of soil organic matter complexes than bacteria. This may allow them to better cope with the various bonding mechanisms involved with adsorbed materials.

Table A.3-2 shows the most favorable conditions for growth of hydrocarbonoclastic fungi for removal of contaminants.

3.1.3 Photosynthetic Microorganisms

The surface soil usually supports large populations of eukaryotic algae and cyanobacteria (blue-green algae) (JRB Associates, Inc., 1984b). Since light cannot penetrate far into the soil, the algal biomass is usually low. These organisms may enhance photodecomposition of hazardous organic compounds at the soil surface.

The photosynthetic microorganisms of interest are the algae, cyanobacteria (blue-green algae), and photosynthetic bacteria (Kobayashi and Rittmann, 1982). These organisms are potentiallly important in situations involving low concentration of nutrients, because they are able to obtain energy from sunlight and carbon by carbon dioxide fixation. Some of the cyanobacteria and photosynthetic bacteria are also able to fix nitrogen; hence, they can survive in situations in which the dissolved nitrogen concentration is inadequate to support bacterial growth.

The ability to oxidize aromatic hydrocarbons is widely distributed among the cyanobacteria and algae. Cyanobacteria, Chlorella (green algae), and especially the Chromatiaceae (photosynthetic bacteria) are pollution tolerant (Kobayashi and Rittmann, 1982; Pfennig, 1978a). Of the latter, the purple sulfur bacteria (Thiorhodaceae) are important. The purple nonsulfur (Rhodospirilliaceae) bacteria are also of interest. For these bacteria, organic compounds serve as the major source of electrons and carbon for cellular components (Berry, Francis, and Bollag, 1987). The photosynthetic bacteria are known to be able to metabolize a wide variety of substances (e.g., simple sugars, alcohols, volatile fatty acids, tricarboxylic acid cycle intermediates, aromatic compounds, benzoates (Dietz, 1980; Stanier, Doudoroff, and Adelberg, 1970)) and to have a wide range of inducible enzymes (Laskin and Lechevalier, 1974).

Rhodopseudomonas capsulata is known to transform nitrosamines (carcinogens) to innocuous compounds. Some of the purple nonsulfur organisms can grow microaerobically and anaerobically as phototrophs, yet live as heterotrophs aerobically in the dark, by respiratory metabolism of organic compounds (Pfennig, 1978a; Stanier, Doudoroff, and Adelberg, 1970). Cyanobacteria and Chlorella are tolerant of low concentrations of dissolved oxygen (Kobayashi and Rittmann, 1982). Dunaliella can tolerate a wide salinity range. See Table A.3-2 for a summary of the most favorable conditions for these organisms for removal of anthropogenic compounds (Kobayashi and Rittmann, 1982).

In general, phototrophs do not promote complete degradation, but only transformation. Table A.3-1 lists some of the compounds that can be degraded by certain phototrophs. The contribution of these microbes to complete degradation requires interactions with other organisms. However, the metabolic products they form stimulate growth of heterotrophic organisms. The proper balance between algae and bacteria can result in extensive biodegradation of anthropogenic compounds.

Phototrophic organisms can not only break down organic compounds, but they can also bioaccumulate hydrophobic compounds (Table A.3-7) (Kobayashi and Rittmann, 1982). This offers the greatest potential for exploitation in treatment processes because such organisms can be self-sustaining without the presence of organic matter in concentrations large enough to serve as carbon and electron donors. Some cyanobacteria and photosynthetic bacteria can also fix nitrogen and, thus, survive in a concentration of dissolved nitrogen that is insufficient to support bacterial growth. On the other hand, when xenobiotics accumulate in the cells of microorganisms they can become concentrated and eventually enter into the food chain (de Klerk and van der Linden, 1974).

Biodegradation of organic compounds by specific phototrophs is discussed in Appendix A, Section 3.1.3.

3.1.4 Higher Life Forms and Predation

The role of the soil macrofauna, such as insects, protozoa, earthworms, and slugs, in the decomposition of organic materials is significant, but predominantly indirect (Parr, Sikora, and Burge, 1983). It is minor compared with microorganisms, but it is still essential. Of the total respiration associated with soils amended with organic material, 10 to 20% could be from macrofauna. Because only a few of these organisms have the ability to produce their own enzymes for the degradation of substrate, their main degradation feature is mechanical.

The gut of most soil animals contains microorganisms, which produce the necessary enzymes for the degradation of a substrate to the point where the animals can absorb the nutrients. The remainder of the substrate passes into the soil where microorganisms complete the degradation.

Earthworms play a prominent role in the degradation of organic materials in soil. With their movement, the soil is aerated and nutrients are carried to deeper soil profiles where these stimulate microbial growth and decomposition. Among the arthropods, the beetles and termites are most correlated with extensive degradation of organic material. Both animals often have a rich microflora in their guts, and these microorganisms produce the enzymes that degrade cellulosic substrates.

Microbial predators also play a role in the degradation process (Texas Research Institute, Inc., 1982). These organisms graze on bacteria and fungi or feed on detrital matter and associated microflora (JRB Associates, Inc., 1984b; Sinclair and Ghiorse, 1985). Protozoa, nematodes, insects, and other worms affect the decomposition process by controlling bacterial or fungal population size through grazing (Bryant, Woods, Coleman, Fairbanks, McClellan, and Cole, 1982), by harboring in their intestinal tract organisms that might decompose a compound of interest, by comminuting plant materials, or by mixing the soil and contributing to its aeration and homogeneity (JRB and Associates, Inc., 1984b). A cyst-forming amoeba was present at 1111/g dry weight and constituted 15% of the total biovolume of sediments in a groundwater interface zone (Sinclair and Ghiorse, 1985). Many species of hydrocarbon utilizers have been found to be ingested by a large number of ciliate and other cytophagic protozoans (Texas Research Institute, Inc. 1982). These higher organisms may reduce the microbial population from 10^7 to 10^2 bacteria/ml (ZoBell, 1973). Protozoan grazing has been shown to be responsible for most of the acclimation period for the mineralization of organic compounds in some sewage (Wiggins and Alexander, 1986).

3.1.5 Cometabolism

Xenobiotic organic compounds are usually transformed or degraded by microorganisms either in a metabolic sequence that provides energy and nutrients (e.g., carbon (C), nitrogen (N), phosphorus (P)) for growth or maintenance of the organism, or by a biochemically mediated reaction that provides neither energy nor nutrients to the cell (JRB and Associates,

Inc., 1984b). The first process usually results in the complete biodegradation of the organic molecules to mineral products (e.g., carbon dioxide, methane, water, ammonia, phosphate) and is called mineralization. The second process usually results in only a minor transformation of the organic molecule and is called cometabolism or cooxidation (Pierce, 1982a; Alexander, 1977; Alexander, 1973). This transformation product is still unusable to the organisms (Hornick, Fisher, and Paolini, 1983).

Two or more substrates are required for cometabolism; one is the nongrowth substrate that is neither essential for, nor sufficient to, support replication of the microorganism (Perry, 1979; Hulbert and Krawiec, 1977), while the other compound(s) does (do) support growth. The nongrowth substrate is only incidentally and incompletely transformed by the microorganism involved, although other microorganisms can often utilize by-products of the cometabolic process (Perry, 1979; de Klerk and van der Linden, 1974).

Cometabolism is the result of a nonspecific enzyme with a broad substrate specificity attacking a recalcitrant molecule and metabolizing it (de Klerk and van der Linden, 1974; Horvath and Alexander, 1970). The organism supplying the enzyme gains nothing from the metabolic transformation. The enzyme is made by the microorganism to metabolize some other organic compound for its energy (de Klerk and van der Linden, 1974; McKenna, 1977). Oxygenases are often involved in cometabolism because they can be induced by, and can attack, a large set of substrates.

Cometabolism probably occurs frequently in natural soil systems, since many genera of bacteria, fungi, and actinomycetes can participate in the process (Alexander, 1977; Horvath and Alexander, 1970). Cometabolism may be important in the biodegradation of complex organics in hazardous waste-contaminated soils. The mixed chemical environment of petroleum permits a variety of microorganisms to enzymatically attack compounds that would not otherwise be degraded, as they grow on the multitude of potential primary substrates in the oil (Atlas, 1981). Many complex branched and cyclic hydrocarbons are removed as environmental contaminants after oil spills, as a result of cooxidation. Structures with four or more condensed rings have been shown to be attacked by cooxidation or as a result of commensalism, where two or more microorganisms are involved in the degradation and one benefits from it.

Since cometabolism rarely results in the complete oxidation of xenobiotics, it may allow accumulation of transformation products, which may be either more or less toxic than the original substances (Perry, 1979; de Klerk and van der Linden, 1974). Incompletely degraded compounds include saturated hydrocarbons, halogenated hydrocarbons, many pesticides, and single-ringed polycyclic aromatic hydrocarbons (Alexander, 1977; Horvath and Alexander, 1970; Horvath, 1972). Cooxidation may be encouraged by adding a more easily degraded compound that is a chemical analog to the hazardous compound that must be decomposed. See Appendix A, Section 5.1.1.2, for examples of compounds that can be used for analog enrichment and organisms that can be added for cooxidation of specific hydrocarbons.

The metabolism of nongrowth substrates in the absence of a true growth substrate is also referred to as stationary metabolism (Foster, 1962). Aside from traditional fermentation and growth of microorganisms on hydrocarbons, cooxidation and stationary transformation techniques and genetic engineering are the most promising areas of hydrocarbon biotechnology for the future (Hou, 1982). These systems are operating in nature.

3.1.6 Microbial Interactions

Microbial interactions in the soil are very complex and undoubtedly play an important role in the transformation or decomposition of hazardous waste components (JRB and Associates, Inc., 1984b). The growth requirements among the organisms lead to intense competition for the available nutrients. As a result of metabolic specialization (e.g., autotrophic nitrification), some microbes are less dependent upon preformed organic substrates or growth factors. The majority of the organisms, however, have acquired antagonistic abilities that help limit growth of competitors. Examples of these antagonistic agents are antibiotics, acids, bases, and other organic and inorganic compounds (Alexander, 1971; Atlas and Bartha, 1981).

True mutualistic or symbiotic relationships also exist among soil organisms (JRB and Associates, Inc., 1984b). It is common for degradation of a xenobiotic compound to involve sequential metabolism by two or more microorganisms (Beam and Perry, 1974) in a relationship that may benefit only one partner (commensalism) or both (protocooperation or synergism) (Atlas and Bartha, 1981). In such a commensalistic relationship, microbes cannot oxidize a given hydrocarbon individually, but collectively they are able to do so.

This form of commensalism may be very widespread in nature with natural mixed populations employing each other's metabolic intermediates as growth substrates (Donoghue, Griffin, Norris, and Trudgill, 1976). Many genera of bacteria, fungi, and actinomycetes can participate in the process (Alexander, 1977; Horvath and Alexander, 1970). Appendix A, Section 5.1.1.1.1, provides examples of organisms involved in this form of commensalism.

More than 100 strains of bacteria were unable to use unsubstituted cycloparaffinic hydrocarbons as their sole source of carbon and energy; however, many could partially oxidize the hydrocarbons when a suitable energy source was present (e.g., n-alkane) (Beam and Perry, 1973; Beam and Perry, 1974). The resulting cycloalkanones were readily oxidized and used as an energy source by other strains of bacteria. Unsubstituted cycloparaffinic hydrocarbons are readily mineralized in natural soil systems, presumably by a process that includes cometabolism.

3.2 MICROORGANISMS IN GROUNDWATER

Surface soils contain many microorganisms, but their numbers decrease rapidly below the root zone (Bitton and Gerba, 1985). Indigenous microbiological activity may extend to a depth of about 2000 m (assuming a normal temperature increase of 3°C/100 m). Water pressures to this depth would allow their survival, and many bacteria can withstand the high osmotic pressures of saline groundwater. Carbonates and other inorganic carbon compounds present in most underground materials can serve as alternate carbon sources to organic compounds. Other essential elements, such as phosphorus (P), sulfur (S), and trace elements are also generally present in underground materials.

Members of the deep subsurface microbial community range from aerobic heterotrophic bacteria capable of mineralizing a variety of organic substrates (including glucose, acetate, indole, phenol, 4-methoxybenzoic acid, styrene, toluene, chlorobenzene, and bromodichloromethane) (Fredrickson and Hicks, 1987; Ward, Tomson, Bedient, and Lee, 1986) to denitrifying, sulfate-reducing, and methanogenic anaerobic bacteria (Fredrickson and Hicks, 1987). The presence of carboxylated metabolites in groundwater suggests, however, that biotransformation of jet fuel hydrocarbons stops short of complete mineralization to inorganic carbon, presumably reflecting oxygen limitation (Ehrlich, Schroeder, and Martin, 1985). Fungi, protozoa, sporeforming bacteria, and autotrophic bacteria have also been found. Contrary to the perception that deep sediments are highly reduced and anaerobic, microbiological activity and pore water chemistry actually indicate a relatively oxidized environment.   Populations of aerobic heterotrophs are dominant.

Table A.3-8 presents some of the growth requirements for microorganisms in the groundwater environment.

3.2.1 Bacteria

Bacteria have been found at significant levels in groundwater samples and have been the predominant species (Larson and Ventullo, 1983). The origin of bacteria in aquifers is unknown (Bitton and Gerba, 1985). They may have been deposited with the sediments millions of years ago, migrated recently into the formations with surface water or through soil fissures, or been introduced during construction of wells.   Many microbes seem to move passively through soil pores larger than their cell size.

Since groundwater from deep, protected aquifers is considered oligotrophic, the bacteria in this region must be capable of metabolism in low nutrient concentrations (Stetzenbach, Kelley, and Sinclair, 1984). Water collected from wells 200 ft deep contained predominantly gram-negative rods, with a limited number of genera.   These were mostly Acinetobacter.   Both gram-negative and gram-positive bacteria were recovered from depths of 1.2, 3.0, and 5.0 m, where the water table was at 3.6 m and the bedrock at 6.0 m (Wilson, McNabb, Balkwill, and Ghiorse, 1983).   Indigenous bacterial populations present in groundwater have the potential to metabolize both natural and xenobiotic substances, depending upon the suitability of the substrate as a carbon and energy source (Larson and Ventullo, 1983).   For degradable substrates, rates of biodegradation are fairly rapid, even in oligotrophic groundwater.

3.2.1.1 Aerobes

Of 500 isolates recovered from groundwater from deep, protected aquifers, more than 80% were gram-negative, nonmotile, rod-shaped bacteria (Stetzenbach, Sinclair, and Kelley, 1983).   Some 59% of the isolates were Acinetobacter spp. Species of Moraxella, Flavobacterium, and an unidentified oxidase-negative, pigmented bacterium were also present.   Corynebacterium species may be major agents for decomposing heterocyclic compounds and hydrocarbons in contaminated aquatic environments (Kobayashi and Rittmann, 1982).

Oxygen levels in groundwater samples contaminated with gasoline and fuel oil ranged from trace amounts to 4 ppm; however, cultures grown under microaerophilic conditions resembled aerobic forms (Litchfield and Clark, 1972). The numbers of hydrocarbon-utilizing bacteria correlated positively with the levels of contamination, suggesting these organisms were growing in the groundwater. Major isolates from a variety of contaminated groundwaters were species of Pseudomonas, Arthrobacter, and Nocardia, and occasionally Achromobacter and Flavobacterium.

3.2.1.2 Anaerobes

As much as 2 to 8 mg/l have been found in water from deep aquifers in Arizona and Nevada (Winograd and Robertson, 1982). Sometimes deep aquifers (several hundred feet) contain high levels of oxygen because of the absence of organics. Except in very deep, isolated formations, oxidation-reduction potentials tend to be within the tolerance range of aerobic bacteria (Bitton and Gerba, 1985). When molecular oxygen is absent in deeper regions, anaerobic bacteria can be found. These include sulfate reducers, denitrifiers, methanogens, and sulfur and hydrocarbon oxidizers (McNabb and Dunlap, 1975). These bacteria use sulfate, carbon dioxide, nitrate, and simple organic compounds rather than oxygen as electron acceptors in their metabolism.

If oxygen, nutrients, or carbon source of energy is depleted in a contaminated aquifer, anaerobic bacteria can become active in biotransformations (Wilson, Leach, Henson, and Jones, 1986). Anaerobic bacteria are generally unable to utilize most of the compounds in crude oil and its refined derivatives as sources of carbon and nutrients (Ehrlich, Schroeder, and Martin, 1985). However, some biotransformation of hydrocarbons appears to be occurring (e.g., conversion of alkanes to fatty acids), even though there is not a complete oxidation of most hydrocarbons to carbon dioxide and water (Widdle and Nobert, 1981).

Methane-producing bacteria can assimilate only a limited range of compounds; they are capable of autotrophic growth with hydrogen and carbon dioxide, and some species can use formate, acetate, or methanol as growth substrates (Ehrlich, Schroeder, and Martin, 1985). Sulfate-reducing bacteria can utilize certain short-chain fatty acids, including acetate (Widdle and Nobert, 1981). Molecular hydrogen can also be used to reduce inorganic carbonate to methane (Wilson, Leach, Henson, and Jones, 1986). The presence of fatty acids and benzoate in water from some wells suggests that an incomplete biotransformation of hydrocarbons is occurring in the shallow zone. This may reflect an oxygen limitation (Widdle and Nobert, 1981).

Methane-producing bacteria and dissolved methane were detected in water samples from two wells from a JP-5-contaminated aquifer, in spite of exposure to oxygen during well recovery following bailing (Ehrlich, Schroeder, and Martin, 1985). Methane-producing bacteria may be less sensitive to oxygen toxicity than is commonly supposed.

It has recently been found that compounds, such as benzene, toluene, the xylenes, and other alkylbenzenes, do not require molecular oxygen to be metabolized (Wilson, Leach, Henson, and Jones, 1986). They could be at least partially metabolized in methanogenic river alluvium to carbon dioxide. Phenolics can also be degraded anaerobically (Rees and King, 1980); however, aromatic compounds are not significantly utilized under anaerobic conditions (Bouwer and McCarty, 1983b). Although strict anaerobes that produce methane can use a very limited range of organic compounds, they can act in combination with other microorganisms to break down more complex organic compounds (Wilson, Leach, Henson, and Jones, 1986). These partnerships can totally degrade a surprising variety of natural and synthetic organic compounds.

The different types of anaerobic degradation are explained in Appendix A, Section 4.1.2.

3.2.1.3 Counts in Uncontaminated Groundwater

Special staining procedures can distinguish cellular material from noncellular subsurface soil particles of the same size and shape (Ghiorse and Balkwill, 1983). This technique was used to count microbes present in core material from shallow water table aquifers and vadose zones. Bacteria, but no yeasts or other fungi, protozoa, or higher animals were found. Tables A.3-9 and A.3-10 show the similarity of counts recovered from different depths.

Appreciable numbers of bacteria (up to 10^8 cells/g of dry soil) have been found to exist in shallow, uncontaminated water table aquifers (Wilson, McNabb, Balkwill, and Ghiorse, 1983; Fredrickson and Hicks, 1987; McNabb and Dunlap, 1975; White, Smith, Gehron, Parker, Findlay, Matz, and Fredrickson, 1982), as well as in deep aquifers (White, Smith, Gehron, Parker, Findlay, Matz, and Fredrickson, 1982). Viable bacteria that degrade naturally occurring compounds range from 0.01 to 10^6/g of soil (Swindoll, Aelion, Dobbins, Jiang, Long, and

Pfaender, 1988) (see Appendix A). The xenobiotic degraders were generally less than 10 g of soil.

3.2.1.4 Counts in Contaminated Groundwater

Numbers of hydrocarbon utilizers were found to increase in sites after contamination. After a gasoline spill in Southern California in 1972, samples from wells that had traces of free gasoline yielded 5×10^4 gasoline-utilizing bacteria/ml or higher; while an uncontaminated well had only 200 organisms/ml (McKee, Laverty, and Hertel, 1972). A hazardous waste site contaminated with high concentrations of jet fuel hydrocarbons, industrial solvents, and heavy metals had bacterial counts of 10^7 cells/wet g sample from both the unsaturated and the saturated zones.

Groundwater samples collected throughout the United States from aquifers contaminated with hydrocarbons were analyzed and found to contain hydrocarbon-utilizing bacteria at levels of up to 10^6 organisms/ml (Litchfield and Clark, 1973). Significant populations of these organisms are present in groundwaters contaminated with gasoline, fuel oil, and other mixed petroleum products. Groundwaters containing less than 10 ppm of gasoline, fuel oil, and other petroleum products generally had populations of less than 10^3 bacteria/ml (Environmental Protection Agency, 1985b). Concentrations of these hydrocarbons in excess of 10 ppm sometimes supported growth of 10^6 bacteria/ml. Acridine orange direct counts, standard methods agar, and JP-5 agar plate counts were higher for samples from contaminated (10^6 to 10^7 cells/ml) than from uncontaminated (2×10^4 cells/ml) wells (Ehrlich, Schroeder, and Martin, 1985). They were highest in contaminated wells with a free fuel layer.

3.3 MICROORGANISMS IN LAKE, ESTUARINE, AND MARINE ENVIRONMENTS

More than 10 million metric tons of petroleum pollutants are estimated to contaminate the world's oceans each year (Atlas, 1978c). Most of these pollutants come from routine operations, such as release of oil from tanker ballast. Relatively little is contributed by catastrophic accidents. Petroleum biodegradation is also a naturally occurring process, which may account for the fact that not all of the world's surface waters are completely covered with a layer of oil (Gholson, Guire, and Friede, 1972). Evaporation and biodegradation of petroleum hydrocarbons are the major processes that remove petroleum pollutants from the contaminated ecosystem (Atlas, 1978c). The biodegradation of the complex mixture of hydrocarbons in petroleum is generally incomplete, leaving a highly asphaltic residue that can form very persistent tar balls.

Microbes are important in degrading petroleum hydrocarbons in ocean waters (Lee and Ryan, 1976). The addition of hydrocarbons to an ecosystem, as occurs from oil spillages, may result in selective increases or decreases in the microbial populations (Bartha and Atlas, 1977). Addition of different oil fractions to seawater results in increases in the populations of different bacterial species (Horowitz, Gutnick, and Rosenberg, 1975). Petroleum hydrocarbons also select for particular species of hydrocarbonoclastic microorganisms (Zajic and Daugulis, 1975). Both oil type and temperature are important in causing differential population increases (Westlake, Robson, Phillippe, and Cook, 1974).

Bacteria and yeasts appear to be the prevalent hydrocarbon degraders in aquatic ecosystems (Bartha and Atlas, 1977). The most important genera (based upon frequency of isolation) of hydrocarbon utilizers in aquatic environments are Pseudomonas, Achromobacter, Arthrobacter, Micrococcus, Nocardia, Vibrio, Acinetobacter, Brevibacterium, Corynebacterium, Flavobacterium, Candida, Rhodotorula, and Sporobolomyces. Corynebacterium species may be major agents for decomposing heterocyclic compounds and hydrocarbons in contaminated aquatic environments (Kobayashi and Rittmann, 1982).

In agitated enrichments, bacteria tend to take over, but some investigators believe that fungal forms can be significant or even predominant in undisturbed surface slicks (Bartha and Atlas, 1977). In polluted freshwater ecosystems, bacteria, yeasts, and filamentous fungi all appear to be important hydrocarbon degraders (Cooney and Summers, 1976). Table A.3-12 lists the predominant hydrocarbon-degrading microorganisms found in aquatic environments.

Marine and estuarine bacteria appear to be susceptible to lysis with fluctuations in salinity and osmotic pressure (Ahearn, Meyers, and Standard, 1971). Increased numbers and localization of strongly hydrocarbonoclastic yeasts in oil-bearing regions suggest their in situ activity in oil biodegradation processes. Vegetative yeast cells are more resistant than those of bacteria to variable salinity, osmotic pressure, and ultraviolet rays. Oil-degrading yeast strains isolated from freshwater are also active in seawater.

Field and laboratory data suggest that yeasts play a role in microbial decomposition of surface oil depositions in the marine environment (Ahearn, Meyers, and Standard, 1971). However, their activity is greatly affected by the concentration of hydrocarbons. Yeasts of the genera Candida, Trichosporon, Rhodosporidium, Rhodotorula, Debaryomyces, Endomycopsis, and Pichia, isolated from all marine environments, readily assimilated hexadecane and kerosene at a concentration of 2%; however, many of the strains did not grow well on 4% hydrocarbon. Such strains are generally concentrated in oil-polluted habitats.

Aquatic algae have also been implicated in degradation of organic chemicals. Algae may degrade organic compounds by mediation of accelerated photolysis or by uptake and degradation of the molecule with intracellular enzymes. Section 3.1.3 discusses these processes. A greater number of heterotrophic microorganisms can be supported in dense algal communities. The achlorophyllous algae Prototheca hydrocarbonea and P. zopfii can degrade hydrocarbons without light, while Scenedesmus strains can utilize n-heptadecane in the light only (Masters and Zajic, 1971).

The spillage of oil in the marine environment is a severe stress to that ecosystem because of the sudden large increase in organic matter (Atlas, 1977). Marine microorganisms are generally adapted to low levels of organic matter. Appendix A, Section 3.3 explains what can happen to a microbial community with the sudden appearance of high concentrations of organics as a result of an oil spill.

Concentrations of microorganisms in marine environments are relatively low compared with other ecosystems. Therefore, seed microorganisms would be able to compete with the naturally occurring microorganisms in an ecosystem very different from the one to which the indigenous microorganisms were adapted.

Temperature and chemical composition of contaminants have a selective influence on the genera of hydrocarbon utilizers (Atlas, 1981). Different combinations of Achromobacter, Alcaligenes, Flavobacterium, Cytophaga, Acinetobacter, Pseudomonas, Arthrobacter, and Xanthomonas were isolated from crude oil samples, depending upon the source of the sample and temperature of incubation (4° or 30°C) (Cook and Westlake, 1974). Several gram-negative and gram-positive thermophilic hydrocarbon-utilizing bacteria have also been recovered, including species of Thermomicrobium (Merkel, Stapleton, and Perry, 1978). Some thermophiles are obligate hydrocarbon utilizers and cannot grow on other carbon sources.

Psychrophilic isolates that utilized paraffinic, aromatic, and asphaltic petroleum components were found in an asphaltic flow near a natural seepage in Alaska (Jensen, 1975).   They belong to the genera Pseudomonas, Brevibacterium, Spirillum, Xanthomonas, Alcaligenes, and Arthrobacter. Hydrocarbon-utilizing Nocardia, Flavobacterium, Vibrio, Pseudomonas, and Achromobacter were present in northwest Atlantic coastal waters and sediment (Walker, Colwell, and Petrakis, 1976).

Chronic exposure to very high concentrations of degradable hydrocarbons can greatly increase the activity and numbers of hydrocarbon-degrading microorganisms (Heitkamp and Cerniglia, 1987). Water samples from oil-contaminated sections of lakes were better able to degrade test hydrocarbons than water from noncontaminated parts of the same lakes, presumably as a result of enrichment of hydrocarbon-using microorganisms in the former (Boyle, Finger, Petty, Smith, and Huckings, 1984).   An aquatic environment exposed to petroleum hydrocarbons had a larger population of hydrocarbon-degrading microorganisms than an uncontaminated environment, even though the latter had higher numbers of heterotrophic microorganisms (Heitkamp and Cerniglia, 1987).   Not all heterotrophs are able to use organic compounds as substrates, and plate counts (or other enumerative procedures that require growth on artificial media) of total heterotrophic microbial populations may not be a reliable indicator of the hydrocarbon-degrading potential of aquatic sediments. PAHs with less than four aromatic rings can be degraded in freshwater and estaurine ecosystems, but higher molecular weight PAHs, such as benzo(a)pyrene, persist for relatively long periods of time, even in ecosystems microbially adapted to aromatic hydrocarbons.

3.3.1 Counts

The number of oil degraders in a natural body of water is chiefly determined by the pollution history (Bartha and Atlas, 1977).   See Appendix A, Section 3.3.1, for the bacterial counts found in aquatic environments.

Microbial Degradation and Transformation of Petroleum Constituents and Related Elements

There are three processes by which microorganisms can break down hydrocarbons: fermentation, aerobic respiration, and anaerobic respiration (Canter and Knox, 1985). In fermentation, the carbon and energy source is broken down by a series of enzyme-mediated reactions that do not involve an electron transport chain. In aerobic respiration, the carbon and energy source is broken down by a series of enzyme-mediated reactions in which oxygen serves as an external electron acceptor. In anaerobic respiration, the carbon and energy source is broken down by a series of enzyme-mediated reactions in which nitrates, sulfates, carbon dioxide, and other oxidized compounds (excluding oxygen), serve as external electron acceptors. These three processes of obtaining energy form the basis for the various biological wastewater treatment processes.

Table A.4-2 summarizes the different ways these processes are used by the organisms and shows the relationship between the processes and redox potential. Appendix A, Section 4, also elaborates on the differences between heterotrophic and autotrophic metabolism.

4.1 ORGANIC COMPOUNDS

The ability of certain microorganisms to oxidize simple aromatic hydrocarbons has been demonstrated (Table A.4-3) (Gibson, 1977). Most of our present knowledge of the microbial degradation of aromatic hydrocarbons has been obtained with single hydrocarbon substrates and pure cultures of different microorganisms (Tables A.4-4 and A.4-5). Little is known about how microorganisms interact with compounds when they are present in a sample of petroleum.

At a given density of oil-degrading microorganisms, their actual contribution to the elimination of oil depends upon their inherent metabolic capability; i.e., "heterotrophic potential," and the degree to which environmental conditions allow this potential to be expressed (Bartha and Atlas, 1977).

Petroleum is chemically very complex, consisting of hundreds of individual organic compounds (Atlas, 1977). For microorganisms to biodegrade petroleum completely or attack even simpler refined oils, thousands of different compounds may be involved, which must be metabolized. The chemical nature of these petroleum components varies from the simple n-paraffin, monoalicyclic, and monoaromatic compounds, to the much more complex branched chains and condensed ring structures (Horowitz, Gutnick, and Rosenberg, 1975). Many different enzymes are necessary to biodegrade these types of compounds.

The principal biochemical reactions associated with the microbial metabolism of xenobiotics include acylation, alkylation, dealkylation, dehalogenation, amide or ester hydrolysis, oxidation, reduction, hydroxylation, aromatic ring cleavage, and condensation and conjugate formation (Kaufman, 1983).

4.1.1 Aerobic Degradation

Aerobic degradation in soil is dominated by a variety of organisms, including bacteria, actinomycetes, and fungi, which require oxygen during chemical degradation (Parr, Sikora, and Burge, 1983). This process involves oxidation-reduction reactions in which molecular oxygen serves as the ultimate electron acceptor, while an organic component of the contaminating substance functions as the electron donor or energy source in heterotrophic metabolism. Appendix A, Section 4.1.1, explains the differences between aerobic oxidation of aromatic hydrocarbons by eukaryotic and procaryotic organisms (Gibson, 1977). Most aerobic bacteria use oxygen to decompose organic compounds into carbon dioxide and other inorganic compounds (Freeze and Cherry, 1979). In soil, oxygen is supplied through diffusion. If the oxygen demand is greater than the rate of oxygen diffused, the soil naturally becomes anaerobic. Maximum degradation rates for petroleum hydrocarbons are dependent upon the availability of molecular oxygen.

Aerobic biodegradation occurs via more efficient and rapid metabolic pathways than anaerobic reactions (Zitrides, 1983). Therefore, most site decontaminations involving refined oils and fuels are conducted under aerobic conditions.

Metabolic pathways have been established for the degradation of a number of simple aliphatic and aromatic structures (Atlas, 1978c). The general degradation pathway for an alkane involves sequential formation of an alcohol, an aldehyde, and a fatty acid. The fatty acid is then cleaved, releasing carbon dioxide and forming a new fatty acid two carbon units shorter than the parent molecule. This process is known as beta oxidation. The initial enzymatic attack involves a class of enzymes called oxygenases. The general pathway for degradation of an aromatic hydrocarbon involves cis-hydroxylation of the ring structure forming a diol, e.g., catechol. (In the cis configuration, the hydroxyl groups are on the same side of the molecule.) The ring is then oxidatively cleaved by oxygenases, forming a dicarboxylic acid, e.g., muconic acid.

Degradation of substituted aromatic compounds generally proceeds by initial beta oxidation of the sidechain, followed by cleavage of the ring structure. Simple alkyl substitution of benzene generally increases its rate of degradation, but extensive alkylation inhibits degradation. Branching generally retards the rate of alkane degradation. It may also change the metabolic pathway for utilization of a hydrocarbon (Pirnik, 1977). Some long chain alkanes may be degraded by different metabolic pathways, such as subterminal oxidation (Markovetz, 1971). The degradative pathway for a highly branched compound, such as pristane or phytane, may proceed by omega oxidation, forming a dicarboxylic acid instead of only monocarboxylic acids, as in normal beta-oxidation. Appendix A, Section 4.1.1, discusses the reactions involved in these processes and describes some of the major pathways microorganisms use for aerobic degradation.

Appendix A presents individual hydrocarbons and some of the organisms that can degrade them (Section 4.1.1) and some end products of the reactions (Section 4.1.3).

4.1.2 Anaerobic Degradation

Anaerobic microbial transformations of organic compounds are important in anoxic environments (Young, 1984; Sleat and Robinson, 1984). Less cell material is formed under anaerobic conditions because of the lower growth yield, but organic fermentation products are likely to accumulate, unless they are converted into methane, or other hydrocarbon gases. Anaerobiosis usually occurs in any habitat in which the oxygen consumption rate exceeds its supply rate and is a common phenomenon in many natural aquatic environments receiving organic materials (Berry, Francis, and Bollag, 1987). Examples include flooded soils and sediments, eutrophic lagoons, stagnant fresh and ocean waters, and some groundwaters.

Petroleum can be microbially degraded anaerobically by the reduction of sulfates and nitrates (Shelton and Hunter, 1975). Aromatic hydrocarbons, common to many fuels, that can be biodegraded without the presence of molecular oxygen, include toluene, xylene, alkylbenzenes, and possibly benzene (Grbic-Galic and Young, 1985; Reinhard, Goodman, and Barker, 1984; Kuhn, Colberg, Schnoor, Wanner, Zehnder, and Schwarzenback, 1985).

Some xenobiotic compounds can be degraded under anaerobic conditions via fermentation, anaerobic respiration, and photometabolism, both by single species of microbes and by microbial communities (Harder, 1981). Single species have been shown to degrade aromatic compounds in the presence of nitrate or light (Evans, 1977), whereas, a number of microbial communities have been described that degrade these compounds in association with methane production (Balba and Evans, 1977) or in the presence of nitrate (Bakker, 1977). Interactions between different microbial populations are potentially important in the degradation of complex molecules (Hornick, Fisher, and Paolini, 1983).

4.1.2.1 Anaerobic Respiration

Depending upon which of the electron acceptors (e.g., NO_3^-, SO_4^{-2}, or CO_2) is dominant in an anoxic environment, anaerobic respiration may be employed to degrade organic compounds (Berry, Francis, and Bollag, 1987). This may be performed by denitrifying bacteria, sulfate reducing bacteria, or methanogens. The processes by which these organisms degrade hydrocarbons are described in Appendix A, Section 4.1.2.1.

4.1.2.2 Fermentation

Many microorganisms that inhabit anoxic environments obtain their energy for growth through fermentation of organic carbon (Schnitzer, 1982). Fermentation is a process that can be carried out in the absence of light by facultative or obligatory anaerobes. In fermentation, organic compounds serve as both electron donors and acceptors. The process is further described in Appendix A, Section 4.1.2.2.

4.1.2.3 Specific Compounds

Appendix A, Section 4.1.2.3, presents specific examples of hydrocarbons and the organisms that can degrade them anaerobically, intermediate and end products of the process, and various factors affecting anaerobic biodegradation.

4.1.3 End Products

In soil, organic chemicals are subject to alteration by biochemical reactions that are catalyzed by enzymes from a wide range of organisms (Kaufman, 1983). In general, metabolites arising from these microbial reactions are usually nontoxic, polar molecules that exhibit little ability to accumulate in food chains. However, the breakdown products of many chemicals can be toxic; sometimes they are even more toxic than the parent compound.

In addition to carbon dioxide and water, the products resulting from complete mineralization of hydrocarbons, there are various hydroperoxides, alcohols, phenols, carbonyls, aldehydes, ketones, and esters that result from incomplete oxidation (ZoBell, 1973). The biodegradation of aromatic hydrocarbons yields phenolics and benzoic acid intermediates (Bartha and Atlas, 1977). Complete oxidation is more likely when a diverse mixture of microbes is available (Texas Research Institute, Inc., 1982). The tendency for primary and intermediate oxidation products to accumulate is much greater for a pure culture than a mixed culture. The biodegradation of higher molecular weight hydrocarbons involves many intermediates, some of which may accumulate to inhibitory levels (Bartha and Atlas, 1977). It was found that C_5 to C_9 alkanes were not toxic to a population of bacteria, but that the alcohols of these hydrocarbons were inhibitory.

As oxygen in the soil is depleted, microbial reactions become anaerobic with the production of malodorous compounds, such as amines, mercaptans, and H_2S, which can be phytotoxic. However, under aerobic conditions, the end products will be inorganic carbon, nitrogen, phosphorus, and sulfur compounds.

Table A.4-7 lists the products formed from the action of various microorganisms on specific hydrocarbons.

4.2 HEAVY METALS AND METALLOIDS

The toxic nature of arsenic (As), barium (Ba), cadmium (Cd), chromium (Cr), lead (Pb), mercury (Hg), nickel (Ni), selenium (Se), silver (Ag), thallium (Tl), and vanadium (V) can adversely affect microbial populations when the soil is contaminated (JRB and Associates, Inc., 1984b). These inorganic elements cannot be destroyed; however, they can be recovered and recycled (Scholze, Wu, Smith, Bandy, and Basilico, 1986). Ag and Hg are the most toxic to microorganisms, followed by Cd, Zn, Cu, Cr, Pb, and Ni at higher concentrations (Josephson, 1983). Some chemicals will be more toxic and biocidal to the soil microflora than others, and they may cause major changes in microbial populations that could persist for weeks or indefinitely. The effect of the metals in waste on the soil microorganisms present should be determined. Industrial wastes are a major source of metals that could be inhibitory to the degradation of petroleum products, such as by inhibition of enzyme activity. This toxicity can be alleviated by limestone application, if the metal loadings are restricted. However, the effects of heavy loadings of metals are difficult to correct, and the soil may have to be removed.

Some microorganisms show a tolerance for heavy metals (Sims and Bass, 1984; Monroe, 1985). They may require low levels of heavy metals for their microbial enzyme systems (Parr, Sikora, and Burge, 1983) and may use redox-sensitive metals and metalloids (e.g., iron, manganese, selenium) as a source of energy and respiration (Monroe, 1985). They may be able to oxidize, reduce, methylate, demethylate, or otherwise transform these elements so their solubility, sorption, or volatility in the soil is greatly affected (JRB and Associates, Inc., 1984b). Microorganisms that play an important role in oil degradation have often been able to utilize some heavy metals (e.g., iron and manganese), even at high concentrations, as energy sources or electron acceptors in their respiratory processes (Hornick, Fisher, and Paolini, 1983). These reactions may involve precipitation, adsorption, or volatilization of the metals, thereby, making the environment more favorable for other microbial species (Ehrich, 1978).

Biodegradation is generally used as a detoxification mechanism for organic contaminants in soil; however, it may also have applications in treating contaminations where heavy metals are present (Sims and Bass, 1984). Some organisms are even employed for extraction and recovery of different metals (Monroe, 1985). The bacteria, fungi, or algae utilized and the metals recovered are shown in Table A.4-8.

Appendix A, Section 4.2, describes the mechanisms used by microorganisms for resistance to metals and the measures that can be taken to increase metal biotransformation. It lists the elements present in the fuels addressed in this review and discusses the potential for using microbes in contamination incidents.

Enhancement of Biodegradation

5.1 OPTIMIZATION OF SOIL BIODEGRADATION

Some of the information in this section may duplicate material covered in Section 5.2, Optimization of Groundwater Biodegradation, and Section 5.3, Optimization of Freshwater, Estaurine, and Marine Biodegradation; however, it is presented here under a separate heading, with other related information, to accommodate those readers who may specifically wish to address treatment of soil contamination only.

An in-depth evaluation of the potential for microorganisms to remove anthropogenic organic compounds from a contaminated environment indicates that use of properly selected populations of microbes, and the maintenance of environmental conditions most conducive to their metabolism, can be an important means of optimizing biological treatment of organic wastes (Kobayashi and Rittmann, 1982).

Most bacteria in the subsurface are firmly attached to soil particles (Wilson, Leach, Henson, and Jones, 1986). As a result, nutrients must be brought to the microbes by advection or diffusion through the mobile phases, i.e., water and soil gas. In the simplest and perhaps most common case, the organic compound to be consumed for energy and cell synthesis is brought into aqueous solution by infiltrating water. At the same time, oxygen, the electron acceptor used to oxidize the carbon source is brought by diffusion through the soil gas. In the unsaturated zone, volatile organic compounds can also move readily as vapors in the soil gas. Below the water table all transport must be through liquid phases, and as a result, the prospects for aerobic metabolism are severely limited by the very low solubility of oxygen in water. In the final analysis, the rate of biological activity is controlled by:

1. The stoichiometry of the metabolic process
2. The concentration of the required nutrients in the mobile phases
3. The advective flow of the mobile phases or the steepness of concentration gradients within the phases
4. Opportunity for colonization in the subsurface by metabolically capable organisms
5. Toxicity exhibited by the waste or a co-occurring material.

Biological restoration of contaminated soils and aquifers has been accomplished using several techniques. Suntech pioneered the use of "bioreclamation" to restore petroleum-contaminated aquifers (Raymond, 1974; Raymond, 1978; Raymond, Jamison, and Hudson, 1976). This process enhances the activity of the natural microbial population in soil and groundwater by providing dissolved oxygen and nutrients to accelerate biodegradation. This technology of enhanced biodegradation has now been significantly modified for use at hazardous waste sites (JRB and Associates, Inc., 1982; Jhaveria and Mazzacca, 1982).

It might be possible to enhance the degradation of petroleum if the microbial population can be altered by seeding with mutant or acclimated organisms in a mixture that collectively can degrade all the components of the contaminant (JRB and Associates, Inc., 1982; Jhaveria and Mazzacca, 1982). The indigenous hydrocarbon-utilizing microflora of a contaminated site has been enhanced by the recirculation of pumped groundwater containing nutrients and oxygen. The groundwater is injected into wells and circulated through the contaminated zone by pumping groundwater from one or more producing wells (Wilson, Leach, Henson, and Jones, 1986). The environment can be modified to better support the natural microbial activity by altering soil factors; or the petroleum can be modified to make it more susceptible to biodegradation. The surface area available for biodegradation may be increased by dispersing the contaminant, or by emulsification (Atlas, 1977). The "ideal" conditions for efficient utilization of hydrocarbons by microbes includes the following (Texas Research Institute, 1982):

1. A large number and variety of hydrocarbon-utilizing microbes
2. A low population of microbial predators
3. Sufficient oxygen (>2 to 3 ppm in water phase)
4. A high degree of dispersal of the hydrocarbon in the aqueous medium

5. A temperature between 20 and 35°C
6. Sources of nitrogen, phosphorus, and trace minerals
7. A source of readily assimilated organic material
8. A means of removing waste products
9. A nontoxic concentration of hydrocarbons in the aqueous medium
10. A pH of 5 to 9.

In order to improve biological treatment by the use of selected organisms, major issues, such as which organism to select, development of the population, and retention of organisms, must be faced (Kobayashi and Rittmann, 1982). Genetic engineering can be used to develop specialized organisms, but control of environmental factors offers a more practical means of encouraging only the desired species. Sometimes it is necessary to maintain selectively a series, or consortium, of microorganisms to achieve complete degradation (Pfennig, 1978a).

Fixed-film processes (bioreactors) may be the best means to treat wastes above ground to assure population retention and avoid total loss of slow-growing microorganisms (Kobayashi and Rittmann, 1982). The cell concentrations in these fixed-film bioreactors are higher than those found in suspended growth systems. Efficient removal of the organic compounds is possible only when the biomass concentration is large (Matter-Muller, Gujer, Giger, and Strumm, 1980). An important goal is to establish which organisms carry out the desired reactions and assure their proliferation.

Another factor to consider before artificially stimulating oil biodegradation in soil is whether the processes could allow growth of microorganisms that might adversely affect nearby plants and crops as well as livestock and other animals (Atlas, 1977). Ultimately, how much fertilizer to add and its precise chemical formulation; how much forced aeration to use; and the extent to which other factors have to be modified are questions that must be answered by measuring levels of success following actual testing of various treatments.

5.1.1 Biological Enhancement

Controlling environmental factors to encourage only the desired species is a practical method for assuring proliferation and maintenance of selected populations. The factors that must be manipulated vary with the physiological needs of the organisms. It must be determined which microorganism carries out the limiting reaction and assure its proliferation, at the same time considering interactions with other organisms.   In some cases, a series of biological reactors--each maintained under different environmental conditions--is required to provide the wide variety of microorganisms necessary for complete biodegradation.

Since the organisms may grow slowly, population retention is important (Kobayashi and Rittmann, 1982). They could be washed out (total loss of the organism from a reactor) or displace through competition from other organisms. Fixed-film processes may be the best mechanism to assure population retention, when appropriate.

5.1.1.1 Seeding of Microorganisms

Biological treatment methods for petroleum hydrocarbon remediation generally rely upon the stimulation and natural selection of indigenous microorganisms in the soil or groundwater (Sims and Bass, 1984). However, the natural soil flora may not have the metabolic capability to readily degrade certain compounds or classes of compounds or to emulsify the water-insoluble components. On the other hand, they may have the ability but not the biomass necessary to degrade the compounds rapidly enough to meet treatment criteria. In such cases, it might be advisable to add large numbers of exogenously grown microorganisms to the soil.

Considering the diversity of enzymatic activities required, it has been assumed by some that the natural microbial communities of many ecosystems would not possess all the needed enzymes, and, therefore, contaminating oil would not be extensively biodegraded unless hydrocarbon-degrading microorganisms were artificially added (Gholson, Guire, and Friede, 1972). In addition, many of the components would probably be recalcitrant to microbial attack.

The addition of cultures of organisms could reduce the lag period required for indigenous populations to respond to petroleum pollutants (Atlas, 1978c). Some workers have suggested that the natural lag or response time is generally short, in which case, seeding would not usually reduce the impact of petroleum pollutants, except in a few open environments with limited microbial populations.

A special culture collection was begun as a depository for "hydrocarbonoclastic" (hydrocarbon-utilizing) microorganisms (Cobet, 1974). "Superbugs" are being sought to solve the world's pollution problems (Gwynne and Bishop, 1975). Several commercial enterprises are marketing microorganism preparations for removing petroleum pollutants (Atlas and Bartha, 1973e; Azarowicz, 1973).   A hydrocarbonoclastic superbug must be able to degrade extensively most of the components found in petroleum. It must be genetically stable, capable of being stored for long periods of time, and able to reproduce rapidly following storage. It must be capable of enzymatic activity and growth in the environment in which it is to be used, and it must be capable of competing with the naturally occurring microorganisms in that environment for a period sufficient to clean up the site. Finally, it must not produce adverse side effects; it must not be pathogenic or produce toxic metabolic end products.

Creation of a single superbug that combines the genetic information from many microorganisms into one could overcome the problem of interference of organisms with the metabolic activities of each other (Friello, Mylroie, and Chakrabarty, 1976). It will be very difficult to engineer a single organism with all these characteristics or even one capable of metabolizing most petroleum components (Atlas and Bartha, 1973e; Azarowicz, 1973). Therefore, it has been proposed that a mixture of microorganisms should be used as the seed culture for oil biodegradation (Kator, 1973). Each organism would be included for its ability to degrade particular petroleum components and would still have to meet the rest of the above criteria.

Microbes can be added to in situ treatment processes to enhance biodegradation by increasing biomass or by reducing the time necessary for acclimation to occur (Lee and Ward, 1986). Usually natural soil microorganisms that have been previously acclimated to degrade the contaminants are used as "seed"; the microbes may have been selected by enrichment culturing, induced mutation, or genetic manipulation (Ward and Lee, 1984). Organisms selected by enrichment culturing have their tolerance and metabolic activity to a particular substance built up over time (Lee and Ward, 1986). The types of microorganisms that are isolated depend upon the source of the inoculum, the conditions used for the enrichment, and the substrate (Atlas, 1977). Microorganisms can become acclimated to specific compounds by repeated exposure to the petroleum product, to certain components, or to related compounds (Zajic and Daugulis, 1975).

Sequential enrichment techniques are a modification of enrichment culturing and can be used to isolate microorganisms capable of degrading most of the components of petroleum (Environmental Protection Agency, 1985b; Horowitz, Gutnick, and Rosenberg, 1975). The substrate is inoculated with a microbial population, and the organisms that can degrade it are isolated. The undegraded substrate is then used to isolate another set of organisms that are capable of degrading the residual components. This continues until none of the substrate remains or no new isolates are recovered. Different mixtures of organisms may be isolated from soils, if the enrichments are carried out at 4° rather than 20°C (Jobson, Cook, and Westlake, 1972).

The resulting mixture of organisms is more effective than the individual isolates. However, organisms isolated individually in the sequential enrichments may not be able to degrade the oil simultaneously; one organism in the mixture may interfere with another. This process may allow isolation of various microorganisms that could degrade the low solubility, high molecular weight, as well as the more soluble and toxic, hydrocarbons and intermediates of hydrocarbon metabolism (Zajic and Daugulis, 1975). Such selective continuous enrichments may be occurring in nature in areas subjected to constant input of petroleum hydrocarbons. Since intermediary metabolites must also be removed for complete oil cleanup, nonhydrocarbon-utilizing microorganisms, such as fatty acid metabolizers, would also be required in the mixture (Atlas, 1977).

Seeding microorganisms onto soil was first tried around 1968 (Gutnick and Rosenberg, 1979). An inoculum of Cellumonas sp. and nutrients was able to degrade hydrocarbon contaminants more effectively than just fertilizer alone (Schwendinger, 1968). Since then, the seeding of microorganisms has been used in a number of different environments to degrade organics (Environmental Protection Agency, 1985b). Varying success has been obtained when seeding microorganisms onto petroleum-contaminated soils (Schwendinger, 1968).

Variations in chemical composition among different oils might require applying unique mixtures of microorganisms (Buckingham, 1981). Different mixtures of seed microorganisms might also be necessary for stimulated oil biodegradation in different soil environments (Westlake, Jobson, Phillippe, and Cook, 1974). Some compounds may be highly resistant to microbial attack, or recalcitrant, including the high molecular weight and highly branched and condensed ring compounds found in many oils (Atlas and Bartha, 1973b). These would be left as a tar-like residue that would not be easily biodegradable.

Microbial inoculants covering a broad range of metabolic capabilities are available commercially, and they are being used increasingly in treating both contaminated soil and aquatic systems (Thibault and Elliott, 1980). Appendix A, Section 5.1.1.1, provides additional information on using seed organisms for biodegradation.

Commercially available mutant bacteria generally require an adaptation period in order to adjust to site conditions (Buckingham,

1981). However, the time required is similar to the adjustment period for natural in situ bacteria. Therefore, additional factors, such as convenience and costs of application and labor must be considered in selecting the most efficient bacterial treatment.

When groundwater is treated with added bacterial strains, the contaminated water is often pumped from the ground from one or more wells to a tank where aeration and nutrient addition occur (Thibault and Elliott, 1979). The organisms can be conveniently added with the nutrient solution. The water is then pumped back into the ground at a point "upstream," relative to the groundwater flow. Thus, a closed loop for biological treatment is developed. Depending upon the tank configuration, a number of mechanical or diffused aeration devices are available commercially.

Inoculation of contaminated soil with microorganisms selected for their ability to degrade or transform hazardous materials is a very attractive treatment concept (JRB and Associates, Inc., 1984b). However, the treatment manager must realize that the soil environment is restrictive and the soil microbial system has a complex ecology. These factors may limit the ability of introduced organisms to become self-perpetuating and carry out their specialized functions for an extended period of time. It may be necessary to reinoculate the soil several times before satisfactory levels of treatment are achieved. Cost must also be considered, and there is no point in adding microorganisms, if seeding is not necessary. Even if microbial seeding is used, it will generally have to be accompanied by environmental modifications (Atlas, 1977). Seeding without simultaneous modification of environmental conditions is not likely to succeed.

In most ecosystems, some environmental factor will limit oil-degrading activity (Atlas, 1977). The application of microbial amendments to the soils is frequently combined with other treatment techniques, such as soil moisture management, aeration, and fertilizer addition (Thibault and Elliott, 1980). This method may be most effective against one compound or closely related compounds. The effectiveness would be limited by toxicity of a contaminant or the inability of an organism to metabolize a wide range of substrates.

5.1.1.1.1 Commensals. When a particular chemical is not easily degraded or is only transformed by microorganisms using it as a sole source of carbon, it is sometimes possible to employ commensalism to encourage complete biodegradation (de Klerk and van der Linden, 1974). Bacterial mixtures, such as Pseudomonas, Nocardia, and Arthrobacter, have been used as part of a "cocktail" that may consist of 12 to 20 strains of microbes for treating different types of wastes (Cooke and Bluestone, 1986). The mixture of organisms depends upon the composition of the waste. Both selective and controlled mixed cultures are currently being used (Pfennig, 1978a). Specific organisms for target chemicals can be chosen. For example, purple nonsulfur bacteria are used to remove organic compounds and green sulfur bacteria to remove hydrogen sulfide.

5.1.1.1.2 Acclimated Microorganisms. The adaptation process during which microorganisms adjust to growing on a new substrate is defined functionally as an increase in rate of degradation with exposure to a compound (Swindoll, Aelion, Dobbins, Jiang, Long, and Pfaender, 1988). It may involve one or more of the following: 1) an induction or derepression of enzymes specific for the degradation pathways of a particular compound, 2) an increase in the number of organisms in the degrading population, 3) a random mutation in which new metabolic pathways are produced that will allow degradation (Spain, Pritchard, and Bourquin, 1980), or 4) adaptation of existing catabolic enzymes (including associated processes, such as transport and regulatory mechanisms) to the degradation of novel compounds (Harder, 1981).

One approach to biodegradation is the addition to the system of microorganisms that have been especially acclimated to degrade a pollutant of concern (Wilson, Leach, Henson, and Jones, 1986). The organisms may be selected by enrichment culturing or genetic manipulation and can become acclimated to the degradation of compounds by repeated exposure to that substance (Lee and Ward, 1986). Inoculating acclimated organisms into a new environment has met with variable success (Lee, Wilson, and Ward, 1986; Goldstein, Mallory, and Alexander, 1985). There is considerable evidence that this technique is successful in degrading a variety of organic contaminants in surface biological waste treatment processes. However, there is less evidence for its application in in situ and groundwater environments, mainly because the cases reported were not managed under controlled conditions to differentiate the effects of the acclimated microbes from those of an enhanced indigenous population.

For this approach to be successful in subsurface environments (Wilson, Leach, Henson, and Jones, 1986; Goldstein, Mallory, and Alexander, 1985):

1. The added microorganisms must be able to survive in what to them is a foreign, hostile environment and compete for nutrients with indigenous organisms.
2. The added microorganisms must be able to move from a point of injection to the location of the contaminant in a medium where bacterial transport rates are normally very low, especially in fine-grained materials.
3. The added microorganisms must be able to retain their selectivity for metabolizing compounds for which they were initially adapted.

Adaptation affects biodegradation rates (Fournier, Codaccioni, and Soulas, 1981). The rate of transformation of a compound is often increased by prior exposure to the chemical. It can involve different mechanisms, such as gene transfer or mutation, enzyme induction, and population changes. There is a high degree of variability in the ability of a microbial community to adapt, depending upon many factors present at the site (Spain and Van Veld, 1983). Adaptation of aquatic microbial communities can last for several weeks after exposure to a xenobiotic compound. Repeated exposure to a contaminant at a site will usually increase the adaptive capabilities of the microorganisms. In fact, organisms from contaminated sites may be able to degrade a wider range of compounds, once they become acclimated to the organic compounds, a process that may require several months. The ability of communities to adapt varies from site to site. It has been suggested that preadaptation of the microbial population is not a major factor in long-term degradation of petroleum in soils, because adaptation will eventually occur naturally. However, the addition of soil containing previously adapted microorganisms may improve the initial degradation rates (Hornick, Fisher, and Paolini, 1983).

Organisms capable of degrading a recalcitrant contaminant are frequently found in a soil that has been exposed to that contaminant for several years (Brubaker and O'Neill, 1982). This local microbial community is often enriched in enzymes, which allows it to degrade the contaminant, although not necessarily as a sole source of carbon and energy.

Often an organism that is adapted to metabolize a member of a homologous series of molecules may be capable of degrading the rest of the members of the series (Kaufman, 1983). For example, phenol and benzoic acid have been degraded rapidly after acclimatization to p-hydroxybenzoic acid (Healy and Young, 1979). Microorganisms adapted to growth on m-xylene in the absence of molecular oxygen, with nitrate as an electron acceptor, are also able to degrade toluene under the denitrifying conditions (Zeyer, Kuhn, and Schwarzenback, 1986). Sometimes the compounds are not similar in structure. For instance, after growth with succinate plus biphenyl, a mutant strain of Beijerinckia could oxidize benzo(a)pyrene and benzo(a)anthracene (Gibson and Mahadevan, 1975). Chemical analog adaptation can reduce the amount of time required for adaptation to a particular chemical.

Appendix A, Section 5.1.1.1.2, suggests procedures that should be used for biodegradation employing acclimated microorganisms and provides examples of the application of this technique in the field.

5.1.1.1.3 Mutant Microorganisms.

The inability of a microorganism to degrade a certain substrate may simply be related to its inability to induce certain enzymes or to transport the compound into the cell (Hornick, Fisher, and Paolini, 1983). The provision of microbial strains that exhibit improved biodegradation capacities is one of the most challenging fields of microbiological research related to pollution control (Leisinger, 1983). Strains with improved degradation rates or with a widened range of degradative ability may be constructed by in vivo or in vitro genetic manipulation.

The design and construction of microbial systems (cellular or extracellular enzymatic) to degrade specific compounds has been made possible by two developments (Pierce, 1982a). First, the ability of microorganisms to degrade some of these compounds has been shown to be encoded on extrachromosomal DNA (plasmids), and, second, recent developments in recombinant DNA technology permit the "engineering" of DNA that codes for desired enzymatic capabilities.

Development of mutant strains through genetic alteration may increase growth rate and/or endow the organism with the desired biochemical capability (Thibault and Elliot, 1980a). The addition of adapted and/or mutant microbes has not been completely successful, but has great potential (Fox, 1985). Researchers at government and industrial laboratories are developing artificially mutated bacteria that will be more active and selective for the chemicals they destroy (Chowdhury, Parkinson, and Rhein, 1986). A number of companies are producing microbial strains to be used to treat abandoned hazardous waste sites and chemical spills (Anonymous, 1981). Adaptation and mutation of microorganisms involve several steps (Zitrides, 1983). Wild strains known to degrade a specific organic chemical or functional group are exposed to successively increasing concentrations of that chemical. Those least inhibited by high concentrations will grow the fastest. These are then irradiated to induce genetic changes for increased growth rate. Although constitutive enzyme systems are being developed to shorten the lag period, both natural in situ bacteria and commercial mutant strains require an adaptation period to adjust to site conditions. Other factors, such as convenience and costs, must be considered in selecting the most efficient bacterial treatment (Amdurer, Fellman, and Abdelhamid, 1985).

Genetic Engineering

Genetic engineering can be used to increase the degradation capacity of microbes by improving the stability of the enzyme systems (located on chromosomes instead of on plasmids), enhancing their activity, providing them with multiple degradative activities, and ensuring that they are safe, both to the environment and human health (Pierce, 1982d).

Since genes for some enzymes involved in degrading alkanes and simple aromatic hydrocarbons can be found on extrachromosomal DNA elements in the bacterial cell (plasmids) (Alexander, 1981; Hou, 1982; Chakrabarty, 1974; Jain and Sayler, 1987), it may be possible to exchange enzymatic activity among closely related microorganisms through plasmid transfer.   Multiple plasmid transfer has been accomplished between Pseudomonas species (Chakrabarty, 1974) to construct a strain that can degrade several hydrocarbons, including octane and naphthalene (Hornick, Fisher, and Paolini, 1983). However, genetic information for enzymes involved in the degradation of many hydrocarbons may not be located on plasmids (Atlas, 1977).

Catabolic plasmids in microorganisms occur naturally and some of the common and well-studied catabolic plasmids are listed in Table A.5-1. By manipulating the exchange of this genetic material, it is possible to develop strains with extended degradative capability; i.e., organisms that can degrade more than one xenobiotic substrate or that can completely mineralize highly recalcitrant molecules (Kamp and Chakrabarty, 1979). The plasmids can be fused together to provide multiple degradative traits or to produce a novel or previously unexpressed degradative pathway (Pierce, 1982d). There is some indirect evidence that this may occur in nature and that plasmid-born genes may be transferred, to some extent, among indigenous bacteria in the soil (Pemberton, Corney, and Don, 1979). It is not known whether introduced microbes would also transfer these genes (Stotsky and Krasovsky, 1981). Stable strains can be engineered that will not pass on their plasmids (plasmids added to stable chromosomes) or will be capable of growth only under restricted conditions; this should limit their potential for escape into the environment. However, environmental conditions and competition with the native microbial population may prevent the genetically engineered organism from reaching its degradative potential.

Genetic engineering can be used to develop organisms with unique metabolic capabilities. Expanding the ability to degrade a wide range of aliphatic and aromatic compounds should give such strains a selective advantage over other bacteria in degrading a mixture of petroleum hydrocarbons (Hornick, Fisher, and Paolini, 1983). However, there is no conclusive evidence that such organisms have been able to establish themselves in aeration basins or in natural environments having an active microbial population (Johnston and Robinson, 1982). The adapted mutant will probably be at a disadvantage in the competition with the native microbial population, and may only be able to proliferate on the substrate upon which it was isolated (Zitrides, 1978; McDowell, Bourgeois, and Zitrides, 1980). Acclimated or genetically engineered organisms may not survive or offer significant advantage in treatment of hazardous wastes unless environmental parameters (oxygen, temperature, nutrients, etc.) can be controlled to promote survival of the added organisms (Fox, 1985).

There are four major areas that may benefit from genetic engineering: stabilization; enhanced activity; multiple degradative activities; and health, safety, and environmental concerns (see Appendix A, Section 5.1.1.1.3) (Pierce, 1982d). Genetic engineering must be used in combination with knowledge of microbial physiology and biochemistry, chemical engineering, and process engineering to develop a successful microbial system that will degrade hazardous organic compounds. Genetic engineering involves implementation of recombinant-DNA procedures and techniques for strain improvement, such as classical mutation of wild type strains and in vivo gene manipulation techniques (e.g., protoplast fusion, transformation, conjugation, and transduction).

It is important to be able to detect and monitor the fate of genes of a released or introduced organism within a microbial community (Jain and Sayler, 1987). Monitoring of microorganisms is required to predict whether the released organisms would disappear quickly, whether they are unable to compete with the natural microbial community, whether they would proliferate and become temporarily or permanently established in the area, or whether they are being transported from the initial site of the release. Because of the concern over release of genetically engineered organisms into the environment, it may be some time before the issue is resolved and the use of these microbes to clean up hazardous waste sites is widespread (Fox, 1985).

Appendix A, Section 5.1.1.1.3, discusses the methods of producing genetic variants for biodegradation, the advantages of using genetically engineered organisms, the properties of strains that have already been developed (including those with multiple degradative abilities), and the plasmids responsible for degradation of specific hydrocarbons.

5.1.1.2 Use of Analog Enrichment for Cometabolism

Cometabolism of a hazardous compound can be brought about by adding a chemical analog of the compound to contaminated soil or to culture media (Sims and Overcash, 1981). This stimulates production of an enzyme that originally targeted an energy-yielding substrate, but which may also be able to transform the recalcitrant molecule (Alexander, 1981). It has been suggested that a concentration threshold exists for some compounds below which the material cannot be reduced by bacterial action (Bouwer and McCarty, 1984). However, in the presence of one compound at a relatively high concentration, termed the primary substrate, another compound present at trace concentrations, termed the secondary substrate, can also be biotransformed, if the organisms are able to transform both substrates.

Incorporating a small amount of an inducer chemical, to which an organism has become adapted, into a solution of a more persistent compound has been shown to facilitate a more rapid degradation of the compound (Kaufman, 1983). Consideration of the use of inducer molecules to enhance microbial degradation must, however, be accompanied by an understanding of both the mechanisms involved and the ecological acceptability of adding the inducing substrate.

An example of analog enrichment is the use of minimally mutagenic phenanthrene to increase the rate of degradation of highly mutagenic benzo(a)pyrene (Sims and Overcash, 1981). Biphenyl has also been used to stimulate cometabolic degradation of PCBs (Furukawa, 1982). Care must be used in selecting such analogs, since they or their degradation products could be hazardous (Sims and Bass, 1984). They should be added in amounts large enough to stimulate microbial activity, but not at levels that are toxic or that adversely affect public health or the environment. Treatability studies are required to determine the feasibility, loading rate, and effectiveness of the analog(s). The analogs may be applied as solids, liquids, or slurries and mixed thoroughly with the contaminated soil, where feasible. Fertilization may be necessary to maintain microbial activity, and controls may have to be implemented to prevent drainage and erosion problems. The reliability of this technology is unknown.

When the transformation products from analog enrichment are not hazardous and are degradable by other organisms, this technique may be an effective treatment for contaminated soil (Alexander, 1981). This subject is further discussed in Appendix A, Section 5.1.1.2. Examples of compounds that can be used as analogs for specific hydrocarbons are given in Appendix A, Section 5.1.1.2. Table A.5-2 summarizes a number of organic compounds, their chemical analogs, the organisms that can perform their cooxidation, and the resulting products of the reactions.

5.1.1.3 Application of Cell-Free Enzymes

There is evidence for the existence in soil of extracellular enzymes capable of degrading xenobiotics or xenobiotic degradation products (Kaufmann, 1983). Enzymatic degradation of organics with cell-free enzymes holds potential as a possible in situ treatment technique (Environmental Protection Agency, 1985b). Purified enzymes, harvested from microbial cells, are commonly used in industry to catalyze a variety of reactions, including the degradation of carbohydrates and proteins.

Microbial enzymes that have the ability to transform hazardous compounds to nonhazardous or more labile products could possibly be harvested from cells grown in mass culture and applied to contaminated soils (Sims and Bass, 1984). Industry commonly employs crude or purified enzyme extracts, in solution or immobilized on glass beads, resins, or fibers, to catalyze a variety of reactions, including the breakdown of carbohydrates and proteins.

There are a number of advantages in using cell-free extracts. Enzyme activity can often be preserved in environments that would be inhospitable to the organisms; e.g., in soils with extremes of pH and temperature, high salinity, or high solvent concentrations. Increased mobility of a compound by extracellular enzymes in sites that inhibit microorganisms may make it more susceptible to decomposition by the soil microflora present in less hostile environments (Munnecke, Johnson, Talbot, and Barik, 1982).

There are also limitations in using the extracts. For an enzyme to function outside the cell in the soil environment, it must not require cofactors or coenzymes. This would limit the application of many enzymes. Most important (especially in situ) is enzyme stability--they must be fairly stable in extracellular environments. Enzymes might be leached out of the treatment zone, and they might be inactivated or have lower activity if they are bound to clay or humus in the soil. Outside of biochemical and environmental constraints, logistics and costs for producing enzymes in effective quantities may limit current use of this concept.

Organic wastes are amenable to this treatment. Theoretically, enzymes would quickly transform hazardous compounds if they remained active in the soil, and the potential level of treatment is high. However, there is little information available on the use of this technique in soil and no information from the field. Its reliability is unknown. Further applications and concerns for use of this method are discussed in Appendix A, Section 5.1.1.3.

5.1.1.4 Addition of Antibiotics

Many species of hydrocarbon utilizers have been found to be ingested by a large number of ciliated and other cytophagic protozoans (Texas Research Institute, Inc., 1982). They have been shown to reduce drastically the microbial population, e.g., from 10^7/ml to 10^2/ml.

Protozoa present in sewage can halt the bacterial action on specific pollutants in treatment plants by devouring the bacteria (Anonymous, 1986). Grazing of hydrocarbon-degrading microorganisms by protozoa can lengthen the acclimation period prior to biodegradation in wastewaters (Alexander, 1986). This problem can be eliminated with the use of antibiotics that target the protozoa. With eukaryotic inhibitors (cycloheximide and nystatin) added to untreated sewage, the counts of protozoa can be lowered to 60/ml and the acclimation period reduced from 11 days to 1 day. With no inhibitors, protozoan counts can increase to 3.5×10^4/ml in one day (Wiggins and Alexander, 1986).

5.1.1.5 Effect of Biostimulation on Counts

Application of oily sludge and fertilizer to soil resulted in an increase in colony-forming units (CFUs) (51% of total number) but without a simultaneous increase in the direct counts (Lode, 1986). This suggests that many bacteria in soil occur in a state of dormancy, waiting for a suitable material for reactivation (Alexander, 1977). Part of these bacteria might be hydrocarbon degraders, but many are likely to be organisms living on intermediary degradation products.

Addition of nutrients and oxygen to a contaminated site has been shown to increase effectively the number of indigenous microorganisms available for degradation of the contaminants. Appendix A, Section 5.1.1.5, presents examples of counts before and after biostimulation.

Stimulation of pleomorphs (bacteria having multiple forms) in response to adding fertilizer suggests that such organisms adapt to oil degradation more easily than others under improved growth conditions (Lode, 1986). The very high stimulation of nonsporeforming rod-shaped bacteria after sludge and fertilizer application supports the assumption that many of these bacteria (particularly the pigmented types) live on degradation products of hydrocarbons.

5.1.2 Optimization of Soil Factors

The various chemical and physical properties of a soil determine the nature of the environment in which microorganisms are found (Parr, Sikora, and Burge, 1983). In turn, the soil environment affects the composition of the microbiological population both qualitatively and quantitatively. The rate of decomposition of an organic waste depends primarily upon its chemical composition and upon those factors that affect the soil environment. Factors having the greatest effect on microbial growth and activity will have the greatest potential for altering the rate of residue decomposition in soil.

The ability of the upper 6 in. of soil to absorb nutrients and hold water depends upon its physical and chemical properties of texture, infiltration and permeability, water-holding capacity, bulk density, organic matter content, cation exchange capacity, macronutrient content, salinity, and micronutrient content (Hornick, 1983). A typical mineral soil is composed of approximately 45% mineral material (varying proportions of sand, silt, and clay), 25% air and 25% water (i.e., 50% pore space, usually half saturated with water), and 5% organic matter, although this is highly variable. Any significant change in the balance of these components could affect the physical and chemical properties of the soil. This may alter the soil's ability to support the chemical and biological reactions necessary to degrade, detoxify, inactivate, or immobilize toxic waste constituents. Appendix A, Section 5.1.2, discusses the effect of waste additions on the physical properties of soil.

Most soils have a tremendous capacity to detoxify organic chemical wastes by diluting the compounds, acting as a buffering system, and decomposing the material through microbial activity (JRB and Associates, Inc., 1984b). The most important soil characteristics for this detoxification are those that affect water movement and contaminant mobility, i.e., infiltration and permeability. Certain waste characteristics can also affect soil infiltration and permeability, and this interaction should be taken into account. Table A.5-3 lists the site/soil properties that should be identified to be able to predict potential migration of the contaminating material and indicate what will be necessary for manipulating the soil characteristics for optimum results. Some of the soil factors, however, can be managed only near the surface for enhancing the soil treatment.

Unless all the proper conditions are met for a given compound, biodegradation is not likely to occur (Bitton and Gerba, 1985). Before in situ remedial actions can be initiated for treating hazardous waste-contaminated soils, both the site and waste characteristics must be evaluated (Solanas and Pares, 1984). These features will help determine whether a biological approach is the most feasible treatment option and, if selected, how biodegradation can be used most effectively with the prevailing conditions.

There are more than a thousand different soil types in the United States alone (Federle, Dobbins, Thornton-Manning, and Jones, 1986). The U.S. Soil Conservation Service has characterized certain chemical and physical parameters for many of them while preparing soil maps. These data are readily available. They would help in predicting biomass and activity in various profiles.

The soil environment affects the qualitative and quantitative microbial population (Parr, Sikora, and Burge, 1983). The factors that enhance growth of the microorganisms will also result in more rapid decomposition of the contaminants in the soil. The most important soil factors that affect degradation are water, temperature, soil pH, aeration or oxygen supply, available nutrients (e.g., nitrogen (N), phosphorus (P), potassium (K), sulfur (S)), and soil texture and structure. Any treatments applied to the soil to enhance contaminant removal processes must not alter the physical and chemical environment in such ways that they would severely restrict microbial growth or biochemical activity (Sims and Bass, 1984). In general, this means that the soil water potential should be greater than -15 bars (Sommers, Gilmore, Wildung, and Beck, 1981); the pH should be between 5 and 9 (Atlas and Bartha, 1981; Sommers, Gilmore, Wildung, and Beck, 1981); and the oxidation-reduction (redox) potential should be between pe + pH of 17.5 to 2.7 (Baas Becking, Kaplan, and Moore, 1960). Soil pH and redox boundaries should be carefully monitored when chemical and biological treatments are combined.

Since the activity of microorganisms is so dependent upon soil conditions, modification of soil properties is a viable method of enhancing the microbial activity in the soil (Sims and Bass, 1984). In order to vary these factors for use as a treatment technology, the following information is required:

- Characterization and concentration of wastes, both organics and inorganics, at the site
- Microorganisms present at the site
- Biodegradability of waste constituents (half-life, rate constant)
- Biodegradation products, particularly hazardous products
- Depth, profile, and areal distribution of constituents
- Soil moisture
- Other soil properties for biological activity (pH, Eh, oxygen content, nutrient content, organic matter, temperature)
- Trafficability of soil and site.

The influence of soil factors, such as temperature and nutrient concentration, on phenol mineralization shows great variability as a function of soil type and horizon (Thornton-Manning, Jones, and Federle, 1987). Most of these factors do not function independently; i.e., a change in one may effect a change in others (Parr, Sikora, and Burge, 1983). While the soil factors play an important role in biodegradation, because of these interactions, it is not always easy to predict a priori how temperature or another environmental variable will affect biodegradation in a given soil environment (Thornton-Manning, Jones, adn Federle, 1987). However, if any of the factors that affect degradation processes in soil are at less than an optimum level, microbial activity will be lowered accordingly and substrate decomposition decreased (Parr, Sikora, and Burge, 1983).

Table A.5-6 lists the soil factors that may have to be modified during the use of different treatment technologies: extraction, immobilization, degradation, attenuation, and reduction of volatiles.

5.1.2.1 Soil Moisture

Biodegradation of waste chemicals in the soil requires water for microbial growth and for diffusion of nutrients and by-products during the breakdown process (JRB and Associates, Inc., 1984b). A typical soil is about 50% pore space and 50% solid matter. Water entering the soil fills the pore spaces until they are full. The water then continues to move down into the subsoil, displacing air as it goes. The soil is saturated when it is at its maximum retentive capacity. When water drains from the pores, the soil becomes unsaturated. Soils with large pores, such as sands, lose water rapidly. If the soil is too impermeable, it will be difficult to circulate treatment agents or to withdraw the polluted water (Nielsen, 1983). Soils with a mixture of pore sizes, such as loamy soils, hold more water at saturation and lose water more slowly. The density and texture of the soil determine the water-holding capacity, which in turn affects the available oxygen, redox potential, and microbial activity (Parr, Sikora, adn Burge, 1983). The actual microbial species composition of a soil is often dependent upon water availability. The migration of organisms in the soil can also be affected by pore size. Small bacteria are on the order of 0.5 to 1.0 μm in diameter (Bitton and Gerba, 1985). Larger bacteria tend to be immobilized in soils by physical straining or filtering.

Field capacity refers to the percentage of water remaining in a soil after having been saturated and free gravitational drainage has ceased (JRB and Associates, Inc., 1984b). Gravitational water movement is important for mobilizing contaminants and nutrients, due to leaching. Slow drainage can reduce microbial activity as a result of poor aeration, change in oxidation-reduction potential, change in nutrient status, and increased concentration of natural minerals or contaminants to toxic levels in the pore water. The amount of water held in a soil between field capacity and the permanent wilting point for plants is known as available water. This is the water available for plants and a similar quantity may be required for optimum soil microbial and chemical reactions.

Where it is necessary to predict and interpret the response of microorganisms in soils to organic wastes, both the water content and water potential should be reported (Parr, Sikora, and Burge, 1983). Water potential is useful for quantifying the energy status of water in soils containing waste chemicals. Generally, with decreasing water potentials, fewer organisms are able to grow and reproduce; and bacterial activity is usually greatest at high water potentials (wet conditions). Species composition of the soil microflora is regulated largely by water availability, which, in turn, is governed essentially by the energy of the water in contact with the soil or waste.

Some fungi can tolerate dry soils and do not grow well if the soil is wet (Clark, 1967). Bacteria may be antagonistic to fungi under moister conditions. At low potentials, bacteria are less active, allowing fungi to predominate (Cook and Papendick, 1970). Microbial decomposition of organic material in dryer soils is probably due primarily to fungi (Gray, 1978; Harris, 1981). When soil becomes too dry, many microorganisms form spores, cysts, or other resistant forms, while many others are killed by desiccation (JRB and Associates, Inc., 1984b). Although fine-textured soils have the maximum total water-holding capacity, medium-textured soils have the maximum available water due to favorable pore size distribution.

A well-drained soil (e.g., a loamy soil) is one in which water is removed readily but not rapidly (JRB and Associates, Inc., 1984b). A poorly drained soil (e.g., a poorly structured fine soil) can remain water-logged for extended periods of time, producing reducing conditions and insufficient oxygen for biological activity; an excessively drained soil (e.g., a sandy soil) is one in which water can be removed readily to the point that drought conditions occur. For in situ treatment of hazardous waste-contaminated soils, the most desirable soil would be one in which permeability is only large enough to maximize soil attenuation processes (e.g., adequate aeration for aerobic microbial degradation) while still minimizing leaching.

Control of moisture content of soils at an in-place treatment site may be essential for control and optimization of some degradative and sorptive processes, as well as for suppression of volatilization of some hazardous constituents (Sims and Bass, 1984). The moisture content of soil may be controlled to immobilize constituents in contaminated soils and to allow additional time for accomplishing biological degradation. When contaminants are immobilized by this technique and anaerobic decomposition is desired, anaerobiosis must be achieved by a means other than flooding, such as soil compaction or organic matter addition. Control of soil moisture may be achieved through irrigation, drainage, or a combination of methods. These techniques are discussed in Appendix A, Section 5.1.2.1.

Sometimes a site with shallow depth contamination may require soil mixing to dilute the wastes and incorporate nutrients and oxygen, as well as to enhance soil drying (Sims and Bass, 1984). It may be necessary to install a drainage system to reduce soil moisture. Increasing soil temperature will enhance surface soil drying; this technology, called "landfarming," is simple and easy to apply. However, drier soil may retard microbial activity, as well as increase volatilization of volatile waste components.

Appendix A, Section 5.1.2.1, supplies additional information on the moisture requirements of organisms and the use of moisture control in biodegradation.

5.1.2.2 Temperature

Soil temperature is one of the more important factors controlling microbiological activity and the rate of organic matter decomposition (Sims and Bass, 1984). Temperatures of both air and soil affect the rate of biological degradation processes in the soil, as well as the soil moisture content (JRB and Associates, Inc., 1984b).

Generally, an increase in temperature increases the rate of degradation of organic compounds in soil (JRB and Associates, Inc., 1962). This rate usually doubles for every 10°C increase in temperature (Thibault and Elliot, 1980a). Conversely, a lowering of the temperature is associated with a slowing of the microbial growth rate. This property is attributed to a decrease in adsorption with rising temperature, which makes more organics available, or to an increase in biological activity, or both (JRB and Associates, Inc., 1984b). Biological activity has an optimum temperature, beyond which biological activity often rapidly decreases, thus displaying a growth curve that is skewed to the right.

Microbial utilization of hydrocarbons has been shown to occur at temperatures ranging from -2° to 70°C (Texas Research Institute, 1982). Biodegradation was found to occur at a temperature of 5°C, but hydrocarbons were degraded more slowly at lower temperatures (Parr, Sikora, and Burge, 1983). Most soils, especially those in cold climates, contain psychrophilic microorganisms that grow best at temperatures below 20°C (JRB and Associates, Inc., 1984b) and are effective at temperatures below 0°C. Soils in hot environments usually support many thermophilic microorganisms that are effective at temperatures above 60°C. However, most soil microorganisms are mesophiles and exhibit maximum growth in the range of 20° to 35°C (Parr, Sikora, and Burge, 1983). The majority of hydrocarbon utilizers are most active in this range.

Temperatures in the thermophilic range (50 to 60°C) were shown to greatly accelerate decomposition of organic matter, in general (Parr, Sikora, and Burge, 1983). At these temperatures, actinomycetes will be naturally predominant over fungi and bacteria. Therefore, in certain situations, composting may offer potential for maximizing the biodegradation rate of waste industrial chemicals. It should be noted, however, that in another investigation in a test treatment facility, it was found that several aromatic hydrocarbons were not metabolized at 55°C, but were metabolized at 30°C (Phillips and Brown, 1975), while other researchers reported a leveling-off of the hydrocarbon biodegradation rate in soil above 20°C (Dibble and Bartha, 1979a). Although elevated temperature has some advantage for potentially limiting the development of pathogenic microorganisms, too high a temperature would not be beneficial to stimulate petroleum biodegradation (Phillips and Brown, 1975). The increased availability of more toxic hydrocarbons at higher temperatures may counteract the stimulation of metabolic processes (Dibble and Bartha, 1979a).

A microbial community will undergo an adaptation or selection process in the mineralization of a compound, which is reflected in a lag period that often increases with decreasing temperature (Thornton-Manning, Jones, and Federle, 1987). Temperature was also found to affect phenol mineralization differently, as a function of soil type.

The effect of temperature on degradation is important in terms of assessment of the seasonal and geographical variation of degradation rates (JRB and Associates, Inc., 1984b). Disposal sites for oil can be chosen in warm areas that receive direct sunlight to assure temperatures suitable for rapid metabolism by mesophilic microorganisms (Atlas, 1977). Even in near-Arctic environments, absorbance of solar energy raises temperatures into a range that allows for mesophilic microbial oil degradation (Atlas and Schofield, 1975).

Soil temperature is difficult to control in a field situation, but can be modified by regulating the incoming and outgoing radiation, or by changing the thermal properties of the soil (Baver, Gardner, and Gardner, 1972). Vegetation plays a significant role in soil temperature because of the insulating properties of plant cover (Sims and Bass, 1984). Bare soil unprotected from the direct rays of the sun becomes very warm during the hottest part of the day, but also loses its heat rapidly at night and during colder seasons. In the winter, the vegetation acts as an insulator to reduce heat lost from the soil. Frost penetration is more rapid and deeper under bare soils than under a vegetative cover. On the other hand, during the summer months, a well-vegetated soil does not become as warm as a bare soil. Fluctuations in soil temperature decrease with increasing depth (Thornton-Manning, Jones, and Federle, 1987).

Appendix A, Section 5.1.2.2, discusses how the thermal properties of soil can be regulated by the use of mulches, irrigation, and compaction.

5.1.2.3 Soil pH

Soil pH contributes to the surface charge on many colloidal-sized soil particles (JRB and Associate, Inc., 1984b). Clays have a permanent negative charge, and it is primarily their coatings of organic and amorphous inorganic materials that change the charge. At high pH values, the surfaces become more negatively charged; at low pH values, they become positively charged. Thus, the pH of groundwater and soil water in the vadose zone determines the degree of anion or cation adsorption by soil particles. In soils with pH-dependent charge, lowering the pH decreases the net negative charge and, thus, decreases anion repulsion or increases anion adsorption (Hornick, 1983).

Biological activity in the soil is greatly affected by the pH, through the availability of nutrients and toxicants and the tolerance of organisms to pH variations. Some microorganisms can survive within a wide pH range, while others can tolerate only small variations. The optimum pH for rapid decomposition of wastes and residues is usually in the range of 6.5 to 8.5. Bacteria and actinomycetes have pH optima near 7.0. A soil pH of 7.8 should be close to the optimum (Dibble and Bartha, 1979a). If the soil is acidic, these organisms often cannot compete effectively with soil fungi for available nutrients. The pH can influence the solubility or availability of macro- (especially phosphorus) and micronutrients, the mobility of potentially toxic materials, and the reactivity of minerals (e.g., iron or calcium) (Parr, Sikora, and Burge, 1983).

Carbonic acid, organic acid intermediates, and nitrate and sulfate (most important for pH < 5), may accumulate during aerobic degradation of organic molecules (Zitrides, 1983). This can lower the soil pH and inhibit biological activity. The acid conditions can, however, be controlled with reinoculation lime addition, or both. In fact, liming has been found to favor the biodegradation of oil (Dibble and Bartha, 1979a).

Hazardous waste-contaminated soil may contain acidic wastes, which will also decrease the soil pH and change the microorganism distribution (Sims and Bass, 1984; JRB and Associates, Inc., 1984b). Many fungi predominate under acidic conditions (pH < 7). These organisms may transform aromatic hydrocarbons by means of oxygenases into an arene oxide, the mutagenic forms of PAHs (Baver, Gardner, and Gardner, 1972). Bacteria, on the other hand, growing better at a neutral or slightly basic pH, would carry out the dioxygenation of the aromatic nucleus to form a cis-glycol as the first stable intermediate, instead of the arene oxide. Higher organisms (above fungi) do not possess the necessary oxygenases and, thus, form trans aromatic diols, which tend to polymerize.

These differences in the mechanism of aromatic hydrocarbon metabolism by microorganisms have important implications concerning engineering techniques for controlling and possibly detoxifying simple aromatics and PAHs in contaminated soils (Cerniglia, Herbert, Dodge, Szaniszlo, and Gibson, 1979). It appears that selection for dominance of the microbial community by bacteria may avoid the formation of mutagens, and that pH may serve as an important engineering tool to direct the pathway of PAH degradation.

Control of soil pH at an in-place hazardous waste treatment site is a critical factor in several treatment techniques, including metal immobilization and optimum microbial activity (Sims and Bass, 1984). The goal of soil pH adjustment in agricultural application usually is to increase the pH to near neutral values, since most natural soils tend to be slightly acidic. Areas of the country in which the need for increasing soil pH is greatest are the humid regions of the East, South, Middle West, and Northwest States. In areas where rainfall is low and leaching is minimal, such as parts of the Great Plain States and the arid, irrigated saline soils of the Southwest, Intermountain, and Far West States, pH adjustment is usually not necessary but may require reduction. The soil pH may affect the solubility, mobility, and ionized forms of contaminants, (JRB and Associates, Inc., 1984b). The pH of different soil types can vary. A calcareous (containing calcium carbonate) soil can range from pH 7 to 8.3. A sodic (high in sodium carbonate) soil can go as high as pH 8.5 to 10. Saline soils tend to be around pH 7. The soil pH may need to be lowered by adjusting with sulfur or other acid-forming compounds, or raised by adding crushed limestone or lime products to bring it between pH 5.5 and 8.5 to encourage microbial activity. Phosphorus solubility is maximized at pH 6.5; this may be the ideal soil pH. See Appendix A, Section 5.1.2.3, for information on methods for increasing and decreasing soil pH.

5.1.2.4 Oxygen Supply

The degree to which the soil pore space is filled with water affects the exchange of gases through the soil (JRB and Associates, Inc., 1984b). Microbial respiration, plant root respiration, and the respiration of other organisms removes oxygen from the soil and replaces it with carbon dioxide. Gases slowly diffuse into the soil from the air above, and gases in the soil slowly diffuse into the air. However, the oxygen concentration in surface unsaturated soil may be only half that in air, while carbon dioxide concentrations may be many times that of air (Brady, 1974).

As the soil becomes saturated, the diffusion of gases through the soil is severely restricted. In saturated soil, oxygen can be consumed faster than it can be replaced, and the soil becomes anaerobic (JRB and Associates, Inc., 1984b). This drastically alters the composition of the microflora. Facultative anaerobes, which use alternative electron acceptors, such as nitrate (denitrifiers), and strict anaerobic organisms become the dominant species. While many soil bacteria can grow under anaerobic conditions, though less actively, most fungi and actinomycetes do not grow at all (Parr, Sikora, and Burge, 1983). Microbial metabolism shifts from oxidative to fermentative and becomes less efficient in terms of biosynthetic energy production (JRB and Associates, Inc., 1984b). Soil structure and texture primarily determine the size of soil pores, and hence the water content at which gas diffusion is significantly limited in a given soil, and the rate at which anaerobiosis sets in.

As the oxygen is depleted from soils, the reactions become anaerobic with the production of malodorous compounds, such as amines, mercaptans, and H_2S (Parr, Sikora, and Burge, 1983). These can be phytotoxic, and if the soil is heavily overloaded, the soil may remain anaerobic for some time. However, if the oxygen balance is maintained, relative to the amount of contaminants and the soil conditions, rapid aerobic decomposition will occur, and the end products will be inorganic carbon, nitrogen, and sulfur compounds.

Several alternative sources of oxygen have been suggested as a means to increase the degradative activity in contaminated aquifers (Texas Research Institute, 1982) (see Appendix A, Section 5.1.2.4). This section of Appendix A also describes the oxygen requirements for biodegradation of organic compounds and summarizes the advantages and disadvantages of various options for supplying oxygen to the subsurface (Table A.5-7). Oxidizing agents can be used to degrade organic constituents in soil systems, although they may themselves be toxic to microorganisms or may cause the production of more toxic or more mobile oxidation products (Sims and Bass, 1984). Two powerful oxidizing agents for in-place treatment, ozone and hydrogen peroxide, are discussed in depth in Appendix A. Approaches for modifying soil oxygen for both aerobic and anaerobic biodegradation are also described in Appendix A, Section 5.1.2.4.

5.1.2.5 Nutrients

Microbial degradation of hazardous compounds requires the presence of certain nutrients for optimum biological growth (Table A.5-7). Feeding nutrient solutions containing inorganic nutrients, such as soluble nitrogen, phosphorus, and sulfur compounds, to natural soil bacteria often enhances the ability of the microorganisms to degrade organic molecules into carbon dioxide and water (Stotzky and Norman, 1961a; Stotzky and Norman, 1961b).

At one site, decomposition of oil in the soil was shown to proceed at a rate of 0.5 lb/ft^3/month without a nutrient source, and at 1.0 lb/ft^3/month after the addition of fertilizer (Kincannon, 1972). Without added nutrients, aromatic hydrocarbons were noted to be more readily attacked than saturated aliphatic hydrocarbons by the microbes (Atlas, 1981). Addition of nitrogen or phosphorus stimulated degradation of saturated hydrocarbons more than of aromatic hydrocarbons.

These nutrients may be present in contaminating wastes, but may not be readily available or may not supply all that is required (Sims and Bass, 1984). Their supplementation may be necessary. Three of the major nutrients, nitrogen, phosphorus, and potassium, can be supplied with common inorganic fertilizers (JRB and Associates, Inc., 1984b). The carbon, nitrogen, and phosphorus content of bacterial cells is generally in the ratio of 100 parts carbon to 15 parts nitrogen to 3 parts phosphorus (Zitrides, 1983). (Alexander, 1977 specifies the approximate C:N:P: ratio found in bacterial biomass as 120:10:1.) By knowing how much of the carbon in a spilled substance ends up as bacterial cells, it is possible to calculate the amount of nitrogen and phosphorus necessary to equal this ratio for bacterial growth (Thibault and Elliot, 1980a). This procedure is explained in Appendix A, Section 5.1.2.5. Sufficient nitrogen and phosphorus should be applied to ensure that these nutrients do not limit microbial activity (Alexander, 1981).

The main danger at hazardous waste sites may be in overloading the soil with elements that may have been present in the waste (or already in the soil, e.g., phosphate plugging), causing toxicity and leaching problems (JRB and Associates, Inc., 1984).

Appendix A, Section 5.1.2.5, further discusses nutrient requirements for biodegradation in different environments, the effect of various additives, and precautions that should be observed in adding certain substances. An optimum fertilization plan is proposed.

5.1.2.6 Organic Matter

Organic material is very important in the soil matrix (Hornick, 1983). The presence of organic materials may have many effects on soil properties, including degree of structure, water-holding capacity, bulk density, mobilization of nutrients (hindering degradation of organic wastes), reduction in soil erosion potential, and soil temperature (Atlas, 1978c). Naturally occurring organic material can influence the ability of microorganisms to degrade pollutants (Shimp and Pfaender, 1984). Its role in metal reactions or sorption processes that occur in the soil determines the availability of metals and essential nutrients for plants and microorganisms (Hornick, 1983). Sorption of contaminants on soil particles can alter the molecular character and enzymatic attack of a given compound. Humic polymers can act as stabilizing agents, making the compounds less resistant to biodegradation (Verma, Martin, and Haider, 1975). However, the opposite effect has also been observed, with bound material becoming unavailable for biodegradation. After adaptation of a microbial community to four types of compounds, it was found that amino acids, fatty acids, and carbohydrates stimulated biodegradation of monosubstituted phenols, while humics decreased biodegradation rates (Shimp and Pfaender, 1984).

Soil contains organic material in varying stages of decomposition (JRB Associates, Inc. 1984b). Around 65 to 75% of this material usually consists of humic substances; i.e., humic acid, fulvic acid, and humin, which have very large surface areas and high cation exchange capacities (Schnitzer, 1978). The remainder of the organic material consists of polysaccharides and proteins, such as carbohydrates, proteins, peptides, amino acids, fats, waxes, alkanes, and low molecular weight organic acids, which are rapidly decomposed by the soil microorganisms (Schnitzer, 1982). This organic matter can also contribute nitrogen, phosphorus, sulfur, zinc, and boron, all of which add to the nutrient status of the soil (JRB and Associates, Inc., 1984b).

If biodegradable organic materials are added to the soil in order to raise the carbon to nitrogen ratio higher than about 20:1, mineral nitrogen in the soil will be immobilized into microbial biomass, and the decomposition process will be slowed considerably (JRB and Associates, Inc., 1984b). Phosphorus is similarly immobilized when carbon is in excess (Alexander, 1977). If the soil must be managed to decompose organic matter during the treatment of hazardous waste-contaminated soils, nitrogen and phosphorus may be required to bring the C:N:P ratio close to that of the bacterial biomass. However, C:N ratios should be used cautiously, since they do not indicate the availability of the carbon or nitrogen to microorganisms (Taylor, Parr, Sikora, and Willson, 1980).

Organic matter is very important to the microbial ecology and activity of the soil (Sims and Bass; JRB and Associates, Inc., 1984b). Its high cation-exchange capacity and high density of reactive functional groups help to bind both organic and inorganic compounds that may be added to the soil. These properties also help to retain the soil bacteria which can then attack the bound compounds. Thus, the sorbents may immobilize the organic constituents, as well as allow more time for biodegradation. Microbial enzymes can sometimes catalyze the binding of xenobiotics to soil humus, where local conditions may regulate their subsequent release (de Klerk and van der Linden, 1974). If the soil contains cracks and fractures that may increase the potential for mobilization and groundwater contamination, addition of an adsorbent can be useful (Sims and Bass, 1984; JRB and Associates, Inc., 1984b). It is especially important and effective in soils with low organic matter content, such as sandy and strip-mined soils. These sorbants include agricultural products and by-products, sewage sludges, other organic matter, and activated carbon.

Supplemental carbon and energy sources can be used to stimulate the metabolism of even recalcitrant xenobiotics, either through cometabolism (Alexander, 1981) or simply because of the presence of additional carbon and energy (Yagi and Sudo, 1980). The naturally occurring organic components of soil and their involvement in biodegradation are further discussed in Appendix A, Section 5.1.2.6.

5.1.2.7 Oxidation-Reduction Potential

The oxidation-reduction potential, or Eh, of the soil in question basically expresses the electron availability as it affects the oxidation states of hydrogen, carbon, nitrogen, oxygen, sulfur, manganese, iron, cobalt, copper, and other elements with multiple electron states in aqueous systems (Bohn, 1971). This indicates the electron density of a system. As a system becomes reduced, there is an increase in electron density and negative potential (Taylor, Parr, Sikora, and Willson, 1980). Eh decreases during flooding and increases during drying (Bouwer, 1984). The fastest changes occur within the top 2 cm.

The degradative pathways for some hazardous compounds may involve reductive steps (JRB and Associates, Inc., 1984b). This may occur as an initial reaction that requires anaerobiosis, or it may be expressed by more rapid degradation under anaerobic conditions. Many compounds, such as heavily chlorinated compounds, can be transformed under anaerobic or alternating anaerobic-aerobic conditions, but not readily under strict aerobic conditions (Wilson and Wilson, 1985). As previously stated, the initial dechlorinations occur most readily under anaerobic conditions, with chlorine acting probably as an electron acceptor. Some contaminated aquifers will be anaerobic, and if the microbial population is capable of degrading the material, it may be possible to use anaerobic in situ techniques to treat some compounds.

The redox conditions can be controlled to achieve conditions under which specific compounds can be degraded, dehalogenated, or particular organisms or enzyme systems can be selected (Wilson and Wilson, 1985). Alteration of aerobic/anaerobic conditions by adjusting Eh through flooding or cultivation can, therefore, be a useful tool for engineering management to maximize detoxification and degradation of some compounds (Guenther, 1975). Anaerobic conditions can be maintained by keeping the soil saturated with water and limiting aeration. Regular cultivation of soil should maintain aerobic conditions. Table A.5-8 shows the succession of events related to the redox potential, which can occur in poorly aerated soils receiving excessive loadings of organic material.

There are economic advantages of using anaerobic degradative processes. For example, accentuated anaerobic metabolism of fuel oil components would eliminate problems (plugging, intensive management) caused by hydrogen peroxide addition to wells. Although pertinent mainly to chlorinated solvents and PCBs (not really addressed in this report), anaerobic conditions are required to dehalogenate some highly chlorinated materials (e.g., tetrachloroethylene and perhaps higher chlorinated PCBs), while aerobic conditions seem to be required to remove halogens from lower chlorinated compounds (dichloroethylene, vinyl chloride). Thus, anaerobic-aerobic control may be of value here. Stanford University is investigating accentuation of the anaerobic breakdown of aromatic compounds, because the process may be cost effective.

Appendix A, Section 5.1.2.7, presents different ways to express the oxygen levels in soils and aquatic surface and subsurface environments. It also classifies oxygen levels in various soil types, based upon their redox potential.

5.1.2.8 Attenuation

The basic principle of attenuation is the mixing of contaminated soil (or wastes) with clean soil to reduce the concentrations of hazardous compounds to acceptable levels (Sims and Bass, 1984). This is applicable to both inorganics and organics. The mixing of uncontaminated soil with the contaminated increases the extent and effectiveness of immobilization of chemical contaminants and may also aid in decreasing toxicity of the contaminated soil to the soil microorganisms involved in biodegradation. In practice, attenuation systems have been designed, and acceptable concentration limits established, only for heavy metals. However, in principle, this technique should also apply to organic contaminants.

Attenuation is included in Table A.5-4, which lists the soil modification requirements for several treatment technologies. See Appendix A for further discussion of this practice.

5.1.2.9 Texture and Structure

Soil type may be an important determinant of whether groundwater pollutants are biodegraded as they pass through the unsaturated zone of a soil profile (Federle, Dobbins, Thornton-Manning, and Jones, 1986). Microorganisms have been shown to be present, often in large numbers, in the entire vertical profiles of sediments in wells several hundred feet deep. The vertical distribution of microorganisms in a soil profile differs greatly as a function of soil type. In four different soil types, biomass and activity declined with increasing depth; however, the magnitude and pattern of this decline differed for each soil type (see Table A.3-3). The soil type also affected the types of microbial populations present. Horizon and soil type affect the time of transit of a contaminant, as well as the potential for biodegradation. The type of soil will also influence the mobility of microorganisms through the subsurface.   Bacteria generally do not move large distances in fine-textured soil (less than a few meters, for example), but they can travel much larger distances in coarse-textured or fractured materials (Romero, 1970).

The composition of soil influences infiltration rate and permeability, water-holding capacity, and adsorption capacity for waste components (Hornick, 1983). These, in turn, have an effect on the biodegradability of the contaminating wastes and the ability of microorganisms to metabolize the compounds. Appendix A, Section 5.1.2.9, describes how the physical properties of soil (i.e., texture, bulk density, and water-holding capacity) function in the process of biodegradation.

5.1.3 Alteration of Organic Contaminants

When the concentration or availability of organic substrate is low, increasing the surface area-to-liquid volume ratio improves substrate utilization and bacterial growth rates (Heukelekian and Heller, 1940; Characklis, 1973).

5.1.3.1 Addition of Surfactants

Surface active agents are a class of natural and synthetic chemicals that promote the wetting, solubilization, and emulsification of various types of organic chemicals (Amdurer, Fellman, and Abdelhamid, 1985). Surfactants are organic molecules and can be either cationic, anionic, or nonionic (JRB and Associates, Inc., 1984b). These agents would be most effective in promoting the mobilization of organic compounds of relatively low water solubility and high lipid solubility (high K_{ow} values). Surfactants may enhance the recovery of subsurface gasoline contamination by groundwater pumping and promote the mobilization of hydrophobic contaminants from unsaturated soils (Ellis and Payne, 1984).

The physical and biological properties of soil may be affected when surfactants are added (Ellis and Payne, 1984). A concentration of 100 ppm of most synthetic emulsifiers is not substantially deleterious to microorganisms. However, detergents and other strong (especially synthetic) emulsifiers/surfactants may disrupt cell membranes (lipids) or cause toxicity by excessive uptake of toxic hydrophobic contaminants. In addition, surfactants may be precipitated by groundwater with high TDS or alkaline earth cation concentrations (Ca, Mg). They require optimal pH and temperature and can be adsorbed by soil particles, negating their solubilizing properties. Nevertheless, use of surfactants alone to flush otherwise insoluble organics, or in combination with other treatments to solubilize waste materials (thereby, promoting biodegradation), is a promising avenue for further research.

Microbes have been long known to produce surface active agents when grown on specific substrates (Zajic and Panchal, 1976). Many of these surfactants have excellent emulsifying properties (Panchal and Zajic, 1978). Many oil-degrading microorganisms produce emulsifying agents (Reisfeld, Rosenberg, and Gutnick, 1972), and naturally occurring biosurfactants seem to be very important in the elimination of hydrocarbons from polluted biotopes (Rambeloarisoa, Rontani, Giusti, Duvnjak, and Bertrand, 1984). These biosurfactants contain sugars, fatty acids, and lipids (mono- and diglycerides) and have a strong emulsifying effect on petroleum contaminants. Attempts have been made to isolate these emulsifying agents for possible use in dispersing oil. Bioemulsifiers are generally easy and inexpensive to produce and are usually nontoxic (Zajic and Gerson, 1977). Production of these compounds can be easily controlled by media manipulation, and emulsifiers of biological origin are being applied for industrial purposes (Zajic and Panchal, 1976).

Microorganisms generally consume only soluble or solubilized (emulsified) organic molecules; and synthesis of an emulsifier may pseudosolubilize target hydrocarbons (Thibault and Elliot, 1980a; Goma, Al Ani, and Pareilleux, 1976). An <u>Arthrobacter</u> strain was found to emulsify oil extensively when growing on hydrocarbons (Reisfel, Rosenberg, and Gutnick, 1972). However, organisms that can emulsify oil often do not extensively degrade the hydrocarbons in the oil (Atlas, 1981). Where the hydrocarbonoclastic microbes do not have strong emulsification ability, commercially produced, biodegradable emulsifiers could be added to accelerate the biodegradation process. In fact, biodegradation rates could perhaps be enhanced by use of microbes that secrete large quantities of such surface-active agents (Zitrides, 1983). Appendix A, Section 5.1.3.1, describes how microorganism-generated or externally added surface active agents effect the uptake of substances microorganisms could not previously utilize.

There are also many chemical oil dispersants that could be used for this purpose (Canevari, 1971). Successful chemical enhancement of oil biodegradation is dependent upon the particular emulsifying agent (Robichaux and Myrick, 1972). The problem with dispersing oil, even if it accelerates soil biodegradation, is that many oil dispersants can be toxic to soil organisms (Shelton, 1971). On the other hand, some emulsifying agents may not produce these undesirable effects.

Both an oil and a water phase are present in soil contaminated with petroleum products (Genner and Hill, 1981). Nutrients, inhibitors, and metabolic products are always partitioned between these phases. Microorganisms proliferate in the water phase and migrate into the hydrocarbon phase, where they usually die. Hydrocarbon-degrading microorganisms are found mainly at the oil-water interface (Atlas, 1981). In soil, microbes break down oil at oil-water interfaces (Flowers, Pulford, and Duncan, 1984). These organisms can be seen growing over the entire surface of an oil droplet. Growth does not appear to occur within oil droplets in the absence of entrained water. Increasing the oil surface area should accelerate biodegradation. The material is more readily available to microorganisms, and movement of emulsion droplets through the water column increases the availability of oxygen and nutrients.

This led to the commercial development and application of synthetic biodegradable emulsifiers for the cleanup of land-based spills of hydrophobic materials, such as crude and refined petroleum products

(Thibault and Elliott, 1979). These emulsifiers act rapidly on the hydrocarbons and, therefore, assist in their rapid assimilation by the indigenous microorganisms. Use of mutant microorganisms that produce large quantities of biosurfactants have also been shown to be useful in accentuating petroleum hydrocarbon degradation.

Bioemulsifiers are highly substrate specific and are most effective with mixtures of compounds, which may not be emulsified individually (Rosenberg, Zuckerberg, Rubinovitz, and Gutnick, 1979). Bioemulsifiers and biosurfactants may enhance pseudosolubilization of hydrocarbons or direct contact between microorganisms and a hydrocarbon substrate. It should also be noted that emulsification of the hydrocarbons will aid in mobilizing them through the soil (Vanloocke, Verlinde, Verstraete, and DeBurger, 1979).

See Appendix A, Section 5.1.3.1, for a discussion of the application of bioemulsification to petroleum product spills.

5.1.3.2 Supplementing Threshold Concentrations of Contaminants

Laboratory evidence clearly indicates that under proper conditions, many potentially hazardous organic compounds can be biodegraded in the ground, even when present at very low concentrations (Bitton and Gerba, 1985). However, evidence suggests that there may be a threshold concentration for a given contaminant, below which it will not be biodegraded. This would be expected, if the energy derived from low concentrations of the substrate was inadequate for maintenance of the bacterial cell, or if higher concentrations were required to activate the transport and metabolic systems of the cell (Boethling and Alexander, 1979). By adding more of the given contaminant (or other less or nonhazardous analogs) this threshold could perhaps be exceeded, since more inducible enzyme(s) would be produced (for example, adding fulvic acid to stimulate the degradation of benzene, toluene, or phenols).

5.2 OPTIMIZATION OF GROUNDWATER BIODEGRADATION

Some of the information in this section may duplicate material covered in Section 5.1, Optimization of Soil Biodegradation, and Section 5.3, Optimization of Freshwater, Estaurine, and Marine Biodegradation; however, it is presented here under a separate heading, with other related information, to accommodate those readers who may specifically wish to address treatment of groundwater contamination only.

Materials from a spill normally reach a groundwater system either through direct contact with a recharge area or by rainwater percolating through contaminated soil (Buckingham, 1981). Examination of these mechanisms, as well as practical sampling, testing, and management techniques, can make it possible to treat a wide variety of spills through the use of a biodegradation scheme.

Microbial activity in aquifers may be limited by the levels of dissolved oxygen, nutrients, and the numbers of microorganisms capable of degrading the contaminants (Lee and Ward, 1983). In situ treatment processes usually involve the circulation of both oxygen and inorganic nutrients through the aquifer so indigenous organisms can degrade the contaminants (Lee and Ward, 1985). Supplying dissolved oxygen to the groundwater is likely to be the limiting factor in the biostimulation process, especially in low-permeability aquifers.

If a contaminated water table aquifer is oxygenated and contains organisms capable of degrading petroleum hydrocarbons, there is potential for natural remediation of the plumes with time (Lee and Ward, 1984a; Wilson and Rees, 1985). When an aquifer is anaerobic and there is no significant degradation activity after a gasoline or fuel oil contamination, oxygen and nutrients can be added, with a resulting rapid degradation of the hydrocarbons down to approximately 1 ug/l with little lag time (Josephson, 1983b). This, however, requires an initial hydrocarbon concentration below about 10 mg/l. Higher concentrations may produce a considerable adaptation time.

Withdrawal and biological treatment seem to currently be the most effective methods for biological restoration of aquifers contaminated by organic compounds (Knox, Canter, Kincannon, Stover, and Ward, 1968). Biostimulation by the addition of oxygen and nutrients has been chiefly used to reclaim aquifers contaminated by gasoline and has been effective in reducing, although not completely eliminating, the quantity of gasoline. Enhancing the indigenous microbial population has also been fairly effective in aboveground treatment of groundwater contaminated with organic solvents, but augmentation has generally failed to increase petroleum hydrocarbon degradation rates. Seeding contaminated sites with microorganisms and the use of ozone or hydrogen peroxide to increase in situ oxygen levels are promising developments but are, as of yet, unproved.

Biodegradation of contaminants by the indigenous microbial population is one of the most effective methods for removing organic contamination. Manipulating these variables may help improve the biodegradation.

5.2.1 Biological Enhancement

Microorganisms in the groundwater can transform many of the organic contaminants typically found in contamination incidents (Wilson, Leach, Henson, and Jones, 1986). Different isolates are able to utilize certain components of gasoline directly, while cooxidation plays an important role in the biodegradation of the other compounds. Aquifer sediment microorganisms can mineralize trace levels of petroleum hydrocarbon compounds, and the rates can be affected by preexposure of the microbial community to the petroleum product (Dooley, Larson, and Ventullo, 1985).

An approach for in situ treatment of contaminated aquifers is to promote a particular microbial population with specific metabolic capabilities to flourish. (Lee and Ward, 1986).

5.2.1.1 Seeding of Microorganisms

Maintenance of the bacterial population at optimal levels is important, especially for selective mutant organisms, which tend to be more sensitive to environmental variables than naturally occurring species (Environmental Protection Agency, 1985b). In theory, a continuous incubation facility operating at optimal growth temperatures and under more controlled conditions could be used to maintain the microbial population. The high biomass-containing stream produced from such a facility could then be reinjected via wells or trenches so as to reinoculate the subsurface continuously with microorganisms, although clogging of the soil and transport could be serious problems. It has been suggested that seed bacteria in an aqueous environment be encapsulated to ensure that they adhere to and remain with the contaminating oil (Gholson, Guire, and Friede, 1972).

5.2.1.2 Acclimation

Organisms from contaminated sites may be capable of degrading a wider range of compounds once they become acclimated to the organic compounds, a process that may require several months (Wilson, McNabb, Cochran, Wang, Tomson, and Bedient, 1986) or even years (Mackay, Roberts, and Cherry, 1985). Other authors reported a lag period of one to five days, depending upon temperature, before the groundwater microbial flora are able to degrade dissolved hydrocarbons measurably (Kappler and Wuhrman, 1978a; Kappler and Wuhrman, 1978b). The biotransformation rates of trace organic contaminants are highly variable, with half-lives ranging from a few days to many years. Use of microorganisms preadapted to degrade specific chemicals can reduce the lag period normally required for acclimation to the substrates (Mackay, Roberts, and Cherry, 1985). Previously acclimated organisms have been employed in remedial actions to clean up contaminated groundwater (Lee and Ward, 1986).

Withdrawal and biological treatment of contaminated groundwater has been shown to be an effective, albeit slow, method for restoration of aquifers and improvement of degradation rates (Ward and Lee, 1984). Activated sludge treatment of pumped groundwater, often in conjunction with physical treatment processes (such as activated carbon adsorption or air stripping), has been demonstrated to build up an acclimated microbial population that can degrade most petroleum contaminants. See Appendix A, Section 5.2.1.2, for a further discussion of acclimation of microorganisms in aquifer restoration.

5.2.2 Optimization of Groundwater Factors

The biodegradation of a particular class of organic contaminants depends upon the physiological capabilities of the organisms in the aquifer (Wilson, Leach, Henson, and Jones, 1986). These capabilities depend, in turn, upon the geochemical environment of the organisms. Essentially, degradation of organic compounds in groundwater depends upon the geochemical properties of the groundwaters that receive them.

Many of the same procedures for supplementing aquatic and soil systems can be used to restore contaminated aquifers (Ward and Lee, 1984). Contaminated groundwater can be pumped to the surface where aeration, nutrients, microorganisms, and surfactants can be added before it is reinjected (Zitrides, 1983). In addition, it is possible to modify the groundwater (Lee and Ward, 1986). A pilot study where nutrients and oxygen were supplied to a contaminated aquifer showed that the microbial population increased from 10^3/ml to 10^6/ml after seven days and remained constant at that level (Knox, Canter, Kincannon, Stover, and Ward, 1968). Pumping and treating water from a contaminated aquifer, by using activating tanks, maintaining the temperature at 20°C, adding air and nutrients, and supplying an acclimated bacterial culture, resulted in reducing the concentration of the contaminants to an acceptable level in a year.

5.2.2.1 Temperature

As in soil, temperature plays a significant role in aquatic systems. The temperature range for optimal organism growth in aerobic biological wastewater treatment processes has been found to range from 20° to 37°C (68° to 99°F) (Environmental Protection Agency, 1985b). According to the "Q-10" rule, for every 10°C decrease in temperature in a specific system, enzyme activity is halved.   Figure A.5-3 provides typical groundwater temperatures throughout the United States.

Raising the temperature of a contaminated zone by pumping in heated water or recirculating groundwater through a surface heating unit may be feasible under conditions of low groundwater flow (Environmental Protection Agency, 1985b).

5.2.2.2 Oxygen Supply

At many sites contaminated with petroleum hydrocarbon contaminants, the levels of dissolved oxygen may control biodegradation (Lee, Thomas, and Ward, 1985). For biodegradation of most organic contaminants, roughly two parts of oxygen are required to metabolize one part of organic compound completely (Wilson, Leach, Henson, and Jones, 1986). For example, microorganisms in a well-oxygenated groundwater containing 4 mg/l of molecular oxygen can degrade only 2 mg/l benzene. Dissolved oxygen should be maintained above the critical concentration for the promotion of aerobic activity, which ranges from 0.2 to 2.0 mg/l, with the most common being 0.5 mg/l (Environmental Protection Agency, 1985b). The concentration of the contaminants affects degradation. Concentrated plumes cannot be degraded aerobically until dispersion or other processes dilute the plume with oxygenated water (Wilson, Leach, Henson, and Jones, 1986).

Many methods have been developed for aerating fluid and semisolid systems above ground (Texas Research Institute, 1982). These include pumps, propellers, stirrers, spargers, sprayers, and cascades--all designed to bring more of the oxygen-deficient material in contact with oxygen in the air. These are discussed in Appendix A, Section 5.2.2.2. Diffusers that sparge compressed air into the groundwater cannot exceed the solubility of oxygen in water, 8 to 10 ppm. Use of pure oxygen can increase the dissolved oxygen content to 40 to 50 ppm, but pure oxygen is expensive and the supersaturated oxygen is likely to degas before the microbes can utilize it (Brown, Norris, and Raymond, 1984). Oxygen levels required for biodegradation and methods for attaining these levels are presented in Appendix A, Section 5.2.2.2.

Alternative sources of oxygen are being considered for increasing levels of dissolved oxygen in a biodegradative system. Hydrogen peroxide, which decomposes to form water and oxygen, can supply much greater oxygen levels (Westlake, Robson, Phillippe, and Cook, 1974). However, concentrations of hydrogen peroxide as low as 200 ppm may be toxic to microbes and levels above 100 ppm may degas to form air bubbles that block the subsurface formations. Ozone can also be used as a source of oxygen, but with the same limitation as hydrogen peroxide in that it is toxic to bacteria and may generate gas bubbles that block the pores in the formation (Lee and Ward, 1984). Use of ozone and hydrogen peroxide for increasing oxygen levels in biodegradation and methods of overcoming their limitations are discussed in depth in Appendix A, Sections 5.1.2.4 and 5.2.2.2.

Appendix A, Section 5.2.2.2, also further discusses the importance of supplemental oxygen for biodegradation in groundwaters and describes possible sources of oxygen and methods of introducing oxygen into contaminated sites.

5.2.2.3 Nutrients

Most groundwater supplies contain a microbial population capable of oil biodegradation (Jamison, Raymond, and Hudson, 1975; Jamison, Raymond, and Hudson, 1976). However, conditions for growth and metabolism of oil may be limited in oxygen, nitrogen, sulfur, phosphorus, and certain trace minerals, although it has been suggested that the levels of most trace elements in aquifers are usually sufficient to support microbial growth (Yang and Bye, 1979). If nutrients are added to the subsurface, there can be a problem with well plugging, from excessive growth at the nutrient release point. An example of this occurred when diammonium phosphate was added to supply nitrogen and phosphorus, which caused excessive precipitation and the procedure had to be discontinued (Raymond, Jamison, and Hudson, 1976).

5.2.2.4 Oxygen-Reduction Potential

Degradation of high concentrations of organic compounds may lead to a rapid depletion of the dissolved molecular oxygen and to an eventual decrease in the redox potential in an aquifer (Zeyer, Kuhn, and Schwarzenback, 1986). Aerobic microorganisms that predominate at high redox potentials are replaced by denitrifying, sulfate-reducing, or even methanogenic populations. The redox potential also largely determines the metabolic diversity of the microorganisms in the aquifer.

Many contaminated aquifers will be anaerobic if the microbial population is capable of degrading the material, and it may be possible to use anaerobic in situ techniques to treat some compounds (Lee and Ward, 1986), although hydrocarbons degrade slowly under anaerobic conditions. The redox conditions can be controlled to achieve conditions under which specific compounds can be degraded, dehalogenation can be promoted, and particular organisms or enzyme systems can be selected (Appendix A, Section 5.2.2.4).

5.2.3 Alteration of Organic Contaminants

5.2.3.1 Addition of Surfactants

Microbial activity in contaminated aquifers can be enhanced by altering the contaminant chemically or physically to make it more degradable (Ward and Lee, 1984). The dispersal of hydrocarbons in an aqueous system renders them more susceptible to enzymatic attack (Texas Research Institute, 1982). One way to do that is to increase the contaminant surface area and mobility by adding surfactants.

Use of a surfactant can be helpful, but it can also be a mixed blessing (Texas Research Institute, 1982). Since a surfactant will increase the concentrations of both aliphatic and aromatic fractions in the water, the toxicity of the water could possibly increase because the aliphatic molecules may be preferentially degraded. The surfactant itself may also be toxic. The n-alkylpyridinium bromides were found to be more toxic than n-alkylcarboxylates of identical chain length, confirming that the head group of the amphiphiles plays an important role in the microbial toxicity of surfactants (Beaubien, Keita, and Jolicoeur, 1987). The n-alkylcarboxylates appear to be rapidly metabolized by microorganisms in a mixed culture, at least for monologs lower than C_{10}. Cellular toxicity is also dependent upon surfactant hydrophobicity.

See Appendix A, Section 5.1.3.1, for more information on the use of surfactants.

5.2.3.2 Photolysis

Photochemical reactions may be used for the enhancement of compound biodegradation at hazardous waste sites (Sims and Bass, 1984). Photolysis reactions are oxidative and should aid microbial degradation through the oxidation of resistant complex structures (Sims and Overcash, 1981; Crosby, 1971). Such reactions are limited to the surface of the soil or surface treatment of groundwater, but when coupled with soil mixing, may prove to be effective for treating relatively immobile chemicals.

Photolysis can be due either to direct light absorbed by the substrate molecule (direct) or to reactions mitigated by an energy-transferring sensitizer molecule (sensitized photooxidation) (Sims and Bass, 1984). Sensitized reactions result in substrate molecule oxidation rather than substrate isomerism, dehalogenation, or dissociation characteristic of direct photolysis. The reaction rates and breakdown products are only crudely understood. Photolysis can be affected by soil characteristics, such as soil organic content (Spencer, Adam, Shoup, and Spear, 1980), transition metal content (Nilles and Zabik, 1975), soil pigment content (Burkhard and Guth, 1979), and soil moisture content (Burkhard and Guth, 1979).

Of all atmospheric removal mechanisms, including physical, chemical, and photochemical, the photochemical reactions are the most significant for most classes of hazardous compounds and should be investigated further as a viable treatment option (Cupitt, 1980).

Appendix A describes how to enhance photodegradation and how to assess the potential for use of photodegradation for specific organic compounds.

5.3 OPTIMIZATION OF FRESHWATER, ESTAURINE, AND MARINE BIODEGRADATION

Some of the information in this section may duplicate material covered in Section 5.1, Optimization of Soil Biodegradation, and Section 5.2, Optimization of Groundwater Biodegradation; however, it is presented here under a separate heading, with other related information, to accommodate those readers who may specifically wish to address treatment of freshwater, estaurine, and marine contamination only.

There are considerable differences in the environmental and nutritional parameters, as well as in the microbial communities of aquatic and soil environments (Bartha and Atlas, 1977). The physical behavior of oil pollutants in the two environments is also quite distinct. In surface aquatic environments, horizontal spreading is very rapid and extremely hard to control, while vertical movement is initially restricted. Consequently, the bulk of an oil spill in a surface water body is subject to extensive weathering changes that influence its composition and ultimate fate. A quantitative recovery of an oil slick is almost never feasible, and for some or most of the spill, microbial degradation remains the principal mechanism of removal. However, the degradation rate can be vastly improved by accentuating microbial processes or by adding preacclimated hydrocarbon-degrading microorganisms.

Chronic inputs allow for adaptation of the biological community under the stress of polluting petroleum hydrocarbons, which is different from the stress of sudden inputs occurring from accidental spills (Bartha and Atlas, 1977). Sudden catastrophic spillages often occur in near-shore regions, involving relatively large amounts of contaminant in a small region, and affecting both marine and coastal ecosystems.

The limitations, side effects, and high expense of traditional cleanup techniques have stimulated interest in unconventional alternatives (Bartha and Atlas, 1977).

5.3.1 Biological Enhancement

5.3.1.1 Seeding of Microorganisms

Removal of stranded oil from beach sand or intertidal areas is a use for which seeding for accelerated biodegradation has been considered (Guard and Cobet, 1972). Microorganisms seeded into intertidal areas would have to survive alternate periods of desiccation and exposure to high concentrations of sodium chloride. However, the weathered and partially biodegraded oil on beaches leaves an oil or tar composed of few easily biodegradable components. Since there is a limited oxygen exchange in a sand column, the rates of biodegradation here would be slow.

It is possible to inoculate oil slicks with highly efficient hydrocarbon degraders (Bartha and Atlas, 1977). Some commercial mixtures of microorganisms have been marketed for use in degrading oil in lagoons and other situations (Atlas, 1977). The applicability of seeding selected bacteria and fungi to oil spills has been patented by Azarowicz and Bioteknika International, Inc. (Azarowicz, 1973). However, the full claims of the effectiveness of such organisms remain to be proved.

Organisms used for seeding and the conflicting results obtained with this approach are described in Appendix A.

5.3.1.2 Acclimation

Adaptation capabilities of a microbial community are highly variable depending upon site and other factors (Spain and Van Veld, 1983). Adaptation of aquatic microbial communities can last for several weeks after exposure to a xenobiotic compound.

5.3.2 Optimization of Aquatic Factors

5.3.2.1 Temperature

Temperature is a major factor regulating biodegradation of hydrocarbons in seawater (Mulkins-Phillips and Stewart, 1974b). Hydrocarbon degradation is limited in the winter, except at sites previously contaminated and chronically receiving inputs of hydrocarbons and nutrients. Hydrocarbon biodegradation can occur at the low temperatures ($<5°C$) that characterize most of the ecosystems likely to be contaminated by oil spills (Atlas, 1981). Ninety percent of the oceanic water mass has a temperature of 4°C or below (Morita, 1966). However, the temperature of the water should be 10°C or above to maintain near-maximum bacterial heterotrophic potential.

Low temperatures tend to decrease rates of metabolic oil degradation (Atlas, 1977). Other conditions being equal, moderate to high water temperatures favor the biodegradation approach, while very cold water temperatures would generally contraindicate the use of stimulated biodegradation (Bartha and Atlas, 1977).

In practical terms, little can be done about the temperature of the ocean water (Bartha and Atlas, 1977). In large, open environments, such as oceans, it is not feasible to alter temperatures in an effort to stimulate biodegradation (Atlas, 1977). However, understanding the effects of temperature can allow for optimum conditions for biodegradation to occur.

Seasonal and climatic variability, the types of hydrocarbons involved, and the nature of the site are also important (Colwell and Walker, 1977). In colder climates, there appears to be a seasonal shift to a microbial community capable of low-temperature hydrocarbon degradation (Atlas and Bartha, 1973c). For instance, a higher number of hydrocarbon utilizers capable of growth at 5°C were present in Raritan Bay, NJ, during winter than during other seasons. In this situation, rates of hydrocarbon mineralization measured at 5°C were significantly higher in water samples collected in winter than in summer.

Temperature effects depend upon the length of time involved (Nedwell and Floodgate, 1971). The short term reflects the response of a given population to the change of temperature, while the long term may see the development of a different population that is adapted to the new conditions. The temperature optima for the heterotrophic utilization of glucose and amino acids in lake sediments decrease with decreasing ambient temperature, suggesting adaptation by or selection for psychrotrophic populations (Bartholomew and Pfaender, 1983). Although microbes in colder waters are adapted to lower temperatures, biodegradation rates can be expected to be much slower, and bioreclamation will take longer in the extreme North. Moderate degradation rates have been observed even at Arctic temperatures (Traxler, 1972), but the rates may not be adequate to rapidly remove hydrocarbon contaminants (Atlas, 1981). Seed microorganisms for such environments will have to be psychrophilic or psychrotrophic, if they they are to be capable of growth and active metabolic degradation of oil (Atlas and Bartha, 1972a).

The effects of temperature differ, depending upon the hydrocarbon composition of a petroleum mixture (Atlas and Bartha, 1972a) (see Appendix A, Section 5.3.2.1). Low temperatures retard the rates of volatilization of low-molecular weight hydrocarbons, some of which are toxic at low concentrations to microorganisms. The presence of such toxic components has been found to delay the onset of oil biodegradation at low temperatures. There is some preference for paraffin degradation, especially at low temperatures. At low temperatures, cometabolism appears to play an important role in determining the rates of disappearance of hydrocarbons (Horowitz and Atlas, 1977a).

Temperate zone lakes have, in general, more extreme water temperature variations than oceans (Bartha and Atlas, 1977). In temperate lakes of Wisconsin, water temperature was the predominant limiting factor throughout the fall, winter, and spring, and only during the summer did nutrient limitation take precedence (Ward and Brock, 1976).

5.3.2.2 Oxygen Supply

Aquatic environments with low concentrations of dissolved oxygen may have lower degradation potentials. Limitation by oxygen deficiency is least likely to occur in cases of thin, floating oil slicks, but the interior of floating water-in-oil emulsions (chocolate mousse) and tar globules of substantial size may well become anaerobic (Bartha and Atlas, 1977).

At pH values between 5 and 8, mineralization of hydrocarbons in estuarine sediments is highly dependent upon oxygen availability (Hambrick, DeLaune, and Patrick, 1980). Rates of hydrocarbon degradation decrease with decreasing oxygen reduction potential. Therefore, hydrocarbons would persist in reduced sediments for longer periods of time than would hydrocarbon contaminants in aerated surface layers. Low mineralization (about 10 to 20 percent) of some alkanes, but not of naphthalene or hexadecane, occurred under anaerobic conditions. Oxygen was not limiting in the water column of three lakes tested; however, it most probably was a limiting factor for utilization of hydrocarbons in the lake sediments, where most of the hydrocarbons were concentrated. Sinking the oil into anoxic sediments is an undesirable treatment method. Many hydrocarbons that enter anaerobic environments, such as reduced sediments, could persist there indefinitely as environmental contaminants (Atlas, 1981).

Forced aeration can be used to supply oxygen in lagoon and many freshwater environments (Rosenberg, Englander, Horowitz, and Gutnick, 1975). Oil biodegradation in the water column of deep lakes is often oxygen limited at various times of the year (Ward and Brock, 1974). Oxygen is rarely a limiting factor in the decomposition of oil in the open ocean, and the rate of oxygen uptake is linear to the rate of nitrogen uptake (Floodgate, 1976). It would generally be impractical to use forced aeration here (Ward and Brock, 1974). Marine oil spills must generally rely on wave action to achieve the necessary aeration. Oxygen might be supplied by employing algae in open, as well as contained, systems (McLean, 1971). Besides supplying oxygen to the water column, some algae have been found to be capable of heterotrophic metabolism of hydrocarbons (Walker, Colwell, and Petrakis, 1975).

The theoretical oxygen demand is 3.5 1 g of oil oxidized per g of oxygen (Floodgate, 1979; ZoBell, 1969). The dissolved oxygen in 3.2 x 10^5 1 of seawater would, therefore, be required for the complete oxidation of 1 1 of oil (ZoBell, 1969).

5.3.2.3 Nutrients

Although temperature is the main limiting factor much of the year, nutrient deficiencies can limit oil biodegradation in temperate lakes during the summer (Ward and Brock, 1976). One of the most important factors that limit the rate of decomposition of oil in the marine environment is the concentration and rate of supply of inorganic nutrients, especially nitrogen and, to a lesser extent, phosphorus (Atlas and Bartha, 1972b). Marine microbial populations are also limited by the levels of organic matter present (Atlas, 1977).

In marine environments where available phosphorus and nitrogen are plentiful, oxidation of oil might be relatively rapid (Floodgate, 1976). Whether nitrogen and phosphorus are going to be a limiting factor in seawater may depend upon whether the contaminant is an oil slick or composed of soluble hydrocarbons (Atlas, 1981). In an oil slick, nitrogen and phosphorus must be able to diffuse to the slick, and rates of diffusion may be inadequate to supply sufficient nutrients to establish optimal C/N and C/P ratios for microbial growth and metabolism. With soluble hydrocarbons, on the other hand, nitrogen and phosphorus are probably not limiting, since the solubility of the hydrocarbons is so low as to preclude establishment of an unfavorable C/N or C/P ratio.

The oxidative state of the inorganic nitrogen is immaterial; both nitrate and ammonia work well (Floodgate, 1976). An oleophilic nitrogen and phosphorus fertilizer is described in Appendix A, Section 5.3.2.3. Some of the nitrogen become part of the bacterial biomass and are incorporated into the cell structure, while some form enzymic molecules needed for oil decomposition (Floodgate, 1976). The amount of nitrogen ("nitrogen demand") required for degradation of a given amount of oil contaminant can be calculated (see Appendix A, Section 5.3.2.3).

Beaches tend to concentrate the above elements to about three times the level found in seawater (Floodgate, 1976). The inorganic nutrients may be less of a limiting factor for oil biodegradation on beaches, since each tide may bring a new supply. An important factor appears to be the way the oil arrives and its degree of division; e.g., lumps are more recalcitrant than a thin skim (Davis and Gibbs, 1975). On a high energy beach, much of the oil will be washed away by wave action (Floodgate, 1976). Otherwise, it can become buried and the conditions anaerobic, with a slowing of the digestion. Ironically, a limiting factor of oil degradation on beaches may be the lack of water for microbial decomposition of the material (Atlas, 1981).

An estuarine area has a much higher bacterial count than offshore areas, possibly because of the availability of more nutrients. Eutrophic lakes and ponds appear to be better at degrading organic compounds than oligotrophic systems.

The need for mineral nutrients is generally recognized, but the addition of nutrients in the form of water-soluble salts restricts the practical application of stimulated oil biodegradation to oily water contained in tanks, bilges, or holding ponds (Rosenberg, Englander, Horowitz, and Gutnick, 1975).   In enclosed systems, addition of mineral-containing water will supply many of the required nutrients (Atlas, 1977). Supplemental nitrogen and phosphorus would still have to be added for extensive oil biodegradation. These nutrients could be added in a water-soluble form, such as ammonium phosphate in these situations (Rosenberg, Englander, Horowitz, and Gutnick, 1975).

Treatment of free-floating oil slicks requires oil- rather than water-soluble nitrogen and phosphorus supplements in a form that will not allow it to dissipate from the oil-water interface, i.e., oleophilic fertilizers.   These fertilizers consist of paraffinized urea and octyl phosphate and are selectively available to oil-degrading microorganisms only (Atlas and Bartha, 1973a). Application of oleophilic fertilizers also prevents algal blooms, which would follow the use of water-soluble formulations.

5.3.3 Alteration of Organic Contaminants

The physical state of petroleum hydrocarbons has a marked effect on their biodegradation (Atlas, 1981). Most petroleum hydrocarbons are soluble in water at very low concentrations, but most oil spill incidents release petroleum hydrocarbons in concentrations far in excess of the solubility limits (Harrison, Winnik, Kwong, and Mackay, 1975). The degree of spreading determines, in part, the surface area of oil available for microbial colonization; in aquatic systems, the oil normally spreads, forming a thin slick (Berridge, Dean, Fallows, and Fish, 1968). The degree of spreading is reduced at low temperatures because of the viscosity of the oil.

Petroleum that is discharged into the sea is acted upon by a variety of processes that cause changes in its physical and chemical characteristics (Tjessem and Aaberg, 1983). Spreading on the surface, followed by evaporation and emulsification appear to be the dominant initial processes (see Appendix A, Section 5.3.3). These processes are highly dependent upon temperature, wind, and sea waves, as well as the exact chemical composition of the particular oil (Payne and Jordan, 1980).

5.3.3.1 Addition of Surfactants

In a two-phase liquid medium where the bulk of the carbon and energy source is water insoluble and all other mineral nutrients are dissolved in the water phase, microbial growth typically occurs at the interface of the two liquids (Bartha and Atlas, 1977). The interfacial tension reduces this interface to a minimal area. Microorganisms can lower the interfacial tension, thereby, increasing the interface and the accessibility of the hydrocarbon substrate. This phenomenon is apparently due to incidental leakage of fatty acids and other metabolic intermediates. In fact, fatty acids are common emulsifying agents. Many oil-degrading microorganisms produce emulsifying agents (Reisfeld, Rosenberg, and Gutnick, 1972), and there is interest in the use of microbes as oil dispersants, since they would have low toxicity for cleanup of oil and cause minimal environmental damage.

In aquatic ecosystems, oil tends to spread naturally, approaching a monolayer in thickness (Fay, 1969). A greater surface area in these ecosystems can also be achieved by emulsification (Berridge, Dean, Fallows, and Fish, 1968). Use of emulsifiers on a slick to increase the spreading of the oil and its exposure to sunlight (causing photolysis) may help increase the rate of degradation (Tissot and Welte, 1978) (see Appendix A, Section 5.3.3.1). If there are no adverse toxic effects, dispersion of oil should also accelerate microbial hydrocarbon degradation (Atlas, 1981). This is an important consideration when deciding whether dispersants should be added to oil spills. However, some dispersants may contain chemicals that are inhibitory to microorganisms, and increased toxicity must also be considered in their use.

It has been found that addition of oil to an aqueous mixed culture system with mild agitation forms an oil-in-water emulsion within 24 hr (Zajic and Supplisson, 1972). On continued exposure this reverts to a water-in-oil emulsion in 48 hr. Functional groups in the oil that are hydrophilic and contribute to these reactions are: COOH (carboxylic), OH (alcohol), CHO (aldehyde), OSO_3 (sulfate), and SO_3H (sulfonate). Normally, the water-in-oil emulsion is associated with petroleum having asphaltenes. The variation in size of the encapsulated water droplets appears to correlate with the stability of the emulsion. This emulsion is very stable with a droplet size of an estimated average diameter of 0.5 u. Microbes are able to exist in the water capsules within the emulsion, where they selectively remove paraffinic hydrocarbons and low molecular weight aromatics.

Appendix A, Section 5.3.3.1, describes some efficient microbial emulsifiers and the result of emulsification of oil spilled on beaches.

5.4 TREATMENT TRAINS WITH CHEMICAL AND BIOLOGICAL PROCESSES

Contaminated water and wastewaters will normally be composed of a complex mixture of compounds varying in concentration (Sutton, 1987; Wilson, Leach, Henson, and Jones, 1986). The compounds may be degradable, inhibitory, or recalcitrant to various degrees. In most contaminated hydrogeologic systems, a remediation process is so complex in terms of contaminant behavior and site characteristics that no one system or unit will usually meet all requirements. Often it is necessary to combine several unit operations, in series or parallel, into one treatment process train to restore groundwater quality to an acceptable level.

Physical-chemical treatment techniques may be required to render the water or wastewater less inhibitory to microbial treatment or to ensure the removal of nonbiodegradable compounds. Physical-chemical treatment will normally be provided in conjunction with the biological step (Wilson, Leach, Henson, and Jones, 1986). Barriers and hydrodynamic controls serve as temporary plume control measures. However, hydrodynamic processes must also be integral parts of any withdrawal and treatment or in situ treatment measures. Heavily contaminated soils may have to be removed or attenuated (Brubaker and O'Neill, 1982). Then, pumping systems can be installed to remove free product floating on the groundwater before biorestoration enhancement measures are initiated to degrade the more diluted portions of the plume (Wilson, Leach, Henson, and Jones, 1986). A limiting factor is moving the contaminated subsurface material to the treatment unit, or getting the treatment process to the contaminated material, in the case of in situ processes. Withdrawal and biological treatment seem to currently be the most effective methods for biological restoration of aquifers contaminated by organic compounds (Knox, Canter, Kincannon, Stover, and Ward, 1968).

Physical processes, implemented quickly and efficiently, can retrieve large quantities of the contaminant (Dietz, 1980). Salvage wells for oil average 30 percent recovery. Pumping of drilled wells recovered two-thirds of a gasoline spill in one study (Raymond, Jamison, and Hudson, 19761). Biodegradation has sometimes been applied as a treatment for spill management after an unsuccessful attempt to recover all of a contaminant by physical means; i.e., pumping, air stripping, or vapor extraction (Raymond, Jamison, and Hudson, 1976; Walton and Dobbs, 1980). These auxiliary physical treatments can also be employed during bioreclamation (Brubaker and O'Neill, 1982). Integration of these removal mechanisms into the biological step will represent a cost-effective alternative, if technically feasible. Biodegradation is an alternative to physical recovery processes once they become nonproductive in terms of cost and effectiveness (Buckingham, 1981). One case study began stimulating soil organisms after estimating that physical recovery methods would require 100 years of operation and maintenance to make the contaminated water potable (Raymond, Jamison, and Hudson, 1976).

A thorough understanding of the hydrogeologic and geochemical characteristics of the area will permit full optimization of all possible remedial actions, maximum predictability of remediation effectiveness, minimum remediation costs, and more reliable cost estimates (Wilson, Leach, Henson, and Jones, 1986). Treatment techniques that can be used in combination with bioreclamation to reduce toxic concentrations to tolerable levels are discussed in Appendix A, Section 5.4.

5.5 BIODEGRADATION IMPLEMENTATION PLAN

Before a biodegradation program is begun, laboratory and field investigations would have to support the hypothesis that a significant number of the toxicants present in the soil could be degraded by natural soil processes (Buckingham, 1981; Sommers, Gilmore, Wildung, and Beck, 1981; Arthur D. Little, Inc., 1976; Quince and Gardner, 1982). The biodegradability of the contaminants must be assessed; the source, quantity, and nature of the spilled material must be determined; and the environmental conditions of the site must be considered.

A qualitative and quantitative analysis of the contaminants present would have to be made. The rates and products of degradation of each contaminant, as a function of manipulable soil parameters, would need to be established. The rates and products of degradation of mixtures of contaminants, as a function of soil parameters, would also have to be known. The geology and hydrogeology of the site and the extent of contamination must be assessed, including formation porosity, hydraulic gradient, depth to water, permeability, groundwater velocity, and direction and recharge/discharge quantification. A laboratory investigation on the potential for inhibition of the microorganisms, their oxygen and nutrient requirements, and the effects of temperature is necessary. Finally, the extent of degradation achieved under field conditions must be determined.

The optimum conditions would then be established for each of the contaminants. There would have to be an integration of each set of optimum conditions into an overall plan of action, considering sequencing of treatment processes and the particular toxicants found together in given areas. The quantity and character of the contaminants and their location in the aquifer will determine what containment technique, recovery method, and treatment system(s) should be used. The environmental conditions would be manipulated according to the plan of action, and the progress of toxicant degradation would be monitored.

Detailed procedures for a biodegradation implementation plan are presented in Appendix A, Section 5.5.

Potential Limitations of Biodegradation

In situ bioreclamation is a versatile tool for treating contaminated groundwater; however, it is not the answer to all contamination problems (Brown, Loper, and McGarvey, 1986). Its applicability must be determined for each site and depends upon local site microbiology, hydrogeology, and chemistry.

There are several important limitations for application of biological treatment (Niaki, Pollock, Medlin, Shealy, and Broscious, Draft). These include: 1) environmental parameters must be appropriate for support of microbial growth (pH, temperature, redox state, and nutrients); 2) some chemicals are nonbiodegradable, according to current knowledge; 3) by-products of biodegradation may be more toxic or persistent than the original compound (e.g., DDT-DDD); 4) substrate concentration may be too high (toxic) or too low (inadequate energy source); and 5) complex mixtures of organics may include inhibitory compounds.

Bioreclamation does not elicit destruction of heavy metals and some recalcitrant organics (Lee, Wilson, and Ward, 1987), although microorganisms can promote change in chemical form and mobility of hazardous materials. Introduction of nutrients and the residues generated by the organisms may adversely affect water quality. The process may not work well for aquifers with low permeabilities that prevent adequate circulation of nutrients and dissolved oxygen. A field demonstration using hydrogen peroxide as a source of oxygen in a very gravely clay loam was not very successful, due partly to the low permeability (1×10^{-6} cm/sec), which made it difficult to inject nutrients and transport oxygenated water (Science Applications International Corporation, 1985a). Other factors contributing to the poor success of this demonstration were the complexities of the site, possible mobilization of lead and antimony by the hydrogen peroxide treatment, and reductions in the permeability of the soil due to precipitation of the nutrients.

Some partial degradation products might be more toxic than the parent compounds (Lee, Wilson, and Ward, 1987). Transformation of a toxic organic solute is no assurance that it has been converted to harmless or even less hazardous products (Mackay, Roberts, and Cherry, 1985). For example, anaerobic biotransformation of common groundwater contaminants, such as PCE and TCE, can result in the formation of intermediates, such as vinyl chloride, which cannot be further transformed under prevailing reducing conditions. Given our limited understanding of transformation processes and the factors influencing them, hazardous contaminants must be assumed, in the absence of site-specific evidence to the contrary, to persist indefinitely.

Bacterial growth can plug the soil and reduce groundwater circulation (Lee, Wilson, and Ward, 1987). The plugging of well screens and the neighboring interstitial zones of the aquifer can be a direct result of biofilm generation (Cullimore, 1983). This can result in reduced flow from the wells, sometimes causing a complete shutdown of the system. Such water supply wells can reduce water quality (through the generation of turbidity, taste, odor, and color) and, eventually, lead to the generation of serious anaerobic corrosion problems. Degeneration in well productivity has, on occasion, been expensive, with an estimated annual cost at between $10 million and $12 million (Canadian). Organisms found to be responsible for this plugging have been Gallionella, other bacteria able to deposit iron or manganese oxides or hydroxides in or around the cell (e.g., Leptothrix, Crenothrix, or Sphaerotilus), and heterotrophic bacteria able to grow in a biofilm. The extensive growth of organisms at the aquifer/well interface is probably due to the increase in oxygen concentration at the site of injection. However, disinfectants and physical techniques have been reported for controlling this problem.

Microorganisms can mobilize hydrocarbons by transforming them to polar compounds, such as alcohols, ketones, and phenols, or to organic acids, such as formate, acetate, proprionate, and benzoate, when a site is contaminated with JP-5 (Perry, 1979; Ehrlich, Schroeder, and Martin, 1985).

It appears that microorganisms that are able to degrade chemicals in culture sometimes may not do so when introduced into natural environments because of improper pH, inability to survive, or use of other substrates (Zaidi, Stucki, and Alexander, 1986). Some Pseudomonas strains were able to mineralize biphenyl or p-nitrophenol in lake water at the natural pH of 8.0, while another strain required the pH to be adjusted to 7.0, and yet another did not mineralize the substrate although its population density rose.

To be a useful pollution abatement method, biodegradation of petroleum pollutants would have to occur rapidly enough to prevent contamination of groundwater or movement of pollutants off site (Atlas, 1977). This may not always be the case. It is likely that the treatment could cost in the range of tens of thousands of dollars for simple treatment programs up to tens of millions of dollars for complex, large sites (Lee, Wilson, and Ward, 1987). This, however, compares favorably with costs of hundreds of millions of dollars that could be incurred by using other methods.

Because of the uncertainty involved in applying innovative remedial approaches, there will always be reluctance to test their feasibility (Alexander, 1980). Uncertainty about subsurface geochemical and contamination conditions makes it difficult to provide specific effluent and other data typically required in a permitting process. The complexity of the technologies and the permitting process itself will cause delays that can hinder cost-effective feasibility demonstrations.

There are few additional safety hazards associated with in situ bioreclamation aside from those hazards normally associated with being on a hazardous waste site or a drill site (Environmental Protection Agency, 1985b). Since wastes are treated in the ground, the danger of exposure to contaminants is minimal during a bioreclamation operation relative to excavation and removal.

SECTION 7

Conclusions and Recommendations

7.1 CONCLUSIONS

1. This country is faced with an enormous task of cleaning up hazardous materials accidently or indiscriminately introduced into the environment. Many of the standard techniques employed for treating the contaminated soil and groundwater have not been totally effective or have not offered a permanant solution to the problem. Some have even created additional hazardous sites.

2. Various biological techniques have been applied to treating this contamination. Among these, in situ biodegradation is receiving growing support and widespread testing in the field. In situ biodegradation practices can be combined with on-site biodegradation techniques, as required, to offer a very efficient and cost-effective treatment system.

3. In situ biodegradation techniques can also be employed in conjunction with chemical and physical procedures, as warranted by a given contamination incident. Treatment trains that combine biodegradation with physical and/or chemical remediation methods should optimize conditions for the microbes to facilitate total destruction of the contaminant materials.

4. In most episodes of soil and aquifer contamination, indigenous microorganisms present at the site appear to be able to degrade the material, if sufficient nutrients and oxygen are supplied and other environmental factors modified, as necessary.

5. <u>In situ</u> biodegradation is still a developing technology. A great deal of effort is being directed toward laboratory investigations of hydrocarbon- degrading microorganisms and their requirements for optimum degradation. Metabolic pathways are being elucidated to be able to predict the performance of individual strains and mixtures of organisms. New microbes are being developed to target a wider range of substrates, as well as the more recalcitrant contaminants. Manipulation of environmental factors is being attempted to improve the conditions for the desired organisms. Anaerobic treatments are also being explored, as well as combination of anaerobic with aerobic measures.

6. <u>In situ</u> biodegradation is a very complex process. The interactions of many factors have yet to be understood before a program of specific measures can be applied with certainty in the field to be able to predict and direct the outcome of the treatment. Although this may appear to be an overwhelming task, much useful information has already been obtained, and research is progressing in promising directions. When there is sufficient evidence that this approach (or its combination with on-site, physical, or chemical procedures) is faster and more economical than other methods, and contaminated sites continue to be discovered, the funding may become available from both government and industry to encourage full development of the technology.

7. Genetic engineering of microorganisms may be able to help concentrate and improve performance of their hydrocarbonoclastic capabilities. Developments in this area will depend upon an increased understanding of microbial physiology, the basic biochemical and genetic basis for microbial detoxifications, and an understanding of microbial resistances to antibiotics and heavy metals. The organisms must also have a competitive advantage over the indigenous organisms to be able to survive long enough to accomplish their task. Therefore, genetic enhancement of ecologically important traits may be required.

8. Development of specialty organisms with unique capabilities and the use of fungi, algae, and actinomycetes are being explored for biodegradation applications. However, the value of seeding contaminated sites with supplemented microorganisms has yet to be substantiated.

9. There is still much to be learned about <u>in situ</u> biodegradation processes. Laboratory results will have to be verified in the field.

Unfortunately, few field applications have been conducted as controlled experiments. Inappropriate or inadequate measurements are often taken or too many variables are introduced, including the use of other restoration measures concurrently with the enhancement techniques. Indirect evidence is sometimes offered as proof that the procedure is responsible for the observed reduction in contaminant levels. More emphasis should be placed on devising methods that will furnish conclusive evidence that the process, or the particular variable being tested, can be linked with disappearance of the contaminant.

7.2 RECOMMENDATIONS

The underlying limitation to using technologies associated with in situ treatment of organic deposits is that these methods are often in an experimental stage of development. This finding frustrates efforts to systematize selection of reagent and delivery and recovery systems, and it emphasizes the need for laboratory simulation and testing prior to implementation. Each site at which in situ treatment is considered becomes, in essence, a research project, but the use of in situ methods on a large scale is hampered by the lack of field experience. There is little incentive for site managers to consider their use.

It became apparent through review of the literature that there are significant gaps in the understanding of the factors involved in the use of bioreclamation for treatment of contaminated soils and groundwaters. The following are suggestions for areas where additional research could contribute useful information toward development of a more effective technology.

7.2.1 Microorganisms

1. The methods for sampling and characterizing indigenous subsurface microorganisms should be improved. This should include methods for assessing the degradative capabilities of the organisms present. Methods for detection, identification, enumeration, and biochemical activity of introduced organisms can help monitor their survival and functioning in the field.
2. Since a variety of methods are being used to investigate in situ biodegradation, such as techniques for microbial enumeration, comparison of results from different studies can be difficult.

Efforts should be made to standardize methods where possible and feasible, even as new and better approaches are developed.

3. Nitrogen-fixing and oxygen-producing hydrocarbonoclastic microorganisms should be sought for possible application to nutrient-deficient environments.

4. In many cases, wastes contain carbon sources that microorganisms may use selectively as nutrient sources instead of the targeted toxic chemical. Development of, or selection for, organisms that can utilize the toxic fractions is necessary.

5. Studies are needed on the population dynamics and survival of new and existing organisms in the environment to permit continuing and long-term biodegradation. This should include requirements of the organisms for growth under field conditions.

6. Before a microorganism can be tested outside a contained environment, determination of its fate after it has done its work is needed. Predictive ecological research is essential to ensure that biodegradation of oily wastes do not unintentionally create new and potentially serious environmental problems.

7. Research should be conducted into the need of microorganisms for acclimation to the compounds they will eventually degrade and, ultimately, lead to shortening the acclimation period.

8. Microorganisms with special capabilities, such as the ability to metabolize contaminants at low temperatures, resistance to high concentrations of pollutants, the ability to adhere to solid surfaces or nonaqueous-phase liquids, resistance to heavy metals, the ability to produce emulsifiers, the capacity to function at high salt concentrations, or are "superbugs," should be sought.

9. While aerobic bacteria and fungi may prove to be the most useful microbes for biodegradation, other organisms, such as anaerobes, algae, and actinomycetes should also be investigated.

7.2.1.1 Seeding

1. The value of seeding contaminated sites with microorganisms has yet to be substantiated. While seeding soil with an acclimated or mutant microbial population to remove contaminants holds a great deal of promise, especially when concerned with recalcitrant substances, results from previous attempts have been inconclusive. Further research is required to determine whether seeding with microbes is an effective and cost-effective technique for the restoration of contaminated aquifers and soils.

2. Methods should be devised to determine whether indigenous or supplemented microorganisms would be preferred in a given contaminant incident.

3. Research is needed to allow prediction of whether introduced organisms require nutrients, surfactants, or other amendments to increase their activity. The appropriate requirements for manipulation of the environment should be established for the intended organisms. Studies should determine the requirements of supplemented organisms to be able to outcompete indigenous microflora.

4. The problem of developing potentially pathogenic microorganisms in the subsurface or long-term environmental effects of seed cultures requires further investigation. Many of the hydrocarbonoclastic microorganisms are of the same genera that contain potentially pathogenic species. For example, many hydrocarbon-degrading bacteria are Pseudomonas species. Establishing that seed microorganisms do not cause a public health threat will be difficult because it must be determined what organism to use in the toxicity testing. This would also require bioassay testing for the different environments into which seeding is considered. The direct toxicity of petroleum and oily wastes must be considered, and it should be determined whether microbial seeding would increase or decrease the undesirable effects of oil pollutants. A consideration of toxicity requires studies on sublethal, as well as lethal effects.

5. The reasons for the frequent failure of introduced microorganisms to survive and degrade contaminants should be determined and overcome.

6. More work should be performed to develop seed inocula consisting of various groups of microorganisms that will interact in a predictable manner to degrade specific mixtures of contaminants.

7. The possibility of seeding with specialty organisms with unique capabilities should be explored for meeting a variety of demands. This could include microbes that are active in cold climates, are salt tolerant, are resistant to heavy metals, are "superbugs," or are able to produce emulsifiers. Greater use could be made of fungi, algae, and actinomycetes.

7.2.1.2 Genetic Engineering

1. Development of genetically engineered microorganisms suitable for rapid and more complete destruction of specific organic compounds is currently an active area of research, which may lead to advances in both aboveground and in situ biological treatment technologies. More information is needed on the ability of genetically engineered organisms to survive, grow, and function in the soil environment, as well as any potential permanent negative influences.

2. The genes responsible for biodegradation should be determined to permit cloning and introduction of the genes into desired recipients and to allow regulation of the function of the introduced gene, as required, without negative effects on the recipient or the environment.

3. Additional research to address the controversy of mutant versus natural bacterial stimulation should lead to selection of the most time-efficient and cost-beneficial procedures.

4. Since many of the waste pool constituents are mutagens and the degradative traits are often associated with nonchromosomal genetic material (plasmids), the organisms could mutate to ineffective strains. This needs to be considered when developing suitable microbial strains for field use, such as utilizing chromosomal genes or more stable plasmids.

5. Genetically engineered microorganisms should be developed that can destroy pollutants that might otherwise not be destroyed or only at a slow rate. Multiple degradative pathways might be incorporated in a single engineered strain.

6. The environmental constraints for the particular engineered organism should be determined.

7. Engineered organisms might be particularly useful in bioreactors, where the environmental conditions can be better controlled.

7.2.1.3 Biochemistry

1. More research is needed on the metabolism and metabolic pathways for the biodegradation of specific compounds by different microorganisms or groups of microorganisms, including algae, fungi, and oligotrophic bacteria.

2. The importance in the environment of the various pathways for hydrocarbon biodegradation should be determined. Various

biological strategies exist for the microbial utilization of petroleum hydrocarbons. The natural environments that receive petrochemical hydrocarbons should be analyzed for intermediate products to determine which pathways are actively used by microbial populations in natural ecosystems. It is likely, but as yet unproven, that different pathways will be active under different conditions, e.g., at different hydrocarbon concentrations.

3. While it has been shown that most PAHs are biodegradable and catabolic pathways for a few have been determined, much more research is needed in this area. Careful attention should be given to the potential carcinogenicity and mutagenicity of intermediate products formed during their biodegradation.

4. Microorganisms may use a water-soluble substrate as it spontaneously dissolves in water, or they may metabolize the compound after a biologically mediated solubilization or by mechanisms involving physical contact with the insoluble phase of the substrate. Former studies have been concerned only with comparisons of dissolution and degradation rates. Additional study is needed to assess the significance of other mechanisms by which microorganisms utilize water-insoluble organic chemicals.

5. The pathways of biodegradation in surface waters, sediments, soils, and subsurface materials should be studied.

6. Research should be conducted into the mechanisms of biodegradation to be able to predict the fate of toxic molecules in the environment and to aid in the design of new and innovative reactor systems that exploit those mechanisms.

7. The kinetics of biodegradation should be investigated, including kinetics of chemicals that are acted on by cometabolism, that are sorbed, and that are present in nonaqueous-phase liquids.

7.2.1.4 Cometabolism

1. Aside from traditional fermentation and growth of microorganisms on hydrocarbons, cooxidation and stationary transformation techniques are promising areas of hydrocarbon biotechnology for the future. These are systems actually operating in nature and need to be more fully explored for application in a biodegradation treatment program.

2. Since soil microorganisms capable of cometabolically degrading a xenobiotic may not proliferate on the xenobiotic substrate without their cometabolic companions, unique techniques may be

necessary to isolate microorganisms with this capability. There is limited information available suggesting what alternate substrates would be suitable for isolating microbial populations capable of metabolizing xenobiotics by cometabolic processes. Such techniques might utilize, as a primary substrate, a chemical analog that would permit enrichment of microbial populations having the cometabolic requirements to degrade the xenobiotic.

3. Evidence is accumulating that indicates that cometabolism may be a particularly important phenomenon in the dissipation of the more refractory xenobiotics. However, considerably more research is needed to understand the importance of this process.

7.2.1.5 Enzymes

1. Theoretically, enzymes would quickly transform hazardous compounds if they remained active in the soil, and oily wastes are amenable to treatment by application of cell-free enzymes. The potential level of treatment is high. However, there is little information available on the use of this technique in soil and no information from the field. Its reliability is unknown. It is even possible that the enzymatic degradation products may not be less hazardous than the parent compounds. All these issues should be investigated to determine whether using enyzmes for treating hazardous contaminants is feasible and, if so, to establish a protocol for their application.

2. Extracellular enzymes capable of degrading xenobiotics or xenobiotic degradation products exist in the soil. The presence of peroxidases in soil should be of great interest to researchers investigating soil xenobiotic transformations. The role of lignin peroxidases in the degradation of polluting chemicals should be further explored. Presumably, these enzymes could be involved in many soil metabolic reactions affecting xenobiotic residues.

7.2.1.6 Interactions

1. Most present knowledge of microbial degradation of different hydrocarbons has been obtained in pure cultures of microorganisms containing single hydrocarbon substrates. Little is known about how natural mixtures of microorganisms degrade each chemical component when they are present in the total petroleum sample.

2. Aged oily wastes usually support large indigenous populations of bacteria, fungi, and actinomycetes. The exact interacting role of these microorganisms in decomposition or the nature of their interactions with the natural soil microflora is unknown. A thorough assessment is needed to establish their potential role in the degradation of chemical constituents.

3. Each bacterial strain will generally degrade only one particular chemical or closely related groups of chemicals. Little research has been conducted on developing protocols for using single species of microorganisms or defined microbial consortia to treat complex waste mixtures.

4. The rates of biodegradation of hydrocarbons from oil spills appear to be highly dependent upon localized environmental conditions. The fate of many components in petroleum, the degradative pathways that are active in the environment, the importance of cooxidation in natural ecosystems, and the role of microorganisms in forming persistent environmental contaminants from hydrocarbons, such as the compounds found in tar balls, are unknown and require future research. Although a number of rate-limiting factors have been elucidated, the interactive nature of microorganisms, oil, and environment needs to be better understood.

7.2.2 Environmental Factors

1. The environmental factors that limit the rate or extent of biodegradation or that prevent successful bioremediation must be established for both indigenous and introduced populations. The limiting factors should be overcome by practical means.

7.2.2.1 Aerobic/Anaerobic Conditions

1. Further biodegradation research is needed in the area of oxygen enhancement to develop more efficient aeration techniques for stimulating aerobic biodegradation in soils and groundwater. Innovative approaches should be sought, such as provision of electron acceptors other than oxygen or the use of very small gas bubbles to provide oxygen.

2. Microbiological plugging in wells usually occurs through biofilm growth constricting the flow of water entering the well through the screen slots and the interstitial zone in the aquifer immediately

surrounding the well, probably as a result of the increase in oxygen concentration there. The mechanism of biofilm generation as the critical component in biological plugging should be studied in the well screen/aquifer environment to find a way to eliminate this problem.

3. There is a lack of data for applying an anaerobic degradation system to treat dilute, low-organic-strength wastewater, such as contaminated groundwater, although it is very promising.

4. There have been a few reports on the anaerobic degradation of some petroleum hydrocarbons in natural ecosystems using nitrate, iron, or sulfate and carbon dioxide as alternate electron acceptors during anaerobic respiration. Nitrate has been reported to support the degradation of xylenes in subsurface material. This approach is still experimental but is believed to offer considerable promise. It should be further studied since nitrate is inexpensive, is very soluble, and is nontoxic to microorganisms (although there are environmental restrictions on it). Degradation of all hydrocarbon groups, however, has not been biochemically confirmed in either pure or mixed cultures, and this needs to be explored. The criteria used for assessing anaerobic hydrocarbon degradation have not been adequate to establish definitive results (e.g., anaerobic biodegradation of benzene has been inferred but not confirmed). Therefore, such criteria must be developed to be able to validate these techniques.

5. The possibility that the biodegradation of polluting oil in nitrogen-deficient natural environments is aided by atmospheric nitrogen fixation needs to be further explored.

7.2.2.2 Temperature

1. The ability to actively metabolize hydrocarbons at low temperatures should be further investigated. Psychrophilic oil-degrading microorganisms should be sought for activity in cold marine and cold soil environments.

2. Composting allows an enrichment of thermophilic microorganisms that have their optimum degradative activity between 50° and 60°C. It may offer considerable potential for maximizing the biodegradation rate of oily wastes. It might also provide for the initial detoxification or degradation of these wastes prior to their application to land treatment systems. Research is needed to fully

assess the potential of composting for detoxification, degradation, and inactivation, especially recalcitrant organic constituents.

3. There is minimal literature that documents attempts, successful or unsuccessful, to increase significantly the temperature of the soil in the field. Fairly simple field tests need to be performed to determine whether application of materials, such as placing black PVC sheeting on the soil surface or using it as a mulch, would increase the temperature enough to cause significant increases in degradation rates.

7.2.2.3 Nutrients

1. While there may be a substantial amount of potentially available nutrients in many wastes, their rate of release from organic material may be too slow to sustain microbial activity. Research is needed to determine the rate and extent to which degradation of petroleum products can be increased by addition of available nutrients for microbial utilization.

2. Nutrient injection into groundwater monitoring wells is an area in need of research. It is probable that subsurface injection is economically feasible when compared with methods of surface application. Methods of fertilizer injection into the subsurface needs to be evaluated.

3. Nutrient requirements of indigenous or introduced organisms should be established to optimize their growth and degradation of the pollutants.

4. Methods are needed for introducing the correct amounts of nutrients to locations where they are required and in the form needed.

7.2.2.4 Water Potential

1. Much more information is needed on the effects of water potential on the ability of different microbial species to degrade or detoxify individual components of hydrocarbons. Such information is important in developing management strategies for efficient and effective land treatment systems for chemical wastes.

7.2.2.5 Photolysis

1. There should be an investigation of photochemical reactions that enhance biodegradation of refractory compounds and those that can produce toxic or undesired breakdown products.

7.2.2.6 Factors Affecting Migration of Contaminants or Their Availability for Biodegradation

1. The mobility of waste organics and their degradation products in soil is not well understood. Evaluation is needed of the extent that petroleum contaminants and their metabolites are sorbed to clay and soil organic matter. Additional information should be developed on the fate and mobility of these wastes in the soil. Information is needed on contaminant sorption and precipitation studies at concentrations in the soil matrix characteristic of those found at remedial action sites. Sorption of contaminants on soil particles can alter their molecular character and enzymatic attack on a given compound. Humic polymers can act as stabilizing agents, making the compounds less resistant to biodegradation. However, the opposite effect has also been observed, with bound material becoming unavailable for biodegradation. More research leading to a better resolution of sorption/biodecomposition interactions is desirable.

2. In addition to descriptions of contaminant sources, prediction of contaminant migration in the saturated zone requires quantitative representations of advection, dispersion, sorption, and transformation that are specific or at least applicable to the site, contaminants, and period of time in question. Available generalized models for the latter three processes have not yet been convincingly validated using field-scale observations or experiments in even simple sand and gravel aquifers. There is a need for continued research to formulate improved process understanding in models that can be tested at both the laboratory and field scale.

3. The role of spontaneous desorption in governing the rate of bioremediation and possible means of enhancing desorption should be examined.

4. There must be a better understanding of contaminant transport and fate in the vadose zone and in heterogeneous hydrogeologic

systems, such as discontinuous and interbedded layers of different geologic media.

5. Evidence from laboratory studies shows that certain organic liquids can cause desiccation and cracking of unweathered clay, which can lead to significant increases in permeability. The significance of this effect in the natural environment is unknown. This should be investigated to see whether it may be an important factor in migration of the contaminants and of any solutions used to treat the site.

6. It is not certain how the rate of dissolution or selective sorption to different soil matrices governs the rate of biodegradation of sparingly soluble organic chemicals in natural ecosystems. Because many synthetic chemicals have low water solubilities, studies should be conducted to assess whether rates of dissolution and desorption govern microbial metabolism in natural conditions.

7. Technologies should be devised to promote bacterial movement through soils or subsurface materials to the site of contamination, and transport of indigenous bacteria with plumes of contamination. Means should be found to overcome the problem of channeling of liquids introduced into the subsurface, whether they contain nutrients, oxygen, or a suspension of bacteria.

7.2.2.7. Effect on Environment

1. More studies are needed to determine the effects of oil on species diversity and the possible influence of chronic or incidental pollution by petroleum hydrocarbons on ecosystem stability. Studies involving the effects of oily wastes on the soil should include studies on soil fauna (e.g., protozoans, worms, insects, and other arthropods), because the overall rate of decomposition can be significantly altered by the reduction in numbers of soil fauna resulting from the addition of toxic materials.

7.2.3 Contaminants

1. Slightly to moderately soluble compounds, such as benzene, will develop a plume in the saturated zone. More information is needed on the interactions among different contaminants in superimposed plumes, the rates and products of biotransformation, the effects of complex contaminant distributions on the activity of

microorganisms, and the flow and dissolution of immiscible organic liquids.

2. New long-chain n-alkanes (C_{25} to C_{45}) appear during biodegradation of some crude oils as a result of microbial activity and contribute to the formation of tar balls. Reactive biodegradation intermediates may undergo spontaneous head-to-head condensation reactions, resulting in these very long n-alkane chains that are largely immune to further enzymic attack. A clear need exists for further experimental work in this area.

3. The fate of many complex hydrocarbons decomposing in the environment remains to be elucidated. The effects of continued accumulation of partially degraded hydrocarbons on normal microbial functions has not been adequately addressed.

4. Little study has been conducted on degradation of mixtures of many compounds, as would occur in wastes. One compound may prevent or promote the degradation of another in a mixture. The occurrence of multiple waste residues in soil provides the opportunity for the formation of complex or hybrid residues. Our knowledge of such interactions occurring in soil is very limited at present and needs to be expanded.

5. The products of biotransformation of organic pollutants should be identified to be able to assess their toxicity and possible deleterious effects on human health or on populations of animals or plants. Some products may persist in, or immediately adjacent to, the site at which the bioremediation is taking place. Some of the products are toxic and are greater hazards than the original compounds. Characterization should be made of these products and their behavior in the environment to assure minimum or acceptable risks from bioremediation.

6. PAHs containing more than three aromatic rings are very difficult to degrade using existing biodegradation approaches. Further research is needed to develop methods that are appropriate for the removal of recalcitrant PAHs from the environment.

7. Methods should be found to reduce the toxicity of organic or inorganic materials to the microorganisms involved in biodegradation.

8. Use of microorganisms for treatment of mixtures of wastes containing organic and inorganic components should be investigated.

9. Research should be conducted on means to biodegrade chemicals present in nonaqueous-phase liquids. A chemical might become

resistant to biodegradation because it is in a nonaqueous phase into which microorganisms or their enzymes may not penetrate. Such compounds appear only slowly in the free water phase and represent long-term pollutants.

10. Special attention should be given to water-insoluble compounds, which may present problems different from those in solution.

11. Research is needed to determine why organic contaminants may not be available for microbial attack. The chemical may be in a form that is not readily utilized.

12. The effects of bioremediation on the partitioning, mobility, and fate of other contaminants present at the same site should be studied.

7.2.3.1 Trace Organics

1. The mechanisms involved and the extent of microbial degradation of trace organics should be further investigated through laboratory and field studies. If the concentration is too low to support growth or may support growth at too slow a rate, methods should be found to bring about a more rapid destruction of the chemical.

7.2.3.2 Metals

1. Despite concern over the environmental impact of heavy metals contained in nearly all major wastes, there are few studies on the effects of these pollutants on microorganisms in their natural environment. Research is needed to evaluate the effects of single and multiple metals (representative of those in petroleum wastes) on microbial processes pertinent to hydrocarbon transformations, such as respiration, nitrogen mineralization, and pertinent enzyme activities. Metal tolerances between different groups of organisms have been observed; however, additional studies are needed to characterize the mechanism by which heavy metals affect metabolism of residues in soils.

2. The nature of trace metal transformations and interactions in microbial degradation processes must be elucidated to prevent or anticipate waste or trace metal overloadings, as could occur during landtreatment, which would inhibit normal soil biodegradation processes.

3. Research is needed to evaluate ways and means of alleviating the toxic effects of metals in soils by addition of organic amendments,

such as crop residues, municipal wastes, composts, and by pH adjustments.

4. Information is needed on behavior of specific metals and metal species, i.e., arsenic, beryllium, silver, selenium, mercury, and chromium, in petroleum-contaminated soil systems.

5. More information concerning the factors governing the competition of clays and organic matter for the binding of metals needs to be obtained through research in order to predict the long-term fate of wastes in soils.

6. Research is needed into bioconcentration of metals by microbial cells.

7. Studies should be conducted to exploit possibly the use of oxidation, reduction, and solubilization reactions to promote bioremediation of sites contaminated with toxic metals.

7.2.4 Additives

1. The use of surfactants either alone (to flush otherwise insoluble organics) or in combination with other treatments (to solubilize the waste materials and, thereby, promote biodegradation) is a promising avenue for further research. Both synthetic and natural surfactants warrant further attention.

2. Information is needed on the effects of chemical addition (oxidants, reductants, and polymerizing agents) on the soil properties affecting biotreatment.

3. The use of chelating agents may be a very effective means of managing oily wastes contaminated with toxic metal concentrations. Depending upon the specific chelating agent, stable metal chelates may be highly mobile or may be strongly sorbed to the soil (e.g., Tetran). Considerable research is needed in this area.

7.2.5 Bioreclamation Monitoring

1. Thousands of contamination events are probably remediated naturally before the contamination reaches a point of detection. There is a major need to have methods to determine when natural biorestoration is occurring, the stage the restoration process is in, whether enhancement of the process is possible or desirable, and what will happen if natural processes are allowed to run their course.

2. Further research is needed to relate laboratory results to field situations, where the physical, chemical, and hydrogeological constraints on the microbial community will be different.
3. Research is needed to verify actual destruction efficiencies of the wide range of hazardous constituents found in mixed oily wastes. This will also require some real world testing and demonstration studies. If biological treatment can be developed for RCRA and CERCLA wastes, the commercial opportunities should be significant.
4. Unless all the proper conditions are met for a given compound, biodegradation may not occur, or only at low rates. Better methods for evaluating when these conditions are met and how they affect the rate of degradation are clearly needed. Monitoring can be achieved with indirect evidence from laboratory studies, hydrogeological investigations, and chemical sampling in combination with microbial sampling in the field.

7.2.5.1 Testing Efficiency of Biodegradation

1. Plate counts usually underestimate the total number of microorganisms in a subsurface sample and seldom correlate to required metabolic rate data. A rapid, reliable, and inexpensive index of overall metabolic activity in the subsurface environment is needed.
2. Many of the bacteria in subsurface aquatic environments may be dormant, which could account for some of the difference between the indirect (plate, Most Probably Number (MPN)) and direct (AODC) count methods for the same sample. Further characterization of the metabolic capabilities of groundwater bacteria, or the use of mixtures of labeled substrates to measure total active cells by MPN is needed to verify this.
3. Indirect evidence, such as by nitrogen depletion in the contaminated zone and high bacterial populations in soils, has been used to suggest a certain volume of gasoline has been removed from contaminated soil and groundwater. Indirect methods have been employed since it was not possible to obtain direct evidence of the magnitude of the gasoline conversion to bacterial cells. A more accurate means of making such a determination would be useful to assess the effectiveness of the treatment method, as well as to provide evidence to link the disappearance of the contaminant with biodegradation.

4. Phospholipid concentration is a good estimate of metabolically active biomass; it is feasible to use fatty acids as an indicator of microbial biomass and community structure in soils. Research is needed to determine the exact phospholipid content of subsurface bacteria growing under in situ nutrient conditions.

5. There is a need to determine how accurately math simulation models can predict transformations of petroleum wastes and nutrients in soils once the biocidal and inhibitory effects of toxic constituents are overcome. It may be possible to adopt some existing models, with or without modification for making predictions concerning the fate of oily waste components.

6. Many studies that have indicated soil degradation of petroleum hydrocarbons have merely tested for the disappearance of the total hydrocarbons and not of the many individual components or even major groups of contaminants. There is not yet a conclusive method for determining total petroleum hydrocarbons. Better methods need to be developed pertinent to each fuel type. More information is needed on the extent of detoxification that occurs.

7. Protocols are needed to evaluate the effectiveness of the bioremediation, the organisms, or the products used to enhance bioremediation.

8. Performance criteria should be established to provide standards for evaluating bioremediation technologies. With no standard of achievement against which to optimize the processes, research and development are impeded. Performance criteria would also allow different products or technologies to be compared.

9. Better protocols are needed for quality assurance/quality control of laboratory, microcosm, and field tests.

10. There should be a method for determining the concentration of chemicals remaining after bioremediation.

7.2.6 Process Applications

1. To date, the application of permeable treatment beds at hazardous waste sites has not been performed. However, bench- and pilot-scale testing has provided preliminary quantification of treatment bed effectiveness. There may also be a possibility of inoculating such beds with organisms to create a biofilm over which the contaminated groundwater would pass. Because of potential saturation of bed material, plugging of bed with precipitates, and the short life of treatment materials, this

treatment might have to be used as a temporary remedial action rather than a permanent one. These possible limitations should be further investigated. The great variety of infill that could be used in these trenches offers the possibility of specifically targeting a particular contaminant or groundwater condition and could be very useful in pretreatment of a contaminated plume.

2. Indigenous microorganisms that secrete oil repelling polymers could be stimulated to grow in situ to create a biological barrier to contaminant movement.

7.2.7 Microcosms

1. Microcosms are needed to evaluate organisms, treatments, or technologies to promote bioremediation. When appropriately designed to simulate field conditions, microcosms will permit a better evaluation of the effectiveness of the different approaches being studied.

2. Microcosms would be beneficial for studying the fate of pathogens and the fate of chemicals in groundwater. Such microcosms should simulate groundwater conditions.

7.2.8 Bioreactors

1. The kinetics of biodegradation in bioreactors should be studied for pollutants that are destroyed as a direct result of microbial growth on the compounds and for pollutants that serve as secondary substrates. Kinetic expressions should also be developed for compounds at trace levels and at levels that are somewhat inhibitory to the biodegradation process.

2. Use of a bioreactor could be convenient for assessing the value of anaerobic composting to destroy compounds that are resistant to aerobic processes. Optimal conditions for this process should be established.

3. The conditions necessary for destruction of toxic chemicals could be more easily established by using a bioreactor. This controlled environment could investigate the effect of microbial mixtures, levels of nutrients, electron acceptors, aeration, pH, etc. Reactor design and operating conditions should allow for changing biological and reactor conditions. The essential engineering factors should be determined, evaluated, optimized, and demonstrated.

4. Anaerobic bioreactors may allow optimization of conditions for degradation of recalcitrant compounds, increase in degradation rates, and prevention of accumulation of products that are environmentally unacceptable. Criteria should be developed to permit evaluation of whether aerobic or anaerobic processes are more appropriate for particular needs.

5. Bioreactors could be used to study sequential treatments, such as aerobic-anaerobic systems for compounds that are not readily or totally degraded under either condition alone. The sequence could also include preceding biodegradation with a physical, chemical, or photochemical treatment.

7.2.9 Field Applications

1. There is a general lack of information on remedial applications. This includes general information about the technologies in any setting, remedial action, land treatment, agriculture, etc., and specific information on individual technologies. Data are needed on the capacity of soil systems to support in-place technologies, as well as data on ways to enhance the natural capacity by the addition of reagents or modification of the soil/waste system.

2. Information is needed from controlled experiments on a field scale, comparing effectiveness of adding treatment agents (e.g., chemicals, microorganisms, adsorbents) to promote biodegradation, with the effectiveness of natural processes (e.g., degradation, transformation, immobilization) for soil treatment of constituents in complex hazardous wastes.

3. Although degradation of petroleum hydrocarbons has been extensively studied in the laboratory over the last 25 years, extensive data on full-scale field studies is seldom available or published, and much of the experience rests with experts who have a financial interest in the technology. Publication of the results of the applied methodologies would allow the procedures to be validated and would benefit further development of the technology.

4. A well-controlled test site for research on in situ methods of groundwater decontamination would be very valuable. Such a site could involve containment of groundwater flow (or careful, known constriction). Facilities could be available for conducting physical and chemical process testing, either within an aquifer or at an aboveground collection point. Such a site could either be at

an actual contaminated aquifer or at a specifically constructed facility. Establishment of an <u>in situ</u> test site would be advantageous to agencies having parallel interests. It would permit investigations and process development under controlled field conditions and over the period of time required for pilot-scale verification of concept.

5. In open (<u>in situ</u>) systems, the effectiveness of the technology is often influenced by poorly controllable environmental conditions, and the supplemented organisms may not survive. Methods must be developed to facilitate control of the optimum conditions required in each remedial situation.

6. The potential for combinations of chemical and biological treatment methods for accelerating treatment of contaminated soil and groundwater should be explored.

7. Information is needed from the controlled application of complex hazardous wastes (including solid, liquid, and semi-solid wastes) to soil systems in laboratory and field experiments. The data must specifically include information on degradation, mobilization, volatilization, photodegradation, and reaction rates.

8. The problems in moving from the laboratory to field testing should be defined. Factors that constrain field demonstration and evaluations must be overcome. After the feasibility of bioremediation has been established in well-characterized microcosms or similar controlled systems that are sufficiently simple to permit definitive conclusions to be drawn, field demonstrations are needed to establish the problems involved in system design and operation.

9. Appropriate monitoring techniques should be employed to evaluate the effectiveness of the bioremediation.

7.2.10 Site Problems

1. Standard procedures are necessary to ensure that meaningful site samples are collected for analysis to assess the effectiveness of bioremediation at different locations in the field.

2. The conditions and methods appropriate for bioremediation of wastes that are deeply buried should be determined.

3. It is necessary to be able to predict the effect of differences among sites on the effectiveness of bioremediation. A practice that works well at one site may not work well or fail at another.

Methods must be established to compare and explain results between different locations.

7.2.11 Case Histories

1. Further examination of case histories is necessary to improve the predictive understanding of the fate of oil pollutants in the environment and the role of microorganisms in biodegradative environmental decontamination.

7.2.12 Funding

1. Neither federal agencies nor the commercial sector has made a substantial long-term investment in biodegradation research. Few biotechnology companies are entering this field and most companies will not fund biotechnology research on waste treatment. Funding has been directed toward the microorganisms, with little emphasis on developing the process technology. More involvement by government agencies and industry would help expand and hasten efforts to produce a viable biodegradation technology.

References

Absalon, J.R. and Starr, R.C. 1980. Practical aspects of ground water monitoring at existing disposal sites. In EPA Nat. Conf. on Mgmt. of Uncontrolled Hazardous Waste Sites. pp. 53-58.

Ahearn, D.G. and Meyers, S.P. (Eds.), 1973. The Microbial Degradation of Oil Pollutants, Atlanta, GA, Dec. 4-6, 1972. Publ. No. LSU-SG-73-01. Louisiana State Univ., Center for Wetland Resources, Baton Rouge, LA. p. 283.

Ahearn, D.G., Meyers, S.P., and Standard, P.G. 1971. The role of yeasts in the decomposition of oils in marine environments. Dev. Ind. Microbiol. 12:126-134.

Ahlert, R.C. and Kosson, D.S. 1983. In-situ and on-site biodegradation of industrial landfill leachate. Report to Department of the Interior, Washington, D.C. on Contract No. 14-34-0001-1132. PB84-136787.

Ahlrichs, J.L. 1972. The soil environment. In Goring, C.A.I. and Hamaker, J.W. (Eds.), Organic Chemicals in the Soil Environment. Vol. 1. Marcel Dekker, Inc., New York, NY.

Alberti, B.N. and Klibanov, A.M. 1981. Enzymatic removal of dissolved aromatics from industrial aqueous effluents. Biotech. Bioeng. Symp. No. 11, John Wiley and Sons, Inc., New York, NY. pp. 373-379.

Alexander, M. 1961. Introduction to Soil Microbiology. John Wiley and Sons, Inc., New York, NY. pp. 139-162, 248-271, 425-441.

Alexander, M. 1971. Microbial Ecology. John Wiley and Sons, Inc., New York. 511 pp.

Alexander, M. 1973. Nonbiodegradable and other recalcitrant molecules. Biotech. Bioeng. 15:611-647.

Alexander, M. 1977. Introduction to Soil Microbiology. 2nd ed. John Wiley and Sons, Inc, New York, NY. 467 p.

Alexander, M. 1980. Biodegradation of toxic chemicals in water and soil. In Haque, R. (Ed.), Dynamics, Exposure and Hazard Assessment of Toxic Chemicals. Ann Arbor Science, Ann Arbor, MI. pp. 179-190.

Alexander, M. 1981. Biodegradation of chemicals of environmental concern. Science. 211:132-138.

Alexander, M. 1985. Biodegradation of organic chemicals. Environ. Sci. Technol. 18:106-111.

Alexander, M. 1986. Biodegradation of chemicals at trace concentrations. Report to Army Research Office on Grant No. DAAG29-83-K-0068.

Amdurer, M., Fellman, R., and Abdelhamid, S. 1985. In situ treatment technologies and superfund. In Proc. Intl. Conf. on New Frontiers for Hazardous Waste Management, Sept. 1985. EPA Report No. EPA-600/9-85/025. Hazardous Waste Eng. Res. Lab., Cincinnati, OH.

American Public Health Assoc. 1955. Standard Methods for the Examination of Water.

American Public Health Assoc. 1971. Standard Methods for the Examination of Water. 13th ed.

Anderson, A.C. and Abdelghani, A.A. 1980. Toxicity of selected arsencial compounds in short term bacterial bioassays. Bull. Environ. Contam. Toxicol. 24:124-127.

Anderson, J.W., Neff, J.M., Cox, B.A., Tatem, H.E., and Hightower, G.M. 1974. Characteristics of dispersions and water-soluble extracts of crude and refined oils and their toxicity to estuarine crustaceans and fish. Mar. Biol. 27:75-88.

Anonymous. 1981. Chemical Week. 128:40-41.

Anonymous. 1982. Chemical Week. 130:49.

Anonymous. 1984. Biostimulation and biodegradation to break down groundwater and soil contaminants. Environ. Sci. Technol. 18:139A.

Anonymous. 1986. Antibiotics speed bacterial attack on pollutants. Chemical Week. p. 29.

Arthur D. Little, Inc. 1976. State-of-the-art survey of land reclamation technology. Report No. EC-CR-76-76 on Contract No. DAAA 15-75-C-0188. Dept. of the Army, Edgewood Arsenal. Aberdeen Proving Ground, MD. 105 p.

Atlas, R.M. 1975a. Microbial degradation of petroleum in marine environments. In Proc. 1st Intersect. Congr. Intl. Assoc. of Microbiol. Studies. Science Council of Japan, Tokyo. p. 527.

Atlas, R.M. 1975b. Effects of temperature and crude oil composition on petroleum biodegradation. Appl. Microbiol. 30:396.

Atlas, R.M. 1977. Stimulated petroleum biodegradation. CRC Crit. Rev. Microbiol. 5:371-386.

Atlas, R.M. 1978a. An assessment of the biodegradation of petroleum in the Arctic. In Loutit, M.W. and Miles, J.A.R. (Eds.), Microbial Ecology. Springer-Verlag, Berlin. pp. 86-90.

Atlas, R.M. 1978b. Measurement of hydrocarbon biodegradation potentials and enumeration of hydrocarbon utilizing microorganisms using 14C radiolabelled spiked crude oil. In Costerton, J.W. and Colwell, R.R. (Eds.), Native Aquatic Bacteria: Enumeration, Activity and Ecology. ASTM-STP 695. Amer. Soc. for Testing and Materials, Philadelphia, PA. pp. 196-204.

Atlas, R.M. 1978c. Microorganisms and petroleum pollutants. BioScience. 28:387-391.

Atlas, R.M. 1981. Microbial degradation of petroleum hydrocarbons: an environmental perspective. Microbiol. Rev. 45:180-209.

Atlas, R.M. and Bartha, R. 1972a. Biodegradation of petroleum in seawater at low temperatures. Can. J. Microbiol. 18:1851-1855.

Atlas, R.M. and Bartha, R. 1972b. Degradation and mineralization of petroleum in sea water: limitation by nitrogen and phosphorus. Biotech. Bioeng. 14:309-318.

Atlas, R.M. and Bartha, R. 1972c. Degradation and mineralization of petroleum by two bacteria isolated from coastal waters. Biotech. Bioeng. 14:297-308.

Atlas, R.M. and Bartha, R. 1972d. Stimulated biodegradation of oil slicks using oleophilic fertilizers. Technical Report No. 3 of Project NR137-843. Sponsored by the Office of Naval Research.

Atlas, R.M. and Bartha, R. 1973a. Stimulated biodegradation of oil slicks using oleophilic fertilizers. Environ. Sci. Technol. 7:538-541.

Atlas, R.M. and Bartha, R. 1973b. Fate and effects of polluting petroleum on the marine environment. Residue Rev. 49:49-85.

Atlas, R.M. and Bartha, R. 1973c. Abundance, distribution and oil-biodegrading potential of microorganisms in Raritan Bay. Environ. Pollut. 4:291-300.

Atlas, R.M. and Bartha, R. 1973d. Effects of some commercial oil herders, dispersants and bacterial inocula on biodegradation of oil in seawater. In Ahearn, D.G. and Meyers, S.P. (Eds.), The Microbial Degradation of Oil Pollutants, Atlanta, GA, Dec. 4-6, 1972. Publ. No. LSU-SG-73-01. Louisiana State Univ., Center for Wetland Resources, Baton Rouge, LA. pp. 283-289.

Atlas, R.M. and Bartha, R. 1981. Microbial Ecology: Fundamentals and Applications. Addison-Wesley Publ. Co., Reading, MA. 560 pp.

Atlas, R.M. and Bronner, A. 1980. Response of intertidal microbial populations to oil from the Amoco Cadiz spillage. Abstr. Ann. Mtg. Amer. Soc. Microbiol. p. 203.

Atlas, R.M. and Busdosh. 1976. Microbial degradation of petroleum in the Arctic. In Sharpley, J.M. and Kaplan, A.M. (Eds.), Proc. 3rd Intl. Biodegradation Symp., Kingston, RI, Aug. 17-23, 1975. Applied Science Publ., London. pp. 79-85.

Atlas, R.M. and Schofield, E.A. 1975. Petroleum biodegradation in the Arctic. In Bourquin, A.W., Ahearn, D.G., and Meyers, S.P. (Eds.), Proc. Impact of the Use of Microorganisms on the Aquatic Environment. EPA Report No. EPA 660-3-75-001. Environmental Protection Agency, Corvallis, OR. p. 185.

Atlas, R.M., Schofield, E.A., Morelli, F.A., and Cameron, R.E. 1976. Effects of petroleum pollutants on Arctic microbial populations. Environ. Poll. 10:35-43.

Austin American Statesman, Nov. 19, 1980.

Azam, F. and Holm-Hansen, O. 1973. Use of tritiated substrates in the study of heterotrophy in sea water. Mar. Biol. 23:191-196.

Azarowicz, E.M. 1973. Microbial degradation of petroleum. Off. Gaz. U.S. Patent Office. 915:1835.

Baas Becking, L.G.M., Kaplan, I.R., and Moore, D. 1960. Limits of the natural environment in terms of pH and oxidation-reduction potentials. J. Geology. 68:243-284.

Babich, H. and Stotzky, G. 1983a. Nickel toxicity to microbes and a bacteriophage in soil and aquatic ecosystems: mediation by environmental characteristics. Abstr. Ann. Mtg. Amer. Soc. Microbiol. p. 261.

Babich, H. and Stotzky, G. 1983b. Abiotic factors affecting nickel toxicity: pure culture studies. Abstr. Ann. Mtg. Amer. Soc. Microbiol. p. 261.

Bache, R. and Pfennig, N. 1981. Selective isolation of Acetobacterium woodii on methoxylated aromatic acids and determination of growth yields. Arch. Microbiol. 130:255-261.

Bae, H.C., Costa-Robles, E.H., and Casida, L.E. 1972. Microflora of soil as viewed by transmission electron microscopy. Appl. Microbiol. 23:637-648.

Baker, M.D. and Mayfield, C.I. 1980. Microbial and nonbiological decomposition of chlorophenols and phenols in soil. Water Air Soil Poll. 13:411-424.

Bakker, G. 1977. Anaerobic degradation of aromatic compounds in the presence of nitrate. FEMS Microbiol. Lett. 1:103-108.

Balba, M.J. and Evans, W.C. 1977. The methanogenic fermentation of aromatic substrates. Biochem. Soc. Trans. 5:302-304.

Balba, M.T. and Evans, W.C. 1980. The anaerobic dissimilation of benzoate by Pseudomonas aeruginosa coupled with Desulfovibrio vulgaris with sulphate as terminal electron acceptor. Biochem. Soc. Trans. 8:624-625.

Balkwill, D.L. and Ghiorse, W.C. 1982. Characterization of subsurface microorganisms. Abstr. Ann. Mtg. Amer. Soc. Microbiol. p. 192.

Ballester, A. and Castelvi, J. 1980. J. Invest. Pes. 44:1.

Bambrick, D.R. 1985. The effect of DTPA on reducing peroxide composition. Tappi J. 68:96-100.

Barker, J.F. and Patrick, G.C. 1985. Natural attenuation of aromatic hydrocarbons in a shallow sand aquifer. Conf. Petrol. Hydro. and Organ. Chemicals in Groundwater. pp. 160-177.

Barkes, L. and Fleming, R.W. 1974. Production of dimethylselenide gas from inorganic selenium by eleven soil fungi. Bull. Environ. Contam. Toxicol. 12:308-311.

Bartha, R. and Atlas, R.M. 1977. The microbiology of aquatic oil spills. Adv. Appl. Microbiol. 22:225-266.

Bartholomew, G.W. and Pfaender, F.K. 1983. Influence of spatial and temporal variations on organic pollutant biodegradation rates in an estuarine environment. Appl. Environ. Microbiol. 45:103-109.

Bassford, P., et al. 1980. In Miller, J.H. and Rezinikoff, W.S. (Eds.), Genetic Fusions of the Lac Operon: A New Approach to the Study of Biological Processes, in the Operon. Cold Spring Harbor Lab., Cold Spring Harbor, New York, NY. pp. 245-262.

Battermann, G. and Werner, P. 1984. Beseitigung einer Untergrundkontamination mit Kohlenwasserstoffen durch mikorbiellen Abbau. Grundwasserforschung-Wasser/Abwasser. 125:366-373.

Batterton, J., Winters, K., and van Baalen, C. 1978. Anilines: selective toxicity to blue-green algae. Science. 199:1068-1070.

Bauer, J.E. and Capone, D.G. 1985. Degradation and mineralization of the polycyclic aromatic hydrocarbons anthracene and napthalene in intertidal marine sediments. Appl. Environ. Microbiol. 50:81-90.

Baver, L.D., Gardner, W.H., and Gardner, W.R. 1972. Soil Physics. 4th ed. John Wiley and Sons, Inc., New York, NY. 498 p.

Beam, H.W. and Perry, J. J. 1973. Co-metabolism as a factor in microbial degradation of cycloparaffinic hydrocarbons. Arch. Mikrobiol. 91:87-90.

Beam, H.W. and Perry, J.J. 1974. Microbial degradation of cycloparaffinic hydrocarbons via cometabolism and commensalism. J. Gen. Microbiol. 82:163-169.

Beaubien, A., Keita, L., and Jolicoeur, C. 1987. Flow microcalorimetry investigation of the influence of surfactants on a heterogeneous aerobic culture. Appl. Environ. Microbiol. 53:2567-2573.

Bedient, P.B., Asce, A.M., Springer, N.K., Baca, E., Bouvette, T.C., Hutchins, S.R., and Tomson, M.B. 1983. J. Environ. Sci. 109:485-501.

Benoit, R.E., Allen, G.C., and Novak, J.T. 1984. Transformation of methanol and t-butyl alcohol by ground water bacteria. Abstr. Ann. Mtg. Amer. Soc. Microbiol. p. 212.

Benoit, R., Novak, J., Goldsmith, C., and Chadduck, J. 1985. Alcohol biodegradation in groundwater microcosms and pure culture systems. Abstr. Ann. Mtg. Amer. Soc. Microbiol. p. 258.

Berridge, S.A., Dean, R.A., Fallows, R.G., and Fish, A.   1968a.   The properties of persistent oils at sea. In Hepple, P. (Ed.), Scientific Aspects of Pollution of the Sea by Oil. Institute of Petroleum, London. pp. 2-11.

Berridge, S.A., Dean, R.A., Fallows, R.G., and Fish, A.   1968b.   The properties of persistent oils at sea. J. Inst. Petroleum, (London). p. 300.

Berry, D.F. and Boyd, S.A.   1984.   Division S-3--Soil Microbiology and Biochemistry. Oxidative coupling of phenols and anilines by peroxidase: structure-activity relationships. Soil Sci. Soc. Amer. J. 48:565-569.

Berry, D.F., Francis, A.J., and Bollag, J.-M. 1987. Microbial metabolism of homocyclic and heterocyclic aromatic compounds under anaerobic conditions. Microbiol. Rev. 51:43-59.

Bertrand, J.C., Esteves, J.L., Mulyono, M., and Mille, G. 1986. Use of continuous flow through systems to determine petroleum hydrocarbon evolution in superficial marine sediments. Chemosphere. 15:205-210.

Biosystems, Inc. May/June 1986. Giving nature a boost. Dupont Magazine, Wilmington, DE.

Bitton, G. and Gerba (Eds.). 1985. Groundwater Pollution Microbiology. John Wiley and Sons, Inc., New York, NY.

Blakebrough, N. 1978. In Charter, K.W.A. and Somerville, H.J. (Eds.), The Oil Industry and Microbial Ecosystems. Institute of Petroleum. London. p. 28.

Blanch, H.W. and Einsele, A. 1973. The kinetics of yeast growth on pure hydrocarbons. Biotechnol. Bioeng. 15:861.

Bland, W.F. and Davidson, R.L. 1967. Petroleum Processing Handbook. McGraw-Hill, New York, NY. pp. 11-54.

Bloom, S.A. 1970. Oil dispersants effect on the microflora of beach sand. J. Mar. Biol. Assoc. U.K. 50:919.

Blumer, M. 1975. Organic compounds in nature: limits of our knowledge. Angewandte Chemie. 14:507-514.

Boehm, D.F. and Pore, R.S. 1984. Studies on hexadecane utilization by Prototheca zopfii. Abstr. Ann. Mtg. Amer. Soc. Microbiol. p. 213.

Boethling, R.S. and Alexander, M. 1979. Effect of concentration of organic chemicals on their biodegradation by natural microbial communities. Appl. Environ. Microbiol. 37:1211-1216.

Bohn, H.L. 1971. Redox potentials. Soil Sci. 112:39-45.

Bollag, J.-M., Czaplicki, E.J., and Minard, R.D. 1975. Agric. Fd. Chem. 23:85-90.

Bollag, J.-M., Sjoblad, R.D., and Minard, R.D. 1977. Polymerization of phenolic intermediates of pesticides by a fungal enzyme. Experientia. 33:1564-1566.

Bollag, J.-M. 1983. Cross-coupling of humus constituents and xenobiotic substances. In Christman, R.F. and Gjessing, E.T. (Eds.), Aquatic and Terrestrial Humic Materials. Ann Arbor Science Publ., Ann Arbor, MI. pp. 127-141.

Boone, D.R., and Bryant, M.P. 1980. Propionate-degrading bacterium, Syntrophobacter wolinii sp. nov. gen. nov. from methanogenic ecosystems. Appl. Environ. Microbiol. 40:626-632.

Bossert, I. 1983. The fate of benzo(a)pyrene and other polynuclear aromatics in soil: effects of structure and cosubstrates. Abstr. Ann. Mtg. Amer. Soc. Microbiol. p. 273.

Bouwer, H. 1984. Elements of soil science and groundwater hydrology. In Groundwater Pollution Microbiology. Bitton, G. and Gerba, C.P. (Eds.). John Wiley and Sons, Inc., New York, NY. pp. 9-38.

Bouwer, E.J. and McCarty, P.L. 1983a. Transformation of halogenated organic compounds under denitrification conditions. Appl. Environ. Microbiol. 45:1295-1299.

Bouwer, E.J. and McCarty, P.L. 1983b. Transformations of 1- and 2-carbon halogenated aliphatic organic compounds under methanogenic conditions. Appl. Environ. Microbiol. 45:1286-1294.

Bouwer, E.J. and McCarty, P.L. 1984. Modeling of trace organics biotransformation in the subsurface. Ground Water. 22:433-440.

Bove, L.J., Lambert, W.P., Lin, L.Y.H., Sullivan, D.E., and Marks, P.J. 1984. Report to U.S. Army Toxic and Hazardous Materials Agency, Aberdeen Proving ground, Maryland, on Contract No. DAAK11-82-C-0017.

Boyd, S.A., Shelton, D.R., Berry, D., and Tiedje, J.M. 1983. Anaerobic biodegradation of phenolic compounds in digested sludge. Appl. Environ. Microbiol. 46:50-54.

Boyle, T.P., Finger, S.E., Petty, J.D., Smith, L.M., and Huckings, J.N. 1984. Distribution and rate of disappearance of fluorene in pond ecosystems. Chemosphere. 13:997-1008.

Boynton, M. 1986. Casebook. Groundwater contamination levels reduced to undetectable levels with new air stripper. Poll. Eng 18(8):52.

Brady, N.C. 1974. The Nature and Properties of Soils. 8th ed. Macmillan Publ. Co., Inc., New York, NY.

Braun, K. and Gibson, D.T. 1984. Anaerobic degradation of 2-aminobenzoate (anthranilic acid) by denitrifying bacteria. Appl. Environ. Microbiol. 48:1187-1192.

Brindley, G.W. and Thompson, T.D. 1966. Clay organic studies. XI. Complexes of benzene, pyridine, piperidine and 1,3-substituted propanes with a synthetic Ca-fluorhectorite. Clay Minerals. 6:345.

Britton, L.N. 1985. API Publication No. 4389. American Petroleum Institute, Washington, D.C.

Brock, T.D. 1978. Thermophilic Microorganisms and Life at High Temperatures. Springer-Verlag, Heidelburg, F.R.G.

Brooks, K. and McGinty, R. 1987. Groundwater treatment know-how comes of age. Chemical Week. May 20. pp. 50-52.

Brown, D.W., Ramos, L.S., Fiedman, A.J., and Macleod, W.D.   1969. Analysis of trace levels of petroleum hydrocarbons in marine sediments using a solvent slurry extraction procedure. In Trace Organic Analysis: a New Frontier in Analytical Chemistry. Special Publication No. 519. National Bureau of Standards, Washington, D.C. pp. 161-167.

Brown, K.W. 1975. An investigation of the feasibility of soil disposal of waste water for the Jefferson Chemical Co., Monroe, Texas. Final Report Phase I. Texas A & M Research Foundation. P.O. Box H. College Station, TX.

Brown, K.W. 1981. In Worm, B.G., Dantin, E.J., and Seals, R.K. (Eds.), Proc. Symp. and Workshop on Hazardous Waste Management Protection of Water Resources, Nov. 16-20, 1981, Louisiana State Univ., Baton Rouge, LA.

Brown, K.W., Deuel, L.E., Jr., and Thomas, J.C. 1983. Land treatability of refinery and petrochemical sludges. EPA Report No. EPA-600/S2-074. Municipal Environmental Research Lab., Cincinnati, OH.

Brown, K.W., Donnelly, K.C., and Deuel, L.E., Jr. 1983. Effects of mineral nutrients, sludge application rate, and application frequency on biodegradation of two oily sludges. Microb. Ecol. 9:363-373.

Brown, R.A., Longfield, J.Y., Norris, R.D., and Wolfe, G.F. 1985. Enhanced bioreclamation: designing complete solution to ground water problems. Proc. Ind. Wastes Symp. Water Poll. Control Fed., Kansas City, KS.

Brown, R.A., Loper, J.R., and McGarvey, D.C.   1986.   Preparedness, prevention, control and cleanup of releases. Proc. 1986 Hazardous Material Spills Conf., May 5-8, 1986, St. Louis, MO.

Brown, R.A., Norris, R.D. and Brubaker, G.R. 1985. Aquifer restoration with enhanced bioreclamation. Poll. Eng. 17:25-28.

Brown, R.A., Norris, R.D., and Raymond, R.L. 1984. Oxygen transport in contaminated aquifers. In Proc. NWWA/API Conf. on Petroleum Hydrocarbons and Organic Chemicals in Groundwater--Prevention, Detection, and Restoration, Houston, TX. National Water Well Association, Worthington, OH. pp. 441-450.

Brubaker, G.R. and O'Neill, E. 1982. Remediation strategies using enhanced bioreclamation. Proc. 5th Nat. Symp. on Aquifer Restoration and Groundwater Monitoring. pp. 1-5.

Brunner, W. and Focht, D.D. 1983. Persistence of polychlorinated biphenyls (PCB) in soils under aerobic and anaerobic conditions. Abstr. Ann. Mtg. Amer. Soc. Microbiol. p. 266.

Bryant, R.J., Woods, L.E., Coleman, D.C., Fairbanks, B.C., McClellan, J.F., and Cole, C.V. 1982. Interactions of bacterial and amoebal populations in soil microcosms with fluctuating moisture content. Appl. Environ. Microbiol. 43:747-752.

Buckingham, B. 1981. Studies in biodegradation. Assoc. Amer. Railroads. Res. and Test Dept., 1920 L St., N.W., Washington, D.C.

Bumpus, J.A., Fernando, T., Mileski, G.J., and Aust, S.D. 1987. Biodegradation of organopollutants by Phanerochaete chrysosporium: practical considerations. Proc. 13th Ann. Res. Symp. on Land Disposal, Remedial Action, Incineration and Treatment of Hazardous Waste, May 6-8, 1987, Cincinnati, OH.

Bumpus, J.A., Tien, M., Wright, D., and Aust, S.D. 1985. Oxidation of persistent environmental pollutants by a white rot fungus. Science. 228:1434-1436.

Burkhard, N. and Guth, J.A. 1979. Photolysis of organophosphorus insecticides on soil surfaces. Pest. Sci. 10:313.

Butkevich, N.V. and Butkevich, V.S. 1936. Microbiology (Moscow). 5:322-343.

Callister, S.M. and Winfrey, M.R. 1983. Microbial mercury resistance and methylation rates in the upper Wisconsin River. Abstr. Ann. Mtg. Amer. Soc. Microbiol. p. 261.

Canevari, G.P. 1971. Oil spill dispersants: current status and future outlook. Proc. Joint Conf. on Prevention and Control of Oil Spills. American Petroleum Institute, Washington, D.C. p. 263.

Canter, L.W. and Knox, R.C. 1985. Ground Water Pollution Control. Lewis Publishers, Chelsea, MI. 526 pp.

CAST. 1976. Application of sewage sludge to cropland. Report No. 64, EPA Report No. EPA-430/9-76-013. Environmental Protection Agency, Washington, D.C.

Cerniglia, C.E. 1984. Microbial metabolism of polycyclic aromatic hydrocarbons. Adv. Appl. Microbiol. 30:31-71.

Cerniglia, C.E. and Crow, S.A. 1981. Metabolism of aromatic hydrocarbons by yeasts. Arch. Microbiol. 129:9-13.

Cerniglia, C.E. and Gibson, D.T. 1979. Oxidation of benzo(a)pyrene by the filamentous fungus Cunninghamella elegans. J. Biol. Chem. 254:12174-12180.

Cerniglia, C.E. and Gibson, D.T. 1980. Fungal oxidation of benzo(a)pyrene: evidence for the formation of a BP-7,8-diol 9,10-epoxide and BP-9,10-diol 7,8-epoxides. Abstr. Ann. Mtg. Amer. Soc. Microbiol. p. 138.

Cerniglia, C.E., van Baalen, C., and Gibson, D.T. 1980a. Metabolism of naphthalene by the cyanobacterium Oscillatoria sp., Strain JCM. J. Gen. Microbiol. 116:485-494.

Cerniglia, C.E., Gibson, D.T., and van Baalen, C. 1980. Oxidation of naphthalene by cyanobacteria and microalgae. J. Gen. Microbiol. 116:495-500.

Cerniglia, C.E., Hebert, R.L., Dodge, R.H., Szaniszlo, P.J., and Gibson, D.T. 1979. Some approaches to studies on the degradation of aromatic hydrocarbons by fungi. In Bourquin, A.L. and Pritshard, H. (Eds.), Microbial Degradation of Pollutants in Marine Environments. EPA Report No. EPA-600/9-79-012.

Cerniglia, C.E., Hebert, R.L., Szaniszlo, P.J., and Gibson, D.T. 1978. Fungal transformation of naphthalene. Arch. Microbiol. 117:135-143.

Cerniglia, C.E. and Perry, J.J. 1973. Oil degradation by microorganisms isolated from the marine environment. Z. Allg. Mikrobiol. 13:299-306.

Cerniglia, C.E., Wyss, O., and van Baalen, C. 1980. Recent studies on the algal oxidation of naphthalene. Abst. Ann. Mtg. Amer. Soc. Microbiol. p. 197.

Chakrabarty, A.M. 1974. Microorganisms having multiple compatible degradative energy-generating plasmids and preparation thereof. Off. Gaz. of the U.S. Patent Office. 922:1224.

Chakrabarty, A.M. 1982. Genetic mechanisms in the dissimilation of chlorinated compounds. In Chakrabarty, A.M. (Ed.), Biodegradation and Detoxification of Environmental Pollutants. CRC Press, Inc., Boca Raton, FL. pp. 127-139.

Challenger, F. and North, H.J. 1933. Production of organo-metalloidal compounds by micro-organisms. II. Dimethyl selenide. J. Chem. Soc. 18:68-71.

Chaney, R.L. and Hornick, S.B. 1978. Accumulation and effects of cadmium on crops. In Proc. 1st Intl. Cadmium Conf. Metals Bulletin, Ltd., London.

Chapin, R.W. 1981. Microbiological degradation of organic hazardous substances: a review of the art and its application to soil and groundwater decontamination at the Goose Farm hazardous waste site, Plumsted Township, New Jersey. U.S. EPA Region II, Hazardous Response Branch, Technical Assistance Team. August, 1981. pp. 1-3.

Characklis, W.G. 1973. Attached microbial growths--I. Attachment and growth. Water Res. 7:1113.

Chen, C., Dey, C.R., Kalb, V.F., Sanglard, D., Sutter, T.R., Turi, T., and Loper, J.C. 1987. Engineering P450 genes in yeast. U.S. EPA 13th Ann. Res. Symp. Land Disposal, Remedial Action, Incineration and Treatment of Hazardous Waste, May 6-8, 1987, Cincinnati, OH.

Chepigo, S.V., Boiko, L.D., Gololobov, A.D., Kryuchkova, A.P., Vorob'era, G.I., Rozhkova, M.I., Fisher, P.N., Pokrovskii, V.K., and Krotchenko, N.I. 1967. Production of fodder yeast from petroleum hydrocarbons. Prikl. Biokim. Mikrobiol. 3:577.

Cherry, J.A., Gillham, R.W. Barker, J.F. 1984. Chapter 3. Groundwater Contamination (Studies in Geophysics). National Academy Press, Washington, D.C. pp. 46-64.

Chibata, I., Tosa, T., and Sato, T. 1979. In Microbial Technology. Peppler, H.J. and Perlman, D. (Eds.). Vol. II. 2nd ed. Academic Press, New York, NY, and London. pp. 433-461.

Chowdhury, J., Parkinson, G., and Rhein, R. 1986. CPI go below to remove groundwater pollutants. Chem. Eng. 93:14-19.

Clark, F.E. 1967. Bacteria in soil. In Burges, A. and Raw, F. (Eds.), Soil Biology. Academic Press, New York, NY. pp. 15-49.

Claus, D. and Walker, N. 1964. The decomposition of toluene by soil bacteria. J. Gen. Microbiol. 36:107-122.

Clewell, H.J., III. 1981. The effect of fuel composition on groundfall from aircraft fuel jettisoning. Report to Eng. and Services Lab., Air Force Eng. and Services Center, Tyndall Air Force Base, FL.

Cobet, A.B. 1974. Hydrocarbonoclastic repository. In Progress Report Abstracts. Microbiology Program, Office of Naval Research, Arlington, VA. p. 131.

Cobet, A.B. and Guard, H.E. 1973. Effect of a bunker fuel on the beach bacterial flora. Proc. Joint Conf. on Prevention and Control of Oil Spills. American Petroleum Institute, Washington, D.C. p. 815.

Cobet, A.B., Guard, H.E., and Chatigny, M.A. 1973. Considerations in application of microorganisms to the environment for degradation of petroleum products. In Ahearn, D.G. and Meyers, S.P. (Eds.), The Microbial Degradation of Oil Pollutants, Atlanta, GA, Dec. 4-6, 1972. Publ. No. LSU-SG-73-01. Louisiana State Univ., Center for Wetland Resources, Baton Rouge, LA. pp. 81- 87.

Coffey, J.C., Ward, C.H., and King, J.M. 1977. Effects of petroleum hydrocarbons on growth of fresh-water algae. Dev. Ind. Microbiol. 18:661-672.

Colwell, R.R. and Walker, J.D. 1977. Ecological aspects of microbial degradation of petroleum in the marine environment. CRC Crit. Rev. Microbiol. 5:423-445.

Colwell, R.R., Walker, J.D., and Nelson, J.D., Jr. 1973. Microbial ecology and the problem of petroleum degradation. In Ahearn, D.G. and Meyers, S.P. (Eds.), The Microbial Degradation Oil Pollutants, Workshop, Atlanta, GA, Dec. 4-6, 1972. Publ. No. LSU-SG-73-01. Louisiana State Univ., Center for Wetland Resources, Baton Rouge, LA. pp. 185-197.

Cook, F.D. and Westlake, D.W.S. 1974. Microbiological degradation of northern crude oils. Environmental-Social Committee, Northern Pipelines, Task Force on Northern Oil Development. Report No. 74-1. Catalog No. R72-12774. Information Canada, Ottawa.

Cook, R.J. and Papendick, R.I. 1970. Soil water potential as a factor in the ecology of _Fusarium_ _roseum_ f. sp. _cerealis_ "colmorum." Plant Soil. 32:131-145.

Cook, W.L. Massey, J.K., and Ahearn, D.G. 1973. The degradation of crude oil by yeasts and its effect on _Lesbistes_ _reticulatus_. In Ahearn, D.G. and Meyers, S.P. (Eds.), The Microbial Degradation of Oil Pollutants, Workshop, Atlanta, GA, Dec. 4-6, 1972. Publ. No. LSU-SG-73-01. Center for Wetland Resources, Louisiana State Univ., Baton Rouge, LA. p. 279.

Cooke, R.C. and Rayner, A.D.M. (Eds.). 1984. Ecology of Saprophytic Fungi. Longman, New York, NY. 415 pp.

Cooke, S. and Bluestone, M. 1986. Invasion from Ireland: grease-eating microbes. Chemical Week. Feb. 5, p. 20.

Cooney, J.J. 1974. Microorganisms Capable of Degrading Refractory Hydrocarbons in Ohio Waters, NTIS PB-237293/6ST. National Technical Information Service. Springfield, VA.

Cooney, J.J., Silver, S.A., and Beck, E.A. 1985. Factors influencing hydrocarbon degradation in three freshwater lakes. Microb. Ecol. 11:127-137.

Cooney, J.J. and Summers, R.J. 1976. Hydrocarbon-using microorganisms in three fresh-water ecosystems. In Sharpley, J.M. and Kaplan, A.M. (Eds.), Proc. 3rd Intl. Biodegradation Symp., Aug. 17-23, 1975, Kingston, RI. pp. 141-156.

Cooper, D.G. 1982. Biosurfactants and enhanced oil recovery. Proc. 1982 Intl. Conf. on Microbiological Enhancement of Oil Recovery, May 16-21, Shangri-La, Afton, Oklahoma. pp. 112-113.

Council on Environmental Quality. 1981. Contamination of groundwater by toxic organic chemicals. Council on Environmental Quality, Washington, D.C.

Cox, D.P. and Williams, A.L. 1980. Biological process for converting naphthalene to cis-1,2-dihydroxy-1,2-dihydronaphthalene. Appl. Environ. Microbiol. 39:320-326.

Craig, P.J. and Wood, J.M. 1981. Environmental Lead. Lynam, D.R., Piantanida, L.E., and Cole, J.F. (Eds.). Academic Press, New York, NY. p. 333.

Crosby, D.G. 1971. Environmental photooxidation of pesticides. In Degradation of Synthetic Organic Molecules in the Biosphere. National Academy of Sciences, Washington, D.C. pp. 260-290.

Crow, S.A., Cook, W.L., Ahearn, D.G., and Bourquin, A.W. 1976. Microbial populations in coastal surface slicks. In Sharpley, J.M. and Kaplan, A.M. (Eds.), Proc. 3rd Intl. Biodegradation Symp. Applied Science Publ., Ltd., London. pp. 93-98.

Cullimore, D.R. 1983. Microbiological parameters controlling plugging of wells and groundwater systems. In Parr, J.F., Marsh, P.B., Kla, J.M. (Eds.), Land Treatment of Hazardous Wastes. Noyes Pub., Park Ridge, NJ. p. 12.

Cupitt, L.T. 1980. Fate of toxic and hazardous materials in the air environment. EPA Report No. EPA-600/53-80-084. Environmental Protection Agency, Athens, GA.

Dagley, S. 1981. New perspectives in aromatic catabolism. In Leisinger, T., Hutter, R., Cook, A.M., and Nuesch, J. (Eds.), FEMS Symposium No. 12. Microbial Degradation of Xenobiotics and Recalcitrant Compounds. 12:141:179.

Dagley, S. and Patel, M.D. 1957. Oxidation of p-cresol and related compounds by a Pseudomonas. Biochem. J. 66:227.

Danielpol, D.L. 1982. Der Einfluss Organischer Verschmutzung auf das Grundwasser-Okosystem der Donau in Raum Weinund Niederosterreich. Limnologisches Institut. Mondsee, Austria.

Davies J.I. and Evans, W.C. 1976. Oxidative metabolism of naphthalene by soil pseudomonads. The ring-fission mechanism. Biochem. J. 91:5988-5996.

Davies, J.S. and Westlake, D.W.S. 1979. Crude oil utilization by fungi. Can. J. Microbiol. 25:146-156.

Davis, B.D., Dulbecco, R., Eisen, H.N., Ginsberg, H.S., and Wood, W.B., Jr. (Eds.). 1970. Microbiology. Harper and Row, Ltd., London. 1464 pp.

Davis, J.B. 1967. Petroleum Microbiology. Elsevier North Holland, New York, NY.

Davis, J.B., et al. 1972. API Report No. 4149. American Petroleum Institute, Washington, D.C.

Davis, S.J. and Gibbs, C.F. 1975. The effect of weathering on a crude oil reside exposed at sea. Water Res. 9:275-288.

Dean-Raymond, D. and Bartha, R. 1975. Biodegradation of some polynuclear aromatic petroleum components by marine bacteria. NTIS No. AD/A-006 346/1ST, National Technical Information Service, Springfield, VA.

de Klerk, H. and van der Linden, A.C. 1974. Bacterial degradation of cyclohexane. Participation of a co-oxidation reaction. Antonie van Leeuwenhoek. 40:7-15.

De Pastrovich, T.L., Baradat, Y., Barthal, R., Chiarelli, A., and Fussel, D.R. 1979. Protection of ground water from oil pollution. CONCAWE Report No. 3179. The Oil Companies' Intl. Study Gp. Den Haag, The Netherlands.

Detox, Inc. Company Sales Literature. Dayton, OH.

Dev, H., Bridges, J.E., and Sresty, G.C. 1984. Decontamination for volatile organics: a case history. In Ludwigson, J. (Ed.), Proc. 1984 Hazardous Material Spills Conf., Nashville, TN. Government Institutes, Inc., Rockville, MD. pp. 57-64.

Dibble, J.T. and Bartha, R. 1976. Effect of iron on the biodegradation of petroleum in seawater. Appl. Environ. Microbiol. 31:544-550.

Dibble, J.T. and Bartha, R. 1979a. Effect of environmental parameters on the biodegradation of oil sludge. Appl. Environ. Microbiol. 37:729-739.

Dibble, J.T. and Bartha, R. 1979b. Leaching aspects of oil sludge biodegradation in soil. Soil Sci. 127:365-370.

Dibble, J.T. and Bartha, R. 1979c. Rehabilitation of oil-inundated agricultural land: a case history. Soil Sci. 128:56-60.

Dietz, D.N. 1980. The intrusion of polluted water into a groundwater body and the biodegradation of a pollutant. Proc. 1980 Conf. Hazardous Material Spills, sponsored by U.S. EPA, et al. pp. 236-244.

Dobbins, D.C., Aelion, M.C., Long, S.C., and Pfaender, F.K. 1986. Methodology for assessing the metabolism of subsurface microbial communities. Abstr. Ann. Mtg. Soc. Microbiol. p. 300.

Dodge, R.H. and Gibson, D.T.   1980.   Fungal metabolism of benzo(a)anthracene. Abstr. Ann. Mtg. Amer. Soc. Microbiol. p. 138.

Dodge, R.H., Cerniglia, C.E., and Gibson, D.T. 1979. Fungal metabolism of biphenyl. Biochem. J. 178:223-230.

Doelle, H.W. 1975. Bacterial Metabolism. 2nd ed. Academic Press, New York, NY. pp. 499-516.

Donnan, W.W. and Schwab, G.O. 1974. Current drainage methods in the U.S.A. In Proc. 1st Intl. Symp. on Acid Precipitation and the Forest Ecosystem. U.S. Forest Service General Technical Report NE-23, USDA--Forest Service, Upper Darby, PA.

Donoghue, N.A., Griffin, M., Norris, D.B., and Trudgill, P.W. 1976. The microbial metabolism of cyclohexane and related compounds. In Sharpley, J. M. and Kaplan, A.M. (Eds.), Proc. 3rd Intl. Biodegradation Symp., Kingston, Rhode Island, August 17-23, 1975. Applied Science Publ., Ltd., London. pp. 43-56.

Dooley, M.A., Larson, R.J., and Ventullo, R.M. 1985. Mineralization of xenobiotic chemicals by aquifer sediments. Abstr. Ann. Mtg. Amer. Soc. Microbiol. p. 258.

Doyle, R.C. 1979. The effect of dairy manure and sewage sludge on pesticide degradation in soil. Ph.D. Dissertation. Univ. of Maryland, College Park, MD.

Drozd, J.W. 1980. In Harrison, D.E.F., Higgins, I.J., and Watkinson, R. (Eds.), Hydrocarbons in Biotechnology. Heyden, London. pp. 75-83.

Dunlap, K.R. and Perry, J.J. 1967. Effect of substrate on the fatty acid composition of hydrocarbon-utilizing microorganisms. J. Bacteriol. 94:1919-1923.

Dunlap, W.J. and McNabb, J.F. 1973. EPA Report No. EPA-660/2-73-014, Environmental Protection Agency, Washington, D.C.

Dunlap, W.J., McNabb, J.F., Scalf, M.R., and Cosby, R.L. 1977. Sampling for organic chemicals and microorganisms in the subsurface. EPA Report No. EPA 600/2-77-176. Robert S. Kerr Environmental Research Lab.

Dutton, P.L. and Evans, W.C. 1969. The metabolism of aromatic compounds by Rhodopseudomonas palustris. Biochem. J. 113:525-535.

Edgehill, R.U. and Finn, R.K. 1983. Microbial treatment of soil to remove pentachlorophenol. Appl. Environ. Microbiol. 45:1122-1125.

Edwards, N.T. 1983. Reviews and analyses. Polycyclic aromatic hydrocarbons (PAH's) in the terrestrial environment--A Review. J. Environ. Qual. 12:427- 441.

Ehrenfeld, J.R. and Bass, J.M. 1983. Handbook for Evaluating Remedial Action Technology Plans. EPA Report No. EPA-600/2-83-076. Municipal Environmental Research Lab., Cincinnati, OH.

Ehrenfeld, J. and Bass, J. 1984. Evaluation of Remedial Action Unit Operations at Hazardous Waste Disposal Sites. Noyes Pub. Park Ridge, NJ. 435 pp.

Ehrich, H.L. 1978. How microbes cope with heavy metals, arsenic and antimony in their environment. In Kushner, J. (Ed.), Microbial Life in Extreme Environments. Academic Press, New York, NY. pp 381-408.

Ehrlich, G.G., Godsy, E.M., Goerlitz, D.F., and Hult, M.F. 1983. Microbial ecology of a creosote-contaminated aquifer at St. Louis Park, MN. Dev. Ind. Microbiol. 24:235-245.

Ehrlich, G.G., Goerlitz, D.F., Godsy, E.M., and Hult, M.F. 1982. Degradation of phenolic contaminants in ground water by anaerobic bacteria: St. Louis Park, MN. Ground Water. 20:703-710.

Ehrlich, G.G., Schroeder, R.A., and Martin, P. 1985. Microbial populations in a jet-fuel-contaminated shallow aquifer at Tustin, California. U.S. Geological Survey. Open-file Report 85-335. Prepared in cooperation with the U.S. Marine Corps.

Ellis, N.D. and Payne, J.R. 1984. EPA Report No. EPA-600/D-84-039.

Ellis, W.D., Payne, J.R., and McNabb, G.D. 1985. Treatment of contaminated soils with aqueous surfactants. EPA Report No. EPA/600/S2-85/129. Hazardous Waste Eng. Res. Lab., Cincinnati, OH.

Ellis, W.D., Payne, J.R., Tafuri, A.N., and Freestone, F.J. 1984. The development of chemical countermeasures for hazardous waste contaminated soil. In Ludwigson, J. (Ed.), Proc. 1984 Hazardous Material Spills Conf., Nashville, TN, 1984. Government Institutes, Inc., Rockville, MD. pp. 116-124.

Environmental Protection Agency. 1971. Impact of oily materials on activated sludge systems. Water Poll. Control Res. Series, 12050 DSH 03/71. Environmental Protection Agency, Washington, D.C.

Environmental Proection Agency. 1974. Manual of Methods for Chemical Analysis of Water and Wastes. EPA Report No. 625/6-74-003a. Environmental Research Center, Cincinnati, OH.

Environmental Protection Agency. 1976. Monitoring groundwater quality: Methods and costs. EPA Report No. 600/4-76-023.

Environmental Protection Agency. 1978. Innovative and Alternative Technology Assessment Manual. EPA Report No. EPA 430-9-78-009.

Environmental Protection Agency. 1979. Water-Related Environmental Fate of 129 Priority Pollutants.

Environmental Protection Agency. 1980. Proposed Ground Water Protection Strategy. Washington, D.C.

Environmental Protection Agency. 1983. Innovative and Alternative Technology Assessment Manual. Draft report.

Environmental Protection Agency. 1984a. Laboratory assessment of potential hydrocarbon emissions from land treatment of refinery oily sludges. EPA Report No. 600/S2-84-108. Environmental Research Laboratory, Ada, OK.

Environmental Protection Agency. 1984b. Remedial Response at Hazardous Waste Sites, Case Studies 1-23. EPA Report No. 540/1-84-002b.

Environmental Protection Agency. 1985a. Land treatment of an oily waste-degradation, immobilization and bioaccumulation. EPA Report No. 600/S2-85-009. Municipal Environmental Research Lab., Ada, OK.

Environmental Protection Agency. 1985b. Handbook Remedial Action at Waste Disposal Sites (Revised). EPA Report No. EPA/625/6-85/006.

Evans, G.B., Jr., Deuel, L.E., Jr., and Brown, R.W. 1980. Mobility of water soluble organic constituents of API separator waste in soils. Report prepared on EPA Grant No. R805474010.

Evans, W.C. 1977. Biochemistry of the bacterial catabolism of aromatic compounds in anaerobic environments. Nature (London). 270:17-22.

Faust, S.J. and Hunter, J.V. 1971. Organic Compounds in Aquatic Environments. Marcel Dekker, NY.

Fay, J.A. 1969. The spread of oil slicks on a calm sea. In Hoult, D.P. (Ed.), Oil on the Sea. Plenum Press, New York, NY. p. 53.

Federle, T.W., Dobbins, D.C., Thornton-Manning, J.R., and Jones, D.D. 1986. Microbial biomass, activity, and community structure in subsurface soils. Ground Water. 24:365-374.

Fedorak, P.M., Semple, K.M., and Westlake, D.W.S. 1984. Oil degrading capabilities of yeasts and fungi isolated from coastal marine environments. Can. J. Microbiol. 30:565-571.

Fenton, H.J. 1894. Oxidation of tartaric acid in presence of iron. J. Chem. Soc. (London). 65:899.

Ferry, J.G. and Wolfe, R.S. 1976. Anaerobic degradation of benzoate to methane by a microbial consortium. Arch. Microbiol. 107:33-40.

Ferry, J.G. and Wolfe. 1977. Nutritional and biochemical characterization of Methanospirillum hungatii. Appl. Environ. Microbiol. 34:371-376.

Fewson, C.A. 1967. The growth and metabolic versatility of the gram-negative bacterium NCIB 8250 ('Vibrio 01'). J. Gen Microbiol. 46:255-266.

Fewson, C.A. 1981. Biodegradation of aromatics with industrial relevance. In Leisinger, T., Hutter, R., Cook, A.M., and Nuesch, J. (Eds.), FEMS Symp. No. 12. Microbial Degradation of Xenobiotics and Recalcitrant Compounds. 12:141-179.

Finnerty, W.R., Kennedy, R.S., Lockwood, P., Spurlock, B.O., and Young, R.A. 1973. In Ahearn, D.G. and Meyers, S.P. (Eds.), Microbial Degradation of Oil Pollutants, Workshop, Atlanta, GA, Dec. 4-6, 1972. Publ. No. LSU-SG-73-01. Louisiana State Univ., Center for Wetland Resources, Baton Rouge, LA. pp. 105-126.

Fisher, P.R., Appleton, J., and Pemberton, J.M. 1978. Isolation and characterization of the pesticide-degrading plasmid pJP1 from Alcaligenes paradoxus. J. Bacteriol. 135:798-804.

Flathman, P.E. and Caplan, J.A. 1985. Biological cleanup of chemical spills. Proc. Hazmacon 85, Apr. 23-25, 1985, Oakland, CA. Preprint.

Floodgate, G.D. 1976. Oil biodegradation in the oceans. In Sharpley, J.M. and Kaplan, A.M. (Eds.), Proc. 3rd Intl. Biodegradation Symp., Kingston, RI, Aug. 17-23, 1975. Applied Sci. Publishers, London. pp. 87-91.

Floodgate, G.D. 1979. Nutrient limitation. In Bourquin, A.W. and Pritchard, P.H. (Eds.), Proc. Workshop, Microbial Degradation of Pollutants in Marine Environments. EPA Report No. EPA-66019-79-012. Environmental Research Lab., Gulf Breeze, FL. pp. 107-118.

Florida Environ. Service Center. 1982. Report No. 4.

Flowers, T.H., Pulford, I.D., and Duncan, H.J. 1984. Studies on the breakdown of oil in soil. Environ. Poll. Series B. 8:71-82.

FMC Aquifer Remediation Systems. Jan. 7, 1986. In situ treatment of ground water: principles, methods, and choices. Report to U.S. Marine Corps.

FMC Corporation. 1979. Industrial waste treatment with hydrogen peroxide. Industrial Chemicals Group, Philadelphia, PA.

Focht, D.D. and Verstraete, W. 1977. Biochemical ecology of nitrification and denitrification. Adv. Microbial Ecol. 1:135-214.

Fogel, S., Lancione, R., Sewall, A., and Boethling, R.S. 1985. Application of biodegradability screening tests to insoluble chemicals: hexadacane. Chemosphere. 14:375-382.

Foght, J.M. and Westlake, D.W.S. 1983. Evidence for plasmid involvement in bacterial degradation of polycyclic aromatic hydrocarbons. Abst. Ann. Mtg. Amer. Soc. Microbiol. p. 275.

Foght, J.M. and Westlake, D.W.S. 1985. Degradation of some polycyclic aromatic hydrocarbons by a plasmid-containing Flavobacterium species. Abstr. Ann. Mtg. of the Amer. Soc. Microbiol. p. 261.

Follett, R.H., Murphy, L.S., and Donahue, R.L. 1981. Fertilizers and Soil Amendments. Prentice-Hall, Inc., Englewood Cliffs, NJ.

Foster, J.W. 1962. Bacterial oxidation of hydrocarbons. In Hayaishi, O. (Ed.), Oxygenases. Academic Press, New York, NY. 241 pp.

Fournier, J.C., Codaccioni, P., and Soulas, G. 1981. Soil adaptation of 2,4-D degradation in relation to the application rates and the metabolic behaviour of the degrading microflora. Chemosphere. 10:977-984.

Fox, J.L. 1985. Fixed up in Philadelphia: genetic engineers meet with ecologists. ASM News. 51:382-386.

Francke, H.C. and Clark, F.E. 1974. Disposal of oil wastes by microbial assimilation. U.S. Atomic Energy Report Y-1934.

Fredericks, K.M. 1966. Adaptation of bacteria from one hydrocarbon to another. Nature (London). 209:1047-1048.

Fredrickson, J.K. and Hicks, R.J. 1987. Probing reveals many microbes beneath earth's surface. ASM News. 53:78-79.

Freedman, W. and Hutchinson, T.C. 1976. Physical and biological effects of experimental crude oil spills on low Arctic tundra in the vicinity of Tuktoyaktut. Can. J. Bot. 54:2219-2230.

Freeze, R.A. and Cherry, J.A.   1979.   Groundwater. Prentice-Hall, Inc., Englewood Cliffs, NJ.   227 p.

Friello, D.A., Chakrabarty, A.M.   1980.   Transposable Mercury Resistance in Pseudomonas putida Plasmids and Transposons. In Suttard, C. and Rozec, K.R. (Eds.), Academic Press, New York, NY.   p. 249.

Friello, D.A., Mylroie, J.R., and Chakrabarty, A.M.   1976.   Use of genetically engineered multi-plasmid microorganisms for rapid degradation of fuel hydrocarbons. In Sharpley J.M. and Kaplan, A.M. (Eds.), Proc. 3rd Intl. Biodegradation Symp. Applied Science Publishers, London.   p. 205.

Frieze, M.P. and Oujesky, H.   1983.   Oil degrading bacterial populations in municipal wastewater. Abstr. Ann. Mtg. Amer. Soc. Microbiol.   p. 269.

Frissel, M.J.   1961.   Verslag. Landbouwk. Onderzoek.   76:3.

Fry, A.W. and Grey, A.S.   1971.   Sprinkler Irrigation Handbook. Rain Bird Sprinkler Mfg. Corp. Glendora, CA.

Fuge, R. and James, K.H.   1973.   Trace metal concentrations in brown seaweeds, Cardigan Bay, Wales. Mar. Chem.   1:281-293.

Furukawa, K.   1982.   Microbial degradation of polychlorinated biphenyls (PCBs). In Chakrabarty, A.M. (Ed.), Biodegradation and Detoxification of Environmental Pollutants. CRC Press, Boca Raton, FL.

Galun, M. Keller, P., Malki, D., Feldstein, H., Galun, E., Siegal, S.M., and Siegal, B.Z.   1983.   Science.   219:285-286.

Gardner, J.J. and Ayres, J.   1980.   Acton, Massachusetts: a groundwater contamination problem. In Proc. 1980 Conf. Hazardous Material Spills. Sponsored by Environmental Protection Agency.   pp. 39-45.

Garvey, K.J., Stewart, M.H., and Yall, I.   1985.   A genetic characterization of hydrocarbon growth substrate utilization by Acinetobacter (Strain P-7): evidence for plasmid mediated decane utilization. Abstr. Ann. Mtg. Amer. Soc. Microbiol.   p. 261.

Gary, J.H. and Hardwork, G.E.   1975.   Petroleum Refining Technology and Economics. Marcel Dekker, New York, NY.

Gatellier, C.R., Oudin, J.L., Fusey, P., Lacase, J., and Priou, M.L.   1973.   Experimental exosystems to measure fate of oil spills dispersed by surface active products. Proc. Joint Conf. on Prevention and Control of Oil Spills. American Petroleum Institute, Washington, D.C.   p. 497.

Gehron, M.J. and White, D.C.   1983.   Sensitive measurements of phospholipid glycerol in environmental samples. J. Microbiol. Methods.   1:23-32.

General Electric Co. Aircraft Engine Business Gp.   1982.   Fuel character effects on USAF gas turbine engine afterburners. Report to Aero Propulsion Lab., Air Force Wright Aeronautical Lab., Air Force Systems Command, Wright-Patterson AFB, OH.   AFWAL-TR-82-2035.

Genner, C. and Hill, E.C.   1981.   In Rose, A.H. (Ed.), Microbial Biodeterioration: Economic Microbiology.   6:260.

Ghassemi, M., Panahloo, A., and Quinlivan, S. 1984a. Comparison of physical and chemical characteristics of shale oil fuels and analogous petroleum products. Environ. Toxicol. Chem. 3:511-535.

Ghassemi, M., Panahloo, A., and Quinlivan, S. 1984b. Physical and chemical characteristics of some widely used petroleum fuels: a reference data base for assessing similarities and differences between synfuel and petrofuel products. Energy Sources. 7:377-401.

Ghiorse, W.C. and Balkwill, D.L. 1983. Enumeration and morphological characterization of bacteria indigenous to subsurface environments. Dev. Ind. Microbiol. 24:213-224.

Ghiorse, W.C. and Balkwill, D.L. 1985a. Microbiological characterization of subsurface environments. In Ward, C.H., McCarty, P.L., and Giger, W. (Eds.), Chemistry and Biology of Groundwater. John Wiley and Sons, Inc., New York, NY.

Ghiorse, W.C. and Balkwill, D.L. 1985b. Microbiological characterization of subsurface environments. In Ward, C.H., Giger, W., and McCarty, P.L. (Eds.), Ground Water Quality. John Wiley and Sons, Inc., New York, NY.

Ghisalba, O. 1983. II. Microbial degradation of chemical waste, an alternative to physical methods of waste disposal. Experientia. 39:1247-1257.

Gholson, R.K., Guire, P., and Friede, J. 1972. Assessment of Biodegradation Potential for Controlling Oil Spills. NTIS Report No. AD-759 848. National Technical Information Service, Springfield, VA.

Gibbs, C.F. 1976. Methods and interpretation in measurement of oil biodegradation rate. In Sharpley J.M. and Kaplan, A.M. (Eds.), Proc. 3rd Intl. Biodegradation Symp. Applied Science Publishers, London. p. 127.

Gibson, D.T. 1971. The microbial oxidation of aromatic hydrocarbons. CRC Crit. Rev. Microbiol. 1:199.

Gibson, D.T. 1976. Microbial degradation of polycyclic aromatic hydrocarbons. In Sharpley, J.M. and Kaplan, A.M. (Eds.), Proc. 3rd Intl. Biodegradation Symp. Applied Science Publishers, London. pp. 57-66.

Gibson, D.T. 1977. Biodegradation of aromatic petroleum hydrocarbons. In Wolfe, D.A. (Ed.), Fate and Effects of Petroleum Hydrocarbons in Marine Organisms and Ecosystems. Pergamon Press, Oxford. pp. 36-46.

Gibson, D.T. 1978. Microbial transformations of aromatic pollutants. In Hutzinger, O., Van Lelyveld, L.H., Zoeteman, B.C.J. (Eds.), Aquatic Pollutants. Pergamon Press, New York, NY.

Gibson, D.T. 1982. Microbial degradation of hydrocarbons. Environ. Toxicol. Chem. 5:237-250.

Gibson, D.T., Koch, J.R., and Kallio, R.E. 1968. Oxidative degradation of aromatic hydrocarbons by microorganisms. I. Enzymatic formation of catechol from benzene. Biochemistry. 7:2653-2662.

Gibson, D.T., Mahadevan, V., Jerina, D.M., Yagi, H., Yeh, H.J.C. 1975. Oxidation of the carcinogens benzo(a)pyrene and benzo(a)anthracene to dihydrodiols by a bacterium. Science. 189:295-297.

Giesy, J.P. (Ed.). 1980. Microcosms in Ecological Research. Dept. of Energy Symp. Series 52 (Conf-781101). National Technical Information Service, Springfield, VA.

Gillan, F.T. 1983. Analysis of complex fatty acid methyl ester mixtures on non-polar capillary GC columns. J. Chromatgr. Sci. 21:293-297.

Gilmore, A.E. 1959. Soil Sci. 87:95.

Girling, C.A. 1984. Selenium in agriculture and the environment. Agriculture, Ecosystems, and Environment. 11:37-65.

Gissel-Nielsen, G. 1971. Selenium content of some fertilisers and their influence on uptake of selenium in plants. J. Agric. Food Chem. 19:565-566.

Gloxhuber, C. 1974. Toxicological properties of surfactants. Arch. Toxicol. 32:245-270.

Gocke, K. 1977. Comparisons of methods for determining the turnover time of dissolved organic compounds. Mar. Biol. 42:131-141.

Godsy, E.M. 1980. Isolation of Methanobacterium bryantii from a deep aquifer by using a novel broth-antibiotic dish method. Appl. Environ. Microbiol. 39:1074-1075.

Goldstein, R.M., Mallory, L.M., and Alexander, M. 1985. Reasons for possible failure of inoculation to enhance biodegradation. Appl. Environ. Microbiol. 50:977-983.

Goma, G., Al Ani, D., and Pareilleux, A. 1976. Hydrocarbon uptake by microorganisms. 5th Intl. Fermentation Symp., Berlin.

Goma, G., Pareilleux, A., and Durand, G. 1974. Aspects physicochemiques de l'assimilation des hydrocarbures par Candida lipolytica. Agric. Biol. Chem. 38:1273-1280.

Gottschalk, G. 1979. Bacterial Metabolism. Springer-Verlag, New York, NY.

Gray, T.R.G. 1978. Microbiological aspects of the soil, plant, aquatic, air, and animal environments: The soil and plant environments. In Hill, I.R. and Wright, S.J.L. (Eds.), Pesticide Microbiology. Academic Press, New York, NY. pp. 19-38.

Grbic-Galic, D. and Vogel, T.M. 1986. Toluene and benzene transformation by ferulate-acclimated methanogenic consortia. Abstr. Ann. Mtg. Amer. Soc. Microbiol. p. 303.

Grbic-Galic, D. and Vogel, T.M. 1987. Transformation of toluene and benzene by mixed methanogenic cultures. Appl. Environ. Microbiol. 53:254-260.

Grbic-Galic, D. and Young, L.Y. 1985. Methane fermentation of ferulate and benzoate: anaerobic degradation pathways. Appl. Environ. Microbiol. 50:292-297.

Green, W.J., Lee, G.F., and Jones, R.A. 1981. Clay-soils permeability and hazardous waste storage. J. Water Poll. Control Fed. 53:1347-1354.

Grimes, D.J. and Colwell, R.R. 1986. Viability and virulence of <u>Escherichia coli</u> suspended by membrane chamber in semitropical ocean water. FEMS Microbiol. Lett. 34:161-165.

Groenewegen, D. and Stolp, H. 1981. Microbial breakdown of polycyclic aromatic hydrocarbons. In Overcash, M.R. (Ed.), Decomposition of Toxic and Nontoxic Organic Compounds in Soils. Ann Arbor Science Publ., Ann Arbor, MI. pp. 233-240.

Groundwater Decontamination Systems, Inc. Company Sales Literature. North Paramus, NJ.

Guard, H.E. and Cobet, A.B. 1972. Fate of petroleum hydrocarbons on beach sands. NTIS Report No. AD-758 740. National Technical Information Service, Springfield, VA.

Guenther, W.B. 1975. Chemical Equilibrium: A Practical Introduction for the Physical and Life Sciences. Plenum Press, New York, NY.

Guidin, C. and Syratt, W.J. 1975. Biological aspects of land rehabilitation following hydrocarbon contamination. Environ. Poll. 8:107-112.

Gutnick, D.L. and Rosenberg, E. 1977. Oil tankers and pollution, a microbiological approach. Ann. Rev. Microbiol. 31:379-396.

Gutnick, D.L. and Rosenberg, E. 1979. Oil tankers and pollution: A microbiological approach. Ann. Rev. Micro. 31:379-396.

Gwynne, P. and Bishop, J., Jr. 1975. Bring on the bugs. Newsweek. 86:116.

Haber, F. and Weiss, J. 1934. Proc. Roy. Soc. London. Ser. A. 147:332.

Hackel, U., Klein, J., Megnet, R., and Wagner, F. 1975. Immobilization of microbial cells on polymeric matrices. Europ. J. Appl. Microbiol. 1:291-293.

Haines, J.R. and Alexander, M. 1974. Microbial degradation of high-molecular weight alkanes. Appl. Microbiol. 28:1084-1085.

Haines, J.R., Pesek, E., Roubal, G., Bronner, A., and Atlas, R. 1981. Microbially mediated chemical evolution of crude oils spilled into differing ecosystems. Abstr. Ann. Mtg. Amer. Soc. Microbiol. p. 213.

Hallberg, R.O. and Martinelli, R. 1976. <u>In-situ</u> purification of ground water. Ground Water. 14:88-93.

Hamaker, J.W. 1966. Mathematical prediction of cumulative levels of pesticides in soil. Advan. Chem. Ser. 60:122-131.

Hamaker, J.W. 1972. Decomposition: quantitative aspects. In Goring, C.A.I. and Hamaker, J.W. (Eds.), Organic Chemicals in Soil Environment. Marcel Dekker, Inc., New York, NY.

Hambrick, G.A., III, DeLaune, R., and Patrick, W.H., Jr. 1980. Effect of estuarine sediment pH and oxidation-reduction potential on microbial hydrocarbon degradation. Appl. Environ. Microbiol. 40:365-369.

Hampton, G.J., Webster, J.J., and Leach, F.R. 1983. The extraction and determination of ATP from soil and subsurface material. In Parr, J.F., March, PB., and Kla, J.M. (Eds.), Land Treatment of Hazrdous Wastes. Noyes Data Corp., Park Ridge, NY. p. 15.

Hance, R.J. and McKone, C.E. 1971. Effect of concentration on the decomposition rates in soil of atrizine, linuron and picloram. Pest. Sci. 3:31-34.

Harder, W. 1981. Enrichment and characterization of degrading organisms. In Leisinger, T., Hutter, R., Cook, A.M., and Nuesch, J. (Eds.), FEMS Symp. No. 12. Microbial Degradation of Xenobiotics and Recalcitrant Compounds. 12:77-96.

Harris, R.F. 1981. Effect of water potential on microbial growth and activity. In Water Potential Relations in Soil Microbiology. SSSA Special Publication No. 9. Soil Sci. Soc. Amer., Madison, WI. pp. 23-95.

Harrison, A.P. and Lawrence, F.R. 1963. J. Bacteriol. 85:742-750.

Harrison, W., Winnik, M.A., Kwong, P.T.Y., and Mackay, D. 1975. Crude oil spills. Disappearance of aromatic and aliphatic components from small sea-surface slicks. Environ. Sci. Technol. 9:231-234.

Hashimoto, Y., Tokura, K., Kishi, H., and Strachan, W.M.J. 1984. Prediction of seawater solubility of aromatic compounds. Chemosphere. 13:881-888.

Healy, J.B., Jr. and Young, L.Y. 1979. Anaerobic biodegradation of eleven aromatic compounds to methane. Appl. Environ. Microbiol. 38:84-89.

Hegeman, G.D. 1972. The evolution of metabolic pathways in bacteria. In Degradation of Synthetic Organic Molecules in the Biosphere. Natural, Pesticidal, and Various Other Man-Made Compounds. Nat. Acad. Sci., Washington, D.C. pp. 56-72.

Heitkamp, M.A. and Cerniglia, C.E. 1986. Microbial degradation of tert-butylphenyl diphenyl phosphate: a comparative microcosm study among five diverse ecosystems. Toxicity Assessment Bull. 1:103-122.

Heitkamp, M.A. and Cerniglia, C.E. 1987. Effects of chemical structure and exposure on the microbial degradation of polycyclic aromatic hydrocarbons in freshwater and estuarine ecosystems. Environ. Toxicol. Chem. 6:535-546.

Heitkamp, M.A., Freeman, J.P., and Cerniglia, C.E. 1987. Naphthalene biodegradation in environmental microcosms: estimates of degradation rates and characterization of metabolites. 53:129-136.

Heringa, J.W., Huybregtse, R., and van der Linden, A.C. 1961. n-Alkane formation by a Pseudomonas. Formation and beta-oxidation of intermediate fatty acids. Antonie van Leeuwenhoek J. Microbiol. Serol. 27:51-58.

Heukelekian, H. and Heller, A. J. 1940. Bacteriol. 40:547.

Heyse, E., James, S.C., and Wetzel, R. 1986. In situ aerobic biodegradation of aquifer contaminants at Kelly Air Force Base. Environ. Progress. 5:207-211.

Higashihara, T., Sato, A., and Simidu, U. 1978. An MPN method for the enumeration of marine hydrocarbon degrading bacteria. Bull. Jpn. Soc. Sci. Fish. 44:1127-1134.

Higgins, I.J., Best, D.J., and Hammond, R.C. 1980. New findings in methane-utilizing bacteria highlight their importance in the biosphere and their commercial potential. Nature (London). 286:561-564.

Hileman, B. 1984. RCRA groundwater protection standards. Environ. Sci. Technol. 18:282A-284A.

Hill, G.D., McGrahen, J.W., Baker, H.M., Finnerty, D.W., and Bingeman, C.W. 1955. The fate of substituted urea herbicides in agricultural soils. Agron. J. 47:93-104.

Hoag, G.E. and Marley, M.C. 1986. Gasoline residual saturation in unsaturated uniform aquifer materials. J. Environ. Eng. 112(3)

Holdeman, L.V. and Moore, W.E.C. 1972. Anaerobe Laboratory Manual. 2nd ed. Blacksburg, VA: Virginia Polytechnic Inst. and State Univ. 130 p.

Holmes, S.H. and Thompson, L.F. 1982. Nitrogen distributions in hydrotreated shale oil products from commercial-scale refining. Review draft. Laramie, WY: Laramie Energy Technology Center. p. 4.

Holt, A.A. 1986. Lignin update. Industrial Chemical News. January: 47-48.

Hornick, S.B. 1983. The interaction of soils with waste constituents. In Parr, J.F., Marsh, P.B., and Kla, J.M. (Eds.), Land Treatment of Hazardous Wastes. Noyes Pub., Park Ridge, NJ. pp. 4-19.

Hornick, S.B., Fisher, R.H., and Paolini, P.A. 1983. Petroleum Wastes. In Parr, J.F., Marsh, P.B., and Kla, J.M. (Eds.), Land Treatment of Hazardous Wastes. Noyes Pub., Park Ridge, NJ. pp. 321-337.

Hornick, S.B., Murray, J.J., and Chaney, R.L. 1979. Overview on utilization of composted municipal sludges. In Proc. Nat. Conf. on Municipal and Industrial Sludge Composting. Information Trans., Inc., Silver Spring, MD.

Horowitz, A. and Atlas, R.M. 1977a. Continuous open flow-through system as a model for oil degradation in the Arctic Ocean. Appl. Environ. Microbiol. 33:647-653.

Horowitz, A. and Atlas, R.M. 1977b. Response of microorganisms to an accidental gasoline spillage in an Arctic freshwater ecosystem. Appl. Environ. Microbiol. 33:1252-1258.

Horowitz, A., Gutnick, D., and Rosenberg, E. 1975a. Sequential growth of bacteria on crude oil. Appl. Microbiol. 30:10-19.

Horowitz, A.D., Gutnick, D., and Rosenberg, E. 1975b. Sequential growth of bacteria on crude oil. Appl. Microbiol. 30:10-19.

Horowitz, A. and Tiedje, J.M. 1980. Anaerobic degradation of substituted monoaromatic compounds. Abstr. Ann. Mtg. Amer. Soc. Microbiol. p. 196.

Horvath, A.L. 1982. Halogenated Hydrocarbons. Marcel Dekker, New York, NY.

Horvath, R.S. 1972. Microbial co-metabolism and the degradation of organic compounds in nature. Bacteriol. Rev. 36:146-155.

Horvath, R.S. and Alexander, M. 1970. Cometabolism of m-chlorobenzoate by an Arthrobacter. Appl. Microbiol. 20:254-258.

Hou, C.T. 1982. Microbial transformation of important industrial hydrocarbons. In Rosazza, J.P. (Ed.), CRC Microbial Transformations of Bioactive Compounds. Vol. I. CRC Press, Inc., Boca Raton, FL. pp. 81-107.

Hsieh, Y.-P., Tomson, M.B., and Ward, C.H. 1980. Toxicity of water-soluble extracts of No. 2 fuel oil to the freshwater alga Selenastrum capricornutum. Dev. Ind. Microbiol. 21:401-409.

Hulbert, M.H. and Krawiec, S. 1977. Cometabolism: a critique. J. Theor. Biol. 69:287-291.

Hunt, P.G., et al. 1973. Proc., Joint Conf. on Prevention and Control of Oil Slicks. American Petroleum Institute, Washington, D.C. pp. 773-740.

Hurle, K. and Walker, A. 1980. Persistence and its prediction. In Hance, R.J. (Ed.), Interaction Between Herbicides and the Soil. Academic Press, London.

Hutchins, S.R., Tomson, M.B., Wilson, J.T., and Ward, C.H. 1984. Microbial removal of wastewater organic compounds as a function of input concentration in soil columns. Appl. Environ. Microbiol. 48:1039-1045.

Hutzinger, O. and Veerkamp, W. 1981. Xenobiotic chemicals with pollution potential. In Leisinger, T., Cook, A.M., Hutter, R., and Nuesch, J., (Eds.), FEMS Symp. No. 12. Microbial Degradation of Xenobiotics and Recalcitrant Compounds. 12:3-45.

Itoh, M., Ohguchi, M., and Doi, S. 1968. Studies on hydrocarbon-utilizing microorganisms: microbial treatment of petroleum waste. J. Ferment. Technol. 46:34.

Jacob, J., Karcher, W., Belliardo, J.J., and Wagstaffe, P.J. 1986. Polycyclic aromatic compounds of environmental and occupational importance. Fresenius. Z. Anal. Chem. 323:1-10.

Jain, R.K. and Sayler, G.S. 1987. Problems and potential for in situ treatment of environmental pollutants by engineered microorganisms. Microbiol. Sci. 4:59-63.

Jamison, V.W., Raymond, R.L., and Hudson, J.O., Jr. 1975. Biodegradation of high-octane gasoline in groundwater. In Underkofler, L.A. and Murray, E.D. (Eds.), Developments in Industrial Microbiology. 16:305-312.

Jamison, V.W., Raymond, R.L., and Hudson, J.O., Jr. 1976. Biodegradation of high-octane gasoline. In Sharpley, J.M. and Kaplan, A.M. (Eds.), Proc. 3rd Intl. Biodegradation Symp. Applied Science Publ., London. pp. 187-196.

Jensen, B., Arvin, E., and Gundersen, A.T. 1985. The degradation of aromatic hydrocarbons with bacteria from oil contaminated aquifers. Proc. Petrol. Hydro. and Organ. Chem. in Groundwater Conf. pp. 421-435.

Jensen, V. 1975. Bacterial flora of soil after application of oily waste. Oikos. 26:152-158.

Jerina, D.M., Selander, H., Yagi, H., Wells, M.C., Davey, J.F., Mahadevan, V., and Gibson, D.T. 1976. Dihydrodiols from anthracene and phenanthrene. J. Amer. Chem. Soc. 98:5988-5996.

Jeter, Capt. 1985. Floating Fuel Recovery with Residual Cleanup. JON: 2103XXX2.

Jhaveria, V. and Mazzacca, A.J. 1982. Bioreclamation of Ground and Groundwater by the GDS Process. Groundwater Decontamination Systems, Inc., Waldwick, NJ.

Jhaveri, V. and Mazzacca, A.J. 1983. 4th Nat. Conf. Mgmt. of Uncontrolled Hazardous Waste Sites, Nov. 1, 1983, Washington, D.C.

Jhaveri, V. and Mazzacca, A.J. 1985. In Case History. Groundwater Decontamination Systems, Inc. Company Literature.

Jobson, A., Cook, F.D., and Westlake, D.W.S. 1972. Microbial utilization of crude oil. Appl. Microbiol. 23:1082-1089.

Jobson, A., McLaughlin, M., Cook, F.D., and Westlake, D.W.S. 1974. Effect of amendments on the microbial utilization of oil applied to soil. Appl. Microbiol. 27:166-171.

Johanides, V. and Hrsak, D. 1976. In Dellweg, H. (Ed.), Abstracts 5th Intl. Symp. on Fermentation. Westkreuz, Berlin. p. 426.

Johnson, B.T., Romanenko, V.I. 1983. Fuel oil #2 in a freshwater microcosm: impact and recovery. Abstr. Ann. Mtg. Amer. Soc. Microbiol. p. 272.

Johnson, L., Kauffman, J., and Krupka, M. 1982. Summer Nat. Mtg. Amer. Inst. Chem. Eng., Aug. 29-Sept. 1, 1982, Cleveland, OH.

Johnson, M.J. 1964. Utilization of hydrocarbons by microorganisms. Chem. Ind. (London). pp. 1532-1537.

Johnson, S.L. 1978. Biological treatment. In Berkowitz, J.B., Funkhouser, J.T., and Stevens, J.I. (Eds.), Unit Operations for Treatment of Hazardous Industrial Wastes. Noyes Data Corp. Park Ridge, NJ. pp. 168-268.

Johnston, J.B. and Robinson, S.G. 1982. Opportunities for development of new detoxification processes through genetic engineering. In Exner, J.H. (Ed.), Detoxification of Hazardous Waste. Ann Arbor Science Pub., Ann Arbor, MI. pp. 301-314.

Jones, J.G. and Eddington, M.A. 1968. An ecological survey of hydrocarbon-oxidizing microorganisms. J. Gen. Microbiol. 52:381-390.

Josephson, J. 1980. Groundwater strategies. Environ. Sci. Technol. 14:1030-1037.

Josephson, J. 1983a. Restoration of aquifers. Environ. Sci. Technol. 17:347A-350A.

Josephson, J. 1983b. Subsurface organic contaminants. Environ. Sci. Tech. 17:518A-521A.

JRB and Associates, Inc. 1982. Final Report, EPA Report No. EPA-625/6-82-006. Environmental Protection Agency, Cincinnati, OH.

JRB and Associates, Inc. 1984a. Remedial response at hazardous waste sites. EPA Report No. EPA-540/2-84-002Aa and b. Environmental Protection Agency, Cincinnati, OH.

JRB and Associates, Inc. 1984b. Review of in-place treatment techniques for contaminated surface soils. Volume 2. Background information for in situ treatment. Prepared for Municipal Environmental Research Lab., Cincinnati, OH.

Kallio, R.E. 1975. Microbial degradation of petroleum. Proc. 3rd Intl. Biodegradation Symp. Univ. of Rhode Island, Kingston, RI. p. I-1.

Kamp, P.F. and Chakrabarty, A.M. 1979. Plasmids specifying p-chlorobiphenyl degradation in enteric bacteria. In Timmis, K.N. and Puhler, A. (Eds.), Plasmids of Medical, Environmental and Commercial Importance. Elsevier/North Holland Biomedical Press, Amsterdam, The Netherlands. pp. 275-285.

Kaplan, D.L. and Kaplan, A.M. 1982. Composting industrial wastes--biochemical consideration. Biocycle. May/June:42-44.

Kappeli, O., Walther, P., Mueller, M., and Fiechter, A. 1984. Structure of the cell surface of the yeast Candida tropicalis and its relation to hydrocarbon transport. Arch. Microbiol. 138:279-282.

Kappler, T. and Wuhrman, K. 1978. Microbial degradation of the water-soluble fraction of gas oil-I. Water Res. 12:327-333.

Kappler, T. and Wuhrman, K. 1978. Microbial degradation of the water-soluble fraction of gas oil-II. Bioassays with pure strains. Water Res. 12:335-342.

Karata, A., Yoichi, Y., and Fumio, T. 1980. Jpn. J. Mar. Chem. 18:1.

Karickhoff, S.W. 1984. Organic pollutant sorption in aquatic systems. J. Hydraul. Eng. 110:707-735.

Karickhoff, W.W., Brown, D.S., and Scott, T.A. 1979. Sorption of hydrophobic pollutants on natural sediments. Water Res. 13:241-248.

Kator, H. 1973. Utilization of crude oil hydrocarbons by mixed cultures of marine bacteria. In Ahearn, D.G. and Meyer, S.P. (Eds.), The Microbial Degradation of Oil Pollutants, Workshop, Atlanta, GA, Dec. 4-6, 1972. Publ. No. LSU-SG-73-01. Louisiana State Univ., Center for Wetland Resources, Baton Rouge, LA. p. 47.

Kator, H., Miget, R., and Oppenheimer, C.H. 1972. Utilization of paraffin hydrocarbons in crude oil by mixed cultures of marine bacteria. Paper No. SPE 4206. Symposium on Environmental Conservation. Society of Petroleum Engineers, Dallas, TX.

Kaufman, D.D. 1983. Fate of toxic organic compounds in land-applied waste. In Parr, J.F., Marsh, P.B., and Kla, J.M. (Eds.), Land Treatment of Hazardous Wastes. Noyes Pub., Park Ridge, NJ. pp. 77-151.

Kaufman, D.D. and Plimmer, J.R. 1972. Approaches to the synthesis of soft pesticides. In Mitchell, R. (Ed.), Water Pollution Microbiology. Wiley: Interscience, New York, NY. pp. 173-203.

Kearney, P.C., Nash, R.G., and Isensee, A.R. 1969. Persistence of pesticide residues in soils. In Miller, M.W. and Berg G.G. (Eds.), Chemical Fallout: Current Research on Persistent Pesticides. Charles C. Thomas, Springfield, IL. pp, 54-67.

Kearney, P.C. and Plimmer, J.R. 1970. Relation of structure to pesticide decomposition. In Intl. Symp. Pesticides in the Soil: Ecology, Degradation and Movement. Michigan State Univ., East Lansing, MI. pp. 65-172.

Keith, L.H. and Telliard, W.A. 1979. Priority pollutants. A perspective view. Environ. Sci. Technol. 13:416-423.

Kellogg, S.J., Chatterjee, D.K., and Chakrabarty, A.M. 1981. Plasmid-assisted molecular breeding: new technique for enhanced biodegradation of persistent toxic chemicals. Science. 214:1113-1135.

Kent, C., Rosevear, A., and Thomson, A.R. 1978. Enzymes immobilised on inorganic supports. In Wiseman, A. (Ed.), Topics in Enzyme and Fermentation Biotechnology. Vol. 2. Ellis Horwood, Ltd., Chichester, England. pp. 12-119.

Kerr, R.P. and Capone, D.G. 1986. Salinity effects on naphthalene and anthracene mineralization by sediment microbes in coastal environments. Abstr. Ann. Mtg. Amer. Soc. Microbiol. p. 303.

Kester, A.S. and Foster, J.W. 1963. Biterminal oxidation of long-chain alkanes by bacteria. J. Bacteriol. 85:859-869.

Khan, K.A., Suidan, M.T., and Cross, W.H. 1981. Anaerobic activated carbon filter for the treatment of phenol-bearing wastewater. J. Water Poll. Control Fed. 53:1519-1532.

Khan, S.U.. 1980. In Dynamics, Exposure and Hazard Assessment of Toxic Chemicals. Haque, R. (Ed.). Ann Arbor Science, Ann Arbor, MI. pp. 215-230.

Khan, S.U. 1982. Studies on bound 14C-prometryn residues in soil and plants. Chemosphere. 11:771-795.

Kilbane, J.J., Chatterjee, D.K., and Chakrabarty, A.M. 1983. Detoxification of 2,4,5-trichlorophenoxyacetic acid from contaminated soil by Pseudomonas cepacia. Appl. Environ. Microbiol. 45:1697-1700.

Kincannon, C.B. 1972. Oily waste disposal by soil cultivation process. EPA-R2-72-110. Office of Research and Monitoring, Environmental Protection Agency, Washington, D.C.

Kirk, T.K., Schultz, E., Connors, W.J., Lorenz, L.F., and Zeikus, J.G. 1978. Influence of culture parameters on lignin metabolism by Phanerochaete chrysosporium. Arch. Microbiol. 117:277-285.

Kitagawa, M. 1956. Studies on the oxidation mechanism of methyl group. J. Biochem., Tokyo. 43:553.

Klein, D.A. and Molise, E.M. 1975. Ecological ramifications of silver iodide nucleating agent accumulation in a semi-arid grassland environment. J. Appl. Meteorology. 14:673-680.

Knackmuss, H.-J. 1981. Degradation of halogenated and sulfonated hydrocarbons. In Leisinger, T., Hutter, R., Cook, A.M., and Nuesch, J. (Eds.), FEMS Symp. No. 12. Microbial Degradation of Xenobiotics and Recalcitrant Compounds. 12:189-212.

Knetting, E. and Zajic, J.E. 1972. Flocculant production from kerosene. Biotechnol. Bioeng. 14:379-390.

Knox, R.C., Canter, L.W., Kincannon, D.F., Stover, E.L., and Ward, C.H. (Eds.). 1968. Aquifer Restoration: State of the Art. Noyes Pub., Park Ridge, NJ.

Knox, R.C., Canter, L.W., Kincannon, D.F., Stover, E.L., and Ward, C.H. 1984. State-of-the-Art of Aquifer Restoration. National Center for Ground Water Research.

Kobayashi, H. and Rittmann, B.F. 1982. Microbial removal of hazardous organic compounds. Environ. Sci. Technol. 16:170A-183A.

Kobayashi, M. and Tchan, Y.T. 1978. Formation of demethylnitrosamine in polluted environment and the role of photosynthetic bacteria. Water Res. 12:199-201.

Kochi, J.D., Tang, R., and Bernath, T. 1972. Mechanism of aromatic substitution: role of cation-radicals in the oxidative substitution of arenes by cobalt (III). J. Amer. Chem. Soc. 95:7114-7123.

Komagata, K., Nakase, T., and Katsu, N. 1964. Assimilation of hydrocarbons by yeast. I. Preliminary screening. J. Gen. Appl. Microbiol. 10:313-321.

Korfhagen, T.R., Sutton, L., and Jacoby, G.A. 1978. Classification and physical properties of Pseudomonas plasmids. In Schlessinger, D. (Ed.), Microbiology--1978. Amer. Soc. Microbiol., Washington, D.C. pp. 221-224.

Kosaric, N. and Zajic, J.E. 1974. Microbial oxidation of methane and methanol. In Ghose, T.K., Fiechter, A., and Blakebrough, N., (Eds.), Advances in Biochemical Engineering. Vol. 3. Springer-Verlag, New York, NY. p. 89.

Koval'skii, S.V. 1968. Geochemical ecology of microorganisms under conditions of different selenium content in soils. Mikrobiologiya. 37:122-130 (in Russian).

Kowalenko, C.G. 1978. Organic nitrogen, phosphorus and sulfur in soils. In Schnitzer, M. and Khan, S.U. (Eds.), Soil Organic Matter. Elsevier Scientific Publ. Co., New York, NY. pp. 95-135.

Krulwick, T.A. and Pellicconi, N.J. 1979. Ann. Rev. Microbiol. 33:94-111.

Kuhn, E.P., Colberg, P.J., Schnoor, J.L., Wanner, O., Zehnder, A.J.B., and Schwarzenbach, R.P. 1985. Microbial transformation of substituted benzenes during infiltration of river water to groundwater: laboratory column studies. Environ. Sci. Technol. 19:961-968.

Kurek, E., Czaban, J., and Bollag, J.-M. 1982. Sorption of cadmium by microorganisms in competition with other soil constituents. Appl. Environ. Microbiol. 43:1011-1015.

Kuznetsov, S.I., Ivanov, M.V., and Lyalikova, N.N. 1963. In Oppenheimer, C.H. (Ed. English edition), Introduction to Geological Microbiology. McGraw-Hill, New York, NY.

Laanbroek, H.J. and Pfennig, N. 1981. Oxidation of short-chain fatty acids by sulfate-reducing bacteria in freshwater and marine sediments. Arch. Microbiol. 128:330-335.

Ladd, T.I., Ventullo, R.M., Wallis, P.M., and Costerton, J.W. 1982. Heterotrophic activity and biodegradation of labile and refractory compounds by groundwater and stream microbial populations. Appl. Environ. Microbiol. 44:321-329.

Lamar, R.T., Larsen, M.J., Kirk, T.K., and Glaser, J.A. 1987. Growth of the white-rot fungus Phanerochaete chrysosporium in soil. Proc. U.S. EPA 13th Ann. Res. Symp. on Land Disposal, Remedial Action, Incineration, and Treatment of Hazardous Waste. Environmental Protection Agency, Cincinnati, OH, May 608, 1987.

Larson, R.J. and Ventullo, R.M. 1983. Biodegradation potential of groundwater bacteria. 3rd Nat. Symp. on Aquifer Restoration and Groundwater Monitoring, May 25-27, 1983, Columbus, OH.

Laskin, A. and Lechevalier, H.A. (Eds.). 1974. Handbook of Microbiology. CRC Press, Cleveland, OH.

Law Engineering Testing Company. 1982. Literature inventory: treatment techniques applicable to gasoline contaminated ground water. American Petroleum Institute, Washington, D.C. p. 60.

Lawlor, G.F., Jr., Shiaris, M.P., and Jambard-Sweet, D. 1986. Phenanthrene metabolites by pure and mixed bacterial cultures as examined by HPLC. Abst. Ann. Mtg. Amer. Soc. Microbiol. p. 302.

Lee, M.D., Thomas, J.M., and Ward, C.H. 1985. Microbially mediated fate of ground water contaminants from abandoned hazardous waste sites. Proc. 5th World Congress on Water Resources. Brussels, Belgium. Int. Water Resources Assoc. pp. 113-122.

Lee, M.D. and Ward, C.H. 17-30.

Lee, M.D. and Ward, C.H. 1983. Microbial ecology of a hazardous waste disposal site: enhancement of biodegradation. In Parr, J.F., Marsh, P.B., and Kla, J.M. (Eds.), Land Treatment of Hazardous Wastes. Noyes Pub., Park Ridge, NJ. p. 25.

Lee, M.D. and Ward, C.H. 1984. Reclamation of contaminated aquifers: Biological techniques. In J. Ludwigson (Ed.), Proc. 1984 Hazardous Material Spills Conf., April 9-12. Nashville, TN. Government Institutes, Inc., Rockville, MD. pp. 98-103.

Lee, M.D. and Ward, C.H. 1985. Environmental and biological methods for the restoration of contaminated aquifers. Environ. Toxic. Chem. 4:743-750.

Lee, M.D. and Ward, C.H. 1986. Ground water restoration.

Lee, M.D., Wilson, J.T., and Ward, C.H. 1984. Microbial degradation of selected aromatics in a hazardous waste site. Dev. Ind. Microbiol. 25:557-565.

Lee, M.D., Wilson, J.T., and Ward, C.H. 1987. In situ restoration techniques for aquifers contaminated with hazardous wastes. J. of Hazardous Materials. 14:71-82.

Lee, R.F. and Ryan, C. 1976. Biodegradation of petroleum hydrocarbons by marine microbes. In Sharpley, J.M. and Kaplan, A.M. (Eds.), Proc. 3rd Intl. Biodeg. Symp., Aug. 17-23, 1975, Kingston, RI. pp. 119-125.

Lehmicke, L.G., Williams, R.T., and Crawford, R.L. 1979. ^{14}C-most-probable-number method for enumeration of active heterotrophic microorganisms in natural waters. Appl. Environ. Microbiol. 38:644-649.

Lehtomakei, M. and Niemela, S. 1975. Improving microbial degradation of oil in soil. Ambio. 4:126-129.

Leisinger, T. 1983. I. General aspects. Microorganisms and xenobiotic compounds. Experientia. 39:1183-1191.

Lemaire, J., Campbell, I., Hulpke, H., Guth, J.A., Merz, W., Philop, L., and Von Waldow, C. 1982. An assessment of test methods for photodegradation of chemicals in the environment. Chemosphere 11:119-164.

Leslie, T.J., Page, M., Reinert, R.H., Moses, C.K., Rodgers, J.H., Jr., Dickson, K.L., and Hinman, M.L. 1983. Biotransformation of organic chemicals--factors important in test design. Abstr. Ann. Mtg. Amer. Soc. Microbiol. p. 275.

Lettinga, G., DeZeeuw, W., and Ouborg, E. 1981. Anaerobic treatment of wastes containing methanol and higher alcohols. Water Res. 15:171-182.

Levin, W., Wood, A.W., Wislocki, P.G., Chang, R.L., Kapitulnik, J., Mah, H.D, Yagi, H., Jerina, D.M., and Conney, A.H. 1978. Mutagenicity and carcinogenicity of benzo(a)pyrene and benzo(a)pyrene derivitives. In Gelboin, H.V. and Ts'o, P.O.P. (Eds.), Polycyclic Hydrocarbons and Cancer: Environment, Chemistry, and Metabolism. Academic Press, New York, NY.

Liang, L.N., Sinclair, J.L., Mallory, L.M., and Alexander, M. 1982. Fate in model ecosystems of microbial species of potential use in genetic engineering. Appl. Environ. Microbiol. 44:708-714.

Lin, W.S. and Kapoor, M. 1979. Induction of aryl hydrocarbon hydroxylase in Neurospora crassa by benzo(a)pyrene. Curr. Microbiol. 3:177-180.

Lindorff, D.E. and Cartwright, K. 1977. Environ. Geol. Notes. No. 81. Illinois State Geological Survey, Urbana, IL.

Litchfield, J.H. and Clark, L.C. 1972. American Petroleum Institute Publ. No. 4211.

Litchfield, J.H. and Clark, L.C. 1973. Bacterial Activity in Ground Waters Containing Petroleum Products. API Publication No. 4211. American Petrolium Institute, Committee on Environmental Affairs, Washington, D.C.

Liu, D. 1980. Enhancement of PCBs biodegradation by sodium ligninsulfonate. Water Res. 14:1467-1475.

Lode, A. 1986. Changes in the bacterial community after application of oily sludge to soil. Appl. Microbiol. Biotechnol. 25:295-299.

Long, S.C., Dobbins, D.C., Aelion, M.C., and Pfaender, F.K. 1986. Metabolism of naturally occurring and xenobiotic compounds by subsurface microbial communities. Abstr. Ann. Mtg. Amer. Soc. Microbiol. p. 285.

Lovelock, J.E. Gaea. 1979. A New Look at Life on Earth. Oxford Univ. Press, Oxford, England.

Lovley, D.R. and Klug, M.J. 1983. Sulfate reducers can outcompete methanogens at freshwater sulfate concentrations. Appl. Environ. Microbiol. 45:187-192.

Lu, J.C.S. and Chen, K.Y. 1977. Migration of trace metals in interfaces of seawater and polluted surficial sediments. Environ. Sci. Technol. 11:174-182.

Ludzack, F.L. and Kinkead, D. 1956. Persistence of oily wastes in polluted water under aerobic conditions: motor oil class of hydrocarbons. Ind. Eng. Chem. 48:263.

Lukins, H.B. 1962. On the Utilization of Hydrocarbons, Methyl Ketones, and Hydrogen by Mycobacteria. Thesis. Univ. of Texas.

Lynch, J.M. and Poole, N.J. (Eds.). 1979. Microbial Ecology, a Conceptual Approach. John Wiley and Sons, Inc., New York, NY.

Mackay, D.M., Roberts, P.V., and Cherry, J.A. 1985. Transport of organic contaminants in groundwater. Environ. Sci. Technol. 19:384-392.

Mackay, D. and Shiu, W.Y. 1977. Aqueous solubility of polynuclear aromatic hydrocarbons. J. Chem. Eng. Data. 22:399-402.

Magor, A.M., Warburton, J., Trower, M.K., and Griffin, M. 1986. Comparative study of the ability of three Xanthobacter species to metabolize cycloalkanes. Appl. Environ. Microbiol. 52:665-671.

Mallon, B.J. and Harrison, F.L. 1984. Octanol-water partition coefficient of benzo(a)pyrene: measurement, calculation, and environmental implications. Bull. Environ. Contam. Toxicol. 32:316-323.

Maloney, S.W., Manem, J., Mallevialle, J., and Fiessinger, F. 1985. The potential use of enzymes for removal of aromatic compounds from water. Water Sci. Tech. 17:273-278.

March, J. 1968. Advanced Organic Chemistry: Reactions, Mechanisms and Structure. McGraw Hill, New York, NY.

Markovetz, A.J. 1971. Subterminal oxidation of aliphatic hydrocarbons by microorganisms. Crit. Rev. Microbiol. 1:225-237.

Marxen, J. 1981. Verh. Internat. Verein. Limnol. 21:1371.

Mason, J.R., Pirt, S.J., and Somerville, H.J. 1978. In Enzyme Engineering. Brown, G.B., Manecke, G., and Wingard, L.B. (Eds.). Vol. 4. Plenum Press, London. pp. 343-344.

Mason, J.W., Anderson, A.C., and Shariat, M. 1979. Role of demethylation of methymercuric chloride by Enterobacter aerogenes and Serratia marcescens. Bull. Environ. Contam. Toxicol. 29:262-268.

Masters, M.J. and Zajic, J.E. 1971. Myxotrophic growth of algae on hydrocarbon substrates. Dev. Ind. Microbiol. 12:77-86.

Matter-Muller, C., Gujer, W., Giger, W. and Strumm, W. 1980. Non-biological elimination mechanisms in a biological sewage treatment plant. Prog. Water Technol. 12:Tor299-314.

Mattingly, G.E.G. 1975. Labile phosphates in soils. Soil Sci. 119:369-375.

McCabe, L.J., Symons, J.M., Lee, R.D., and Robeck, G.G. 1970. Survey of community water supply systems. J. Amer. Water Works Assoc. 62:670:687.

McCarty, P.L. 1980. Organics in water--an engineering challenge. J. Environ. Eng. Div. Amer. Soc. Civil Eng. 106:1.

McCarty, P.L., Reinhard, M., and Rittmann, B.E. 1981. Trace organics in groundwater. Environ. Sci. Tech. 15:40-51.

McCarty, P.L., Rittmann, B.E., and Bouwer, E.J. 1984. In Bitton, G. and Gerba, C.P. (Eds.), Groundwater Pollution Microbiology. John Wiley and Sons, Inc., New York, NY. pp. 89-115.

McDowell, C.S., Bourgeois, H.J., Jr., and Zitrides, T.G. 1980. Biological methods for the in situ cleanup of oil spill residues. Coastal and Off-shore Oil Pollution Conf., The French/American Experience, Sept. 10-12, 1980, New Orleans, LA.

McGill, W.B. 1977. Soil restoration following oil spills--a review. J. Can. Petrol Technol. 16:60-67.

McInerney, M.J. and Bryant, M.P. 1981. Anaerobic degradation of lactate by syntrophic associations of Methanosarcina barkeri and Desulfovibrio species and effect of H_2 on acetate degradation. Appl. Environ. Microbiol. 41:346-354.

McInerney, M.J., Bryant, M.P., Hespell, R.B., and Costerton, J.W. 1981. Syntrophomonas wolfei gen. nov. sp. nov., an anaerobic, syntrophic, fatty acid-oxidizing bacterium. Appl. Environ. Microbiol. 41:1029-1039.

McKee, J.E., Laverty, F.B., and Hertel, R.M. 1972. Gasoline in groundwater. J. Water Poll. Control Fed. 44:293-302.

McKenna, E.J. 1977. Biodegradation of Polynuclear Aromatic Hydrocarbon Pollutants by Soil and Water Microorganisms. 70th Ann. Mtg. Amer. Inst. Chem. Eng., New York, NY.

McKenna, E.J. and Heath, R.D.   1976.   Biodegradation of Polynuclear Aromatic Hydrocarbon Pollutants by Soil and Water Microorganisms. UILU-WRC-76-0113, Univ. of Illinois, Urbana Water Resources Center, Urbana, IL.   p. 25.

McLean, E.O.   1982.   Soil pH and lime requirement.   In Page, A.L. (Ed.), Methods of Soil Analysis. Part 2--Chemical and Microbiological Properties. 2nd ed. American Society of Agronomy, Inc., Madison, WI.

McLean, T.C.   1971.   Sump oil digested by bacteria at Santa Barbara.   Pet. Eng. (Los Angeles).   43:68.

McNabb, J.F. and Dunlap, W.J.   1975.   Ground Water.   13:33-44.

McNabb, J.F. and Mallard, G.E.   1984.   Microbiological sampling in the assessment of groundwater pollution.   In Bitton, G. and Gerba, C.P. (Eds.), Groundwater Pollution Microbiology.   John Wiley and Sons, Inc., New York, NY.   pp. 235-260.

McNabb, J.F., Smith, B.H., and Wilson, J.T. 1981.   Biodegradation of toluene and chlorobenzene in soil and ground water. Abstr. Ann. Mtg. Amer. Soc. Microbiol.   p. 213.

Merkel, G.J., Stapleton, S.S., and Perry, J.J.   1978.   Isolation and peptidoglycan of gram-negative hydrocarbon-utilizing thermophilic bacteria. J. Gen. Microbiol.   109:141-148.

Meyer, C.F. (Ed.).   1973.   Polluted groundwater: some causes, effects, controls and monitoring.   EPA Report No.   EPA-600/4-73-001b.

Miget, R.J.   1973.   Bacterial seeding to enhance biodegradation of oil slicks. In Ahearn, D.G. and Meyers, S.P. (Eds.), The Microbial Degradation of Oil Pollutants, Workshop, Atlanta, GA, Dec. 4-6, 1972.   Publ. No. LSU-SG-73-01.   Louisiana State Univ., Center for Wetland Resources, Baton Rouge, LA.   p. 291.

Mikell, A.T., Jr., Parker, B.C., and Simmons, G.M., Jr.   1984.   Response of the heterotrophic microbial communities of Lake Hoare, Antarctica and Mountain Lake, Virginia to high dissolved oxygen. Abstr. Ann. Mtg. Amer. Soc. Microbiol.   p. 218.

Miller, E.C. and Miller, J.A.   1981.   Searches for ultimate chemical carcinogens and their reactions with cellular macromolecules.   Cancer. 47:2327-2345.

Minugh, E.M., Patry, J.J., Keech, D.A. and Leek, W.R.   1983.   A case history: cleanup of a subsurface leak of refined product.   Proc. 1983 Oil Spill Conf.: Prevention, Behavior, Control and Cleanup, Feb. 28-Mar. 3, 1983, San Antonio, TX.   pp. 397-403.

Mironov, O.G.   1970.   Role of microorganisms growing on oil in self-purification and indication of oil pollution in sea.   Oceanol. Rev. 10:650.

Monroe, D.   1985.   Amer. Biotech. Lab.   3:10-19.

Morita, R.Y.   1966.   Oceanogr. Mar. Biol.   4:105-121.

Morozov, N.V. and Nikolayov, V.N. 1978. Environmental influences on the development of oil-decomposing microorganisms. Hydrobiological J. 14:47-53.

Morrison, S.M. and Cummings, B.A. 1982. Microbiologically-mediated mutagenic activity of crude oil. EPA Report No. EPA-600/S3-81-053. Environmental Research Lab., Corvallis, OR.

Mulder, D. (Ed.). 1979. Soil Disinfection. Elsevier Scientific Publ. Co., Amsterdam.

Mulkins-Phillips, G.J. and Stewart, J.E. 1974a. Distribution of hydrocarbon-utilizing bacteria in northwestern Atlantic waters and coastal sediments. Can. J. Microbiol. 20:955-962.

Mulkins-Phillips, G.J. and Stewart, J.E. 1974b. Effect of environmental parameters on bacterial degradation of bunker C oil, crude oils, and hydrocarbons. Appl. Microbiol. 28:915-922.

Mulkins-Phillips, G.J. and Stewart, J.E. 1974c. Effect of four dispersants on biodegradation and growth of bacteria on crude oil. Appl. Microbiol. 28:547-552.

Munnecke, D.M., Johnson, L.M., Talbot, H.W., and Barik, S. 1982. Microbial metabolism and enzymology of selected pesticides. In Chakrabarty, A.M. (Ed.), Biodegradation and Detoxification of Environmental Pollutants. CRC Press, Inc., Boca Raton, FL. pp. 1-32.

Nagel, G., et al. 1982. Sanitation of groundwater by infiltration of ozone treated water. GWF-Wasser/Abwasser. 123:399-407.

Nakayama, S., et al. 1979. Ozone. Science and Engineering 1:119-132.

Nannipieri, P. 1984. Microbial biomass and activity measurements in soil: ecological significance. In Klug, M.J. and Reddy, C.A. (Eds.), Current Perspectives in Microbial Ecology. Amer. Soc. Microbiol, Washington, D.C. pp. 515-521.

National Academy of Sciences. 1977. Drinking Water and Health. Vol. 1. National Academy of Sciences, Washington, D.C.

Naval Civil Engineering Laboratory. 1986. Plan of action and cost estimate. Demonstration project for on-site remediation of JP-5 and other fuel oil pollution from NAS Patuxent River, Maryland. Report to Chesapeake Division, Naval Facilities Engineering Command.

Nedwell, D.B. and Floodgate, G.D. 1971. The seasonal selection by temperature of heterotrophic bacteria in an intertidal sediment. Marine Biology. II:306-310.

Neely, N.S., Walsh, J.S., Gillespie, D.P., and Schauf, F.J. 1981. Remedial actions at uncontrolled waste sites. In Shultz, D.W. (Ed.), Land Disposal: Hazardous Waste, Proc. 7th Ann. Res. Symp., Philadelphia, 1981. EPA Report No. EPA-600/9-81-002b. Environmental Protection Agency, Cincinnati, 1981. pp. 312-319.

Niaki, S., Pollock, C.R., Medlin, W.E., Shealy, S., and Broscious, J.A. Draft. Underground storage tank technologies--literature review.

Nicholas, R.B. 1987. Biotechnology in hazardous-waste disposal: an unfulfilled promise. Amer. Soc. Microbiol. News. 53:138-142.

Nielsen, D.M. 1983. Remedial methods available in areas of ground water contamination. In Nielsen, D.M. and Aller, L. (Eds.), Ground Water Quality, Proc. 6th Nat. Symp., Atlanta, 1983. National Water Well Association, Worthington, OH. pp. 219-227.

Nilles, G.P. and Zabik, M.J. 1975. Photochemistry of bioactive compounds. Multiphase photodegradation and mass spectral analysis of basagran. J. Agr. Food Chem. 23:410.

Nimah, M.H., Ryan, J., and Chaudhry, M.A. 1983. Effect of synthetic conditioners on soil water retention, hydraulic conductivity, porosity, and aggregation. Soil Sci. Soc. Amer. J. 47:742-745.

Noel, M.R., Benson, R.C., and Beam, P.M. 1983. Advances in mapping organic contamination: alternative solutions to a complex problem. Proc. Nat. Conf. on Mgmt. of Uncontrolled Hazardous Waste Sites. Hazardous Materials Control Research Institute, Silver Spring, MD. pp. 71-75.

Novak, J.T., Goldsmith, C.D., Benoit, R.E., and O'Brien, J.H. 1985. Biodegradation of methanol and tertiary butyl alcohol in subsurface systems. Water Sci. Tech. 17:71-85.

Novick, R.P., Murphy, E., Gryczan, T.J., Baron, E., and Edellman, I. 1979. Penicillinase plasmids of Staphylococcus aureus: restriction-deletion maps. Plasmid. 2:109-129.

Novitsky, J.A. and Morita, R.Y. 1978. Possible strategy for the survival of marine bacteria under starvation conditions. Mar. Biol. 48:289-295.

Nyns, E.J., Auquiere, I.P., and Wiaux, A.L. 1968. Taxonomic value of property of fungi to assimilate hydrocarbons. Antonie van Leeuwenhoek J. Microbiol. Serol. 34:441-457.

Nyns, E.J., Brand, J., and Wiaux, A.L. 1968. Assay of the property of fungi to assimilate hydrocarbons by a simple test-tube technique. Fuel. 47:449-453.

Odu, C.T.I. 1972. Microbiology of soils contaminated with petroleum hydrocarbons. I. Extent of contamination and some soil and microbial properties after contamination. J. Inst. Petrol. 58:201-208.

Ohneck, R.J. and Gardner, G.L. 1982. Restoration of an aquifer contaminated by an accidental spill of organic chemicals. Ground Water Monitor. Rev. 2(4):50-53.

Olivieri, R., Bacchin, P., Robertiello, A., Oddo, N., Degen, L., and Tonolo, A. 1976. Microbial degradation of oil spills enhanced by a slow-release fertilizer. Appl. Environ. Microbiol. 31:629-634.

Olson, B.H., Barkay, T., and Colwell, R.R. 1978. Role of plasmids in mercury transformation by bacteria isolated from the aquatic environment. Appl. Environ. Microbiol. 38:478-485.

Olson, G.J., Iverson, W.P., and Brinckman, F.E. 1981. Volitilization of mercury by Thiobacillus ferrooxidans. Curr. Microbiol. 5:115-118.

Osgood, J.O. 1974. Hydrocarbon dispersion in ground water: significance and characteristics. Ground Water. 12:427-438.

Osman, A., Bull, A.T., and Slater, J.H. 1976. In Dellweg, H. (Ed.), Abstracts 5th Intl. Symp. on Fermentation. Westkreuz, Berlin. p. 124.

Overcash, M.R. and Pal, D. 1979. Design of Land Treatment Systems for Industrial Wastes--Theory and Practice. Ann Arbor Science Publishers, Ann Arbor, MI. 684 p.

Padar, F.V. 1979. Identification and control of organic chemicals in sole source aquifers on Long Island, NY.

Page, G.W. 1981. Comparison of ground water and surface water for patterns and levels of contamination by toxic substances. Environ. Sci. Technol. 15:1475-1481.

Painter, H.A. 1977. Microbial transformation of inorganic nitrogen. Prog. Water Technol. 8:3-29.

Palumbo, A.V., Pfaender, F.K., Paerl, H.W., Bland, P.T., Boyd, P.E., and Cooper, D. 1983. Biodegradation rates in coastal ecosystem: influence of environmental factors. Abstr. Ann. Mtg. Amer. Soc. Microbiol. p. 265.

Panchal, C.J. and Zajic, J.E. 1978. Isolation of emulsifying agents from a species of Corynebacterium. Dev. Ind. Microbiol. 9:569-576.

Panchal, C.J., Zajic. J.E., and Gerson, D.F. 1979. Multiple-phase emulsions using microbial emulsifyers. J. of Colloid Interface Sci. 68:295-307.

Pan-Hou, H.S.K., Hosono, M., and Imura, N. 1980. Plasmid-controlled mercury biotransformation by Clostridium cochlearium T-2. Appl. Environ. Microbiol. 40:1007-1011.

Paris, D.F., Wolfe, N.L., Steen, W.C., and Baughman, G.L. 1983. Effect of phenol molecular structure on bacterial transformation rate constants in pond and river samples. Appl. Environ. Microbiol. 45:1153-1155.

Parr, J.F., Sikora, L.J., and Burge, W.D. 1983. Factors affecting the degradation and inactivation of waste constituents in soils. In Parr, J.E., Marsh, P.B., and Kla, J.M. (Eds.), Land Treatment of Hazardous Wastes. Noyes Pub., Park Ridge, NJ. pp. 20-49, 321-337.

Patrick, W.H., Jr. and Mahapatra, I.C. 1968. Transformation and availability to rice of nitrogen and phosphorus in waterlogged soils. Advan. Agron. 20:323-359.

Payne, J.R. and Jordan, R.E. 1980. In The Fate and Weathering of Petroleum Spilled in the Marine Environment. A Literature Review and Synopsis. Ann Arbor Science Publishers, MI. p. 32.

PEDCO Environmental, Inc. 1978. Assessment of gasoline toxicity. EPA Contract No. 68-02-2515. p. 25.

Pemberton, J.M., Corney, B., and Don, R.H. 1979. Evaluation and spread of pesticide degrading ability among soil microorganisms. In Timmis, K.N. and Puhler, A. (Eds.), Plasmids of Medical, Environmental, and Commercial Importance. Elsevier/North Holland Biomedical Press, Amsterdam, The Netherlands.

Perry, J.J. 1968. Substrate specificity in hydrocarbon utilizing microorganisms. Antonie van Leeuwenhoek. 34:27-36.

Perry, J.J. 1979. Microbial cooxidations involving hydrocarbons. Microbiol. Rev. 43:59-72.

Pettigrew, C. and Sayler, G.S. 1986. The use of DNA-DNA colony hybridization in the rapid isolation of 4-chlorobiphenyl degradative bacteria phenotypes. J. Microbiol. Methods. 5:205-213.

Pettyjohn, W.A. and Hounslow, A.W. 1983. Organic compounds and ground-water pollution. Ground Water Monitoring Rev. 3:41-47.

Pfaender, F.K. and Klump, J.V. 1981. Assessment of environmental biodegradation of organic pollutants. Abstr. Ann. Mtg. Amer. Soc. Microbiol. p. 214.

Pfaender, F.K., Shimp, R.J., Palumbo, A.V., and Bartholomew, G.W. 1985. Comparison of environmental influences on pollutant biodegradation: estuaries, rivers, lakes. Abstr. Ann. Mtg. Amer. Soc. Microbiol. p. 266.

Pfennig, N. 1978a. General physiology and ecology of photosynthetic bacteria. In Clayton, P.K. and Sistrom, W.R. (Eds.), The Photosynthetic Bacteria. Plenum Press, New York. pp. 3-17.

Pfennig, N. 1978b. Rhodocyclus purpureus gen. nov. and sp. nov., a ring-shaped vitamin B_{12}-requiring member of the family Rhodospirillaceae. Intl. J. Syst. Bacteriol. 28:283-288.

Pfennig, N. and Biebl, H. 1976. Arch. Microbiol. 110:3-12.

Pflug, Z. 1980. Effect of humic acids on the activity of two peroxidases. Z. Pflannernechr. Bodenkd. pp. 430-440.

Phillips, W.E., Jr. and Brown, L.R. 1975. The effect of elevated temperature on the aerobic microbiological treatment of a petroleum refinery wastewater. In Underkofler, L.A., Cooney, J.J., and Walker, J.D. (Eds.), Developments in Industrial Microbiology. Amer. Inst. Biol. Sci., Washington, D.C. 16:296.

Pierce, G.E. 1982a. Diversity of microbial degradation and its implications in genetic engineering. In C.F. Kulpa, R.L. Irvine, and S.A. Sojka (Eds.), Symposium at Univ. of Notre Dame on Impact of Applied Genetics in Pollution Control, May 24-26, 1982.

Pierce, G.E. 1982b. Summer Nat. Mtg. of the Amer. Inst. of Chem. Eng., Aug. 29-Sept. 1, 1982.

Pierce, G.E. 1982c. Potential role of genetically engineered microorganisms to degrade toxic chlorinated bydrocarbons. In Exner, J.H. (Ed.), Detoxication of Hazardous Waste. Ann Arbor Science Publishers, Ann Arbor, MI. pp. 315-322.

Pierce, G.E. 1982d. Development of genetically engineered microorganisms to degrade hazardous organic compounds. In Sweeney, T.L., Bhat, H.G., Sykes, R.N., and Sproul, O.J. (Eds.), Hazardous Waste Management for the 1980's. Ann Arbor Science Publishers, Ann Arbor, MI. pp. 431-439.

Pierce, G.E., Facklam, T.J., and Rice, J.M.   1980.   Isolation and characterization of plasmids from environmental strains of bacteria capable of degrading the herbicide 2,4-D. In Underkofler, L.A.,(Ed.), Dev. Ind. Microbiol. Soc. Indust. Microbiol., Arlington, VA.

Pirnik, M.P. 1977. Microbial oxidation of methyl branched alkanes. Crit. Rev. Microbiol. 5:413-422.

Pirnik, M.P., Atlas, R.M., and Bartha, R. 1974. Hydrocarbon metabolism by Brevibacterium erythrogenes: normal and branched alkenes. J. Bacteriol. 119:868-878.

Plumb, R.H., Jr. 1985. Disposal site monitoring data: observations and strategy implications. In Hitchon, B. and Trudell, M. (Eds.), Proc. 2nd Canadian/American Conf. on Hydrogeology, June 25-29, 1985. Banff, Alberta, Canada.

Poglazova, M.M., Fedoseeva, G.E., Khesina, A.J., Meissel, M.N., and Shabad, L.M. 1967. Destruction of benzo(a)pyrene by soil bacteria. Life Sciences. 6:1053-1062.

Poindexter, J. 1981. Workshop on trace organic contaminant removal, July 27-28, 1981. Advanced Environmental Control Technology Res. Center. Univ. of Illinois, Urbana, IL.

Polyakova, I.N. 1962. Mikrobiologiya. 31:1076-1981.

Polybac Corp. 1983. Product Information Packet. Allentown, PA.

Pramer, D. and Bartha, R. 1972. Preparation and processing of soil samples for biodegradation studies. Environ. Lett. 2:217-224.

Pramer, D. and Schmidt, E.L. 1964. Experimental Soil Microbiology. Burgess Publ. Co., Minneapolis, MN.

Prill, R.C., Oaksford, E.T., and Potorti, J.E. 1979. U.S. Geological Survey, Water-Resources Investigation 79-48. Washington, D.C.

Prins, R.A. 1978. Chapter 7. Nutritional impact of intestinal drug-microbe interations. In Hathcock, J. and Coon, J. (Eds.), Intl. Symp. on Nutrition and Drug Interrelations. Academic Press, New York, NY. pp. 189-251.

Pritchard, P.H., Van Veld, P.A., and Cooper, W.P. 1981. Biodegradation of p-cresol in artificial stream channels. Abstr. Ann. Mtg. Amer. Soc. Microbiol. p. 210.

Proctor, M.H. and Scher, S. 1960. Decomposition of benzoate by a photosynthetic bacterium. Biochem. J. 76:33.

Public Law 99-499. Oct. 17, 1986.

Pye, V.I. and Patrick, R. 1983. Ground water contamination in the United States. Science. 221:713-718.

Pye, V.I., Patrick, R., and Quarles, J. 1983. Groundwater Contamination in the United States. Univ. of Pennsylvania Press, Philadelphia, PA.

Quince, J.R. and Gardner, G.L. 1982. Recovery and treatment of contaminated ground water: parts 1 and 2. Ground Water Monitoring Rev. 2:Summer 18-22; 2:Fall 18-25.

Radford, A.J., Oliver, J., Kelly, W.J., and Reanney, D.C. 1981. Translocatable resistance to mercuric and phenylmercuric ions in soil bacteria. J. Bacteriol. 147:1110-1116.

Rambeloarisoa, E., Rontani, J.F., Giusti, G., Duvnjak, Z., and Bertrand, J.C. 1984. Degradation of crude oil by a mixed population of bacteria isolated from sea-surface foams. Mar. Biol. 83:69-81.

Rand, M.C., et al. 1975. Standard methods for the examination of water and wastewater. APHA-AWWA-WPCF, Washington, D.C.

Raymond, R.L. 1974. U.S. Patent Office 3,846,290. Patented Nov. 5, 1974.

Raymond, R.L. 1978. Environmental bioreclamation. Proc. Control of Chemicals and Oil Spills. 1978 Mid-Continent Conference and Exhibition on Control of Chemicals and Oil Spills, Sept. 1978.

Raymond, R.L., Hudson, J.O., and Jamison, V.W. 1976. Oil degradation in soil. Appl. Environ. Microbiol. 31:522-535.

Raymond, R.L., Jamison, V.W., and Hudson, J.O. 1975. Report on Beneficial Stimulation of Bacterial Activity in Ground Waters Containing Petroleum Products. Committee on Environmental Affairs, American Petroleum Institute, Washington, D.C.

Raymond, R.L., Jamison, V.W., and Hudson, J.O. 1976. Beneficial stimulation of bacterial activity in groundwaters containing petroleum products. AIChE Symposium Series. Water--1976. 73:390-404.

Raymond, R.L., Jamison, V.W., and Hudson, J.O. 1978. Report on American Petroleum Project No. 307-77.

Raymond, R.L., Jamison, V.W., Hudson, J.O., Mitchell, R.E., and Farmer, V.E. 1978. Field application of subsurface biodegradation of gasoline in a sand formation. Report on API Project No. 307-77. American Petroleum Institute, Washington, D.C.

Reasoner, D.J. and Geldreich, E.E. 1985. A new medium for the enumeration and subculture of bacteria from potable water. Appl. Environ. Microbiol. 49:1-7.

Rees, J.F. and King, J.W. 1981. The dynamics of anaerobic phenol biodegradation in Lower Greensand. J. Chem. Tech. Biotech. 31:306-310.

Rees, J.F., Wilson, B.H., and Wilson, J.T. 1985. Biotransformation of toluene in methanogenic subsurface material. Abstr. Ann. Mtg. Amer. Soc. Microbiol. p. 258.

Reinhard, M., Goodman, N.L., and Barker, J.F. 1984. Occurrence and distribution of organic chemicals in two landfill leachate plumes. Environ. Sci. Technol. 18:953-961.

Reinjhart, R. and Rose, R. 1982. Evaporation of crude oil at sea. Water Res. 16:1319-1325.

Reisfeld, A., Rosenberg, E., and Gutnick, D. 1972. Microbial degradation of crude oil: factors affecting oil dispersion in sea water by mixed and pure cultures. Appl. Microbiol. 24:363-368.

Rice, R.G. 1984. 5th Nat. Conf. on Mgmt. of Uncontrolled Hazardous Waste Sites, Nov. 7-9, 1984, Washington, D.C.

Rich, L.A., Bluestone, M., and Cannon, D.R. 1986. Technology: In situ bioreclamation. Chem. Week. 139(8):62-64.

Richards, D.J. and Shieh, W.K. 1986. Biological fate of organic priority pollutants in the aquatic environment. Water Res. 20:1077-1090.

Rickabaugh, J. Clement, S., Martin, J., and Sunderhaus, M. 1986. Chemical and microbial stabilization techniques for remedial action sites. 12th Ann. Research Symp., Apr. 21-23, 1986. Sponsored by the Hazardous Waste Engineering Res. Lab., Environmental Protection Agency, Cincinnati, OH.

Rittmann, B.E., Bouwer, E.J., Schreiner, J.E., and McCarty, P.L. 1980. Technical Report No. 255. Stanford Univ., Dept. of Civil Eng.

Roberts, P.V., McCarty, P.L., Reinhard, M., and Schreiner, J. 1980. Organic contaminant behavior during groundwater recharge. J. Water Poll. Control Fed. 52:161-172.

Roberts, P.V. and Valocchi, A.J. 1981. Principles of organic contaminant behavior in groundwater. Sci. Total Envir. 21:161.

Roberts, P.V. and Valocchi, A.J. 1982. J. Amer. Water Works Assn. 74:408-413.

Roberts, R.M., Koff, J.L., and Karr, L.A. 1988. Enzyme and microbe immobilization, toxic and hazardous waste handling. Hazardous Waste Minimization Initiation Decision Report. TN-1787. pp. J-1 - J-8.

Robertson, B., Arhelger, S., Kinney, P.J., and Button, D.K. 1973. Hydrocarbon biodegradation in Alaskan waters. Microbial Degradation of Oil Pollutants, Workshop, Atlanta, GA, Dec. 4-6, 1972. Publ. No. LSU-SG-73-01. Louisiana State Univ., Center for Wetland Resources, Baton Rouge, LA. pp. 171-184.

Robichaux, T.J. and Myrick, H.N. 1972. Chemical enhancement of the biodegradation of crude-oil pollutants. J. Pet. Technol. 24:16.

Robison, R. 1987. Regulations target underground tanks. Civil Eng. February, pp. 72-74.

Romero, J.C. 1970. The movement of bacteria and viruses through porous media. Ground Water. 8(2):37-48.

Rosenberg, E., Englander, E., Horowitz, A., and Gutnick, D. 1975. Bacterial growth and dispersion of crude oil in an oil tanker during its ballast voyage. In Bourquin, A.W., Ahearn, D.G., and Meyers, S.P. (Eds.), Proc. Impact of the Use of Microorganisms on the Aquatic Environment. EPA Report No. EPA 660-3-75-001. Environmental Protection Agency Research Center, Corvallis, OR. p. 157-167.

Rosenberg, E., Perry, A., Gibson, D.F., and Gutnick, D.L. 1979. Emulsifier of Arthrobacter RAG-1: specificity of hydrocarbon substrate. Appl. Environ. Microbiol. 37:409-413.

Rosenberg, E., Zuckerberg, A., Rubinovitz, C., and Gutnick, D.L.   1979. Emulsifier of <u>Arthrobacter</u> RAG-1:   isolation and emulsifying properties. Appl. Environ. Microbiol.   37:402-408.

Rosenfeld, W.D.   1947.   Anaerobic   oxidation   of   hydrocarbons   by sulfate-reducing bacteria. J. Bacteriol.   54:664-665.

Ross, D.S., Sjogren, R.E., and Bartlett, R.J.   1981.   Behavior of chromium in soils.   IV. Toxicity to microorganisms. J. Environ. Qual.   10:145-148.

Roubal, G. and Atlas, R.M.   1980.   Biodegradation of crude oil-mousse in the Gulf of Mecico from the Ixtoc I well blowout.   Abstr. Ann. Mtg. Amer. Soc. Microbiol.   p. 203.

Rubin, H.E. and Alexander, M.   1983.   Effect of nutrients on the rates of mineralization of trace concentrations of phenol and p-nitrophenol. Environ. Sci. Technol.   17:104-107.

Rubin, J.   1983.   Transport of reacting solutes in porous media: relation between mathematical nature of problem formulation and chemical nature of reactions. J. Water Resour. Res.   19:1231-1252.

Sanning, D.E. and Olfenbuttel, R.   1987.   NATO/CCMS pilot study on demonstration of remedial action technologies for contaminated land and groundwater.   Proc. 13th Ann. Res. Symp. on Land Disposal, Remedial Action, Incineration and Treatment of Hazardous Waste, May 6-8, 1987, Cincinnati, OH.

Sawhney, B.L. and Kozloski, R.P.   1984.   Organic pollutants in leachates from landfill sites. J. Environ. Qual.   13:349-352.

Saxena, A. and Bartha, R.   1983.   Microbial mineralization of humic acid--3,4-dichloroaniline complexes.   Soil Biol. Biochem.   15:59-62.

Sayer, J.M., Yagi, H., Wood, A.W., Conney, A.H., and Jerina, D.M.   1982. Extremely   facile   reaction   between   the   ultimate   carcinogen benzo(a)pyrene-7,8-diol 9,10-epoxide and ellagic acid. J. Amer. Chem. Soc. 104:5562-5564.

Sayler, G.S., Harris, C., Pettigrew, C., Pacia, D., Breen, A., and Sirotken, K.M.   1986.   Evaluating the maintenance and effects of genetically engineered microorganisms. Dev. in Ind. Microbiol.   27:

Sayler, G.S., Shields, M.S., Tedford, E.T., Breen, A., Hooper, S.W., Sirotkin,K.M., and Davis, J.W.   1985.   Application of DNA-DNA colony hybridization to the detection of catabolic genotypes in environmental samples. Appl. Environ. Microbiol.   49:1295-1303.

Scalf, M.R., McNabb, J.F., Dunlap, W.J., Cosby, R.L., and Fryberger, J. 1981. Manual of Ground Water Sampling Procedures.   National Water Well Association.   Worthington, OH.

Scheda, S. and Bos, P. 1966.   Hydrocarbons as substrates for yeasts.   Nature. 211:660.

Schink, B.   1985.   Degradation of unsaturated hydrocarbons by methanogenic enrichment cultures.   FEMS Microbiol. Ecol.   31:69-77.

Schink, B. and Pfennig, N. 1982. Fermentation of trihydroxybenzenes by Pelobacter acidigallici gen. nov. sp. nov., a new strictly anaerobic non-sporeforming bacterium. Arch. Microbiol. 133:195-201.

Schmidt, S.K. and Alexander, M. 1985. Effects of dissolved organic carbon and second substrates on the biodegradation of organic compounds at low concentrations. Appl. Environ. Microbiol. 49:822-827.

Schmidt, S.K., Scow, K.M., and Alexander, M. 1985. The kinetics of simultaneous mineralization of two substrates in soil, lake water, and pure cultures of bacteria. Abstr. Ann. Mtg. Amer. Soc. Microbiol. p. 263.

Schmidt-Bleek, F., and Wagenknecht, P. 1979. Umweltchemikalien. Chemosphere. 8:583-721.

Schnitzer, M. 1978. Humic substances: chemistry and reactions. In Schnitzer, M. and Khan, S.U. (Eds.), Soil Organic Matter. Elsevier/North-Holland, Inc., New York, NY. pp. 1-64.

Schnitzer, M. 1982. Organic matter characterization. In Page, A.L., Miller, R.H., and Keeney, D.R. (Eds.), Methods of Soil Analysis. Part 2. Chemical and Microbiological Properties. 2nd ed. Agronomy Monograph No. 9(2). American Society of Agronomy, Inc., Madison, WI. pp. 581-594.

Schnitzer, M. and Khan, S.U. 1978. Soil Organic Matter. Developments in Soil Science 8. Elsevier Scientific Publishing Co., Amsterdam.

Scholze, R.J., Jr., Wu, Y.C., Smith, E.D., Bandy, J.T., and Basilico, J.V. (Eds.). 1986. Proc. Intl. Conf. on Innovative Biological Treatment of Toxic Wastewaters, June 24-26, 1986, Arlington, VA.

Schwab, G.O., Frevert, R.K., Edminster, T.W., and Barhes, K.K. 1981. Soil and Water Conservation Engineering. 3rd ed. John Wiley and Sons, Inc., New York, NY.

Schwartz, R.D. and Hutchinson, D.B. 1981. Microbial and enzymatic production of 4,4' dihydroxybiphenyl via phenol coupling. Enzyme Microbiol. Technol. 3:361.

Schweizer, E.E. 1976. Persistence and movement of ethofumesate in soil. Weed Sci. 14:22-26.

Schwendinger, R.B. 1968. Reclamation of soil contaminated with oil. J. Inst. Pet. 54:182-197.

Science Applications International Corporation. 1985a. Interim Report: Field Demonstration of In Situ Biological Degradation. For Air Force Engineering and Services Center, Tyndall AFB, FA and EPA Hazardous Waste Eng. Res. Lab., Cincinnati, OH.

Science Applications International Corporation. 1985b. Company sales information.

Scow, K.M., Simkins, S., and Alexander, M. 1986. Kinetics of mineralization of organic compounds at low concentrations in soil. Appl. Environ. Microbiol. 51:1028-1035.

Seki, H. 1976. Method for estimating the decomposition of hexadecane in the marine environment. Appl. Environ. Microbiol. 31:439-441.

Shabad, L.M., Cohan, Y.L., Ilnitsky, A.P., Khesina, A. Ya., Shcherbak, N.P., and Smirnov., G.A. 1971. The carcinogenic hydrocarbon beno(a)pyrene in the soil. J. Nat. Cancer Inst. 47:1179-1191.

Shackelford, W.M. and Keith, L.H. 1977. Frequency of organic compounds identified in water. Report NTIS PB-265-470. Environmental Protection Agency, Athens, GA.

Shanley, E.S. and Edwards, J.O. Peroxides and Peroxy Compounds, Kirk-Othner Encyclopedia of Chemical Technology, 2nd Ed. Vol. 14.

Shelley, P.E. 1977. Sampling of water and wastewater. EPA Report No. EPA 600/4-77-039.

Shelton, R.G.J. 1971. Effects of oil and oil dispersants on the marine environment. Proc. Royal Soc. London Series B. 177:411.

Shelton, T.B. and Hunter, J.V. 1975. Anaerobic decomposition of oil in bottom sediments. J. Water Pollut. Control Fed. 47:2256-2270.

Sherrard, J.H. and Schroeder, E.D. 1975. Stoichiometry of industrial biological wastewater treatment. Proc. 30th Ind. Waste Conf., May 6-8, 1975, Purdue Univ.

Sherill, T.W. and Sayler, G.S. 1980. Phenanthrene biodegradation in freshwater environments. Appl. Environ. Microbiol. 39:172-178.

Sherrill, T.W. and Sayler, G.S. 1982. Enhancement of polyaromatic hydrocarbon mineralization rates by polyaromatic hydrocarbon and synthetic oil contamination of freshwater sediments. Abstr. Ann. Mtg. Amer. Soc. Microbiol. p. 216.

Shiaris, M.P. and Cooney, J.J. 1981. Phenanthrene-degrading and cooxidizing microorganisms in estuaring sediments. Abstr. Ann. Mtg. Amer. Soc. Microbiol. p. 213.

Shiaris, M.P. and Jambard-Sweet, D. 1984. Potential transformation rates for polynuclear aromatic hydrocarbons (PNAHs) in surficial estuarine sediments. Abstr. Ann. Mtg. Amer. Soc. Microbiol. p. 217.

Shimp, R.J. and Pfaender, F.K. 1984. Influence of naturally occurring carbon substrates on the biodegradation of monosubstituted phenols by aquatic bacteria. Abstr. Ann. Mtg. Amer Soc. Microbiol. p. 212.

Shin, Y., Chodan, J.J., and Wolcott, A.R. 1970. Adsorption of DDT by soils, soil fractions, and biological materials. J. Agr. Food Chem. 18:1129-1133.

Shine, J. and Dalgarno, L. 1975. Determinant of cistron specificity on bacterial ribosomes. Nature. 254:34-38.

Shoda, M. and Udaka, S. 1980. Preferential utilizaiton of phenol rather than glucose by Trichosporon cutaneum possessing a partially constitutive catechol 1,2-oxygenase. Appl. Environ. Microbiol. 39:1129-1133.

Short, H. and Parkinson, G. 1983. Chem. Eng. 90:26.

Shrift, A. 1973. Selenium in biomedicine. In Klaymann, D.L. and Gunther, W.H.H. (Eds.), Organic Selenium Compounds: Their Chemistry and Biology. John Wiley & Sons, London. pp. 764-814.

Shuckrow, A.J. and Pajak, A.P. 1981. In Proc. Land Disposal Hazardous Waste, Mar. 16-18, 1981, Philadelphia, PA. EPA Report NO. EPA-600/9-81-992b. Environmental Protection Agency, Cincinnati, OH. pp. 341-351.

Shuckrow, A.J. and Pajak, A.P. 1982. In Proc. Land Disposal of Hazardous Waste, Ft. Mitchell, KY, Mar. 8-10, 1982. Shultz, D.W. and Black, D. (Eds.). EPA Report No. EPA-600/9-82-001. Environmental Protection Agency, Cincinnati, OH. pp. 346-359.

Shuckrow, A.J., Pajak, A.P., and Touhill, C.J. 1982. Management of hazardous waste leachate. Report No. SW-871. Environmental Protection Agency, Washington, D.C.

Sikes, D.J. 1984. The containment and mitigation of a formaldehyde rail car spill using naval chemical and biological in situ treatment techniques. Hazardous Material Spills Conf., Nashville, TN.

Silver, S. 1983. Bacterial interactions with mineral cations and anions: good ions and bad. In Westbroek, P. and deJong, E.W. (Eds.), Biomineralization and Biological Metal Accumulation. D. Reidel Publishing Co., Amsterdam, The Netherlands. p. 439-457.

Silver, S. and Kinscherf, T.G. 1982. In Biodegradation and Detoxification of Environmental Pollutants. Chakrabarty, A.M. (Ed.). CRC Press, Inc., Boca Raton, FL.

Simkins, S., Schmidt, S.K., and Alexander, M. 1984. Kinetics of mineralization by bacteria and in environmental samples fit models incorporating only the variables of substrate concentration and cell density. Abstr. Ann. Mtg. Amer. Soc. Microbiol. p. 212.

Simon, R.D. and Weathers, P. 1976. Determination of the structure of the novel polypeptide containing aspartic acid and arginine which is found in cyanobacteria. Biochim. Biophys. Acta. 420:165-176.

Sims, R. and Bass, J. 1984. Review of in-place treatment techniques for contaminated surface soils. Volume 1: technical evaluation. EPA Report No. EPA-540/2-84-003a.

Sims, R.C. and Overcash, M.R. 1981. Land treatment of coal conversion wastewaters. In Environmental Aspects of Coal Conversion Technology VI. A Symposium on Coal-Based Synfuels. EPA Report No. EPA-600 9-82-017. Environmental Protection Agency, Washington, DC. pp. 218-230.

Sims, R.C. and Overcash, M.R. 1983. Fate of polynuclear aromatic compounds (PNAs) in soil-plant systems. Residue Rev. 88:1-68.

Sinclair, J.L. and Ghiorse, W.C. 1985. Isolation and characterization of a subsurface amoeba. Abst. Ann. Mtg. Amer. Soc. Microbiol. p. 258.

Slavnia, G.P. 1965. Mikrobiologiya. 34:103-106.

Sleat, R. and Robinson, J.P. 1984. The bacteriology of anaerobic degradation of aromatic compounds. J. Appl. Bacteriol. 57:381-394.

Slonim, Z., Lien, L.-T., Eckenfelder, W.W., and Roth, J.A. 1985. Anaerobic-aerobic treatment process for the removal of priority pollutants. EPA Report No. EPA/600/S2-85/077.

Smith, J.H., Harper, J.C., and Jaber, H. 1981. SRI Int'l., Menlo Park, CA. Report No. ESL-TR-81-54. Engineering and Services Lab., Tyndall Air Force Base, FL.

Smith, J.L., McWhorter, D.B., and Ward, R.C. 1977. Continuous subsurface injection of liquid dairy manure. EPA No. 600/2-77-117. Robert S. Kerr Environmental Research Lab.

Soil Conservation Service. 1979. Guide for sediment control on construction sites in North Carolina. U.S. Dept. of Agriculture. Soil Conservation Service, Raleigh, NC.

Soil Science Society of America. 1981. Water potential relations in soil microbiology. SSSA Special Publication No. 9. Soil Sci. Soc. Amer., Madison, WI. 151 p.

Sokol, R.A. and Klein, D.A. 1975. The responses of soils and soil microorganisms to silver iodide weather modification agents. J. Environ. Qual. 4:211-214.

Solanas, A.M., Pares, R., Bayona, J.M., and Albaiges, J. 1984. Degradation of aromatic petroleum hydrocarbons by pure microbial cultures. Chemosphere. 13:593-601.

Soli, G. 1973. Marine hydrocarbonoclastic bacteria: types and range of oil degradation. In Ahearn, D.G. and Meyers, S.P. (Eds.), The Microbial Degradation of Oil Pollutants, Workshop, Atlanta, GA, Dec. 4-6, 1972. Publ. No. LSU-SG-73-01. Louisiana State Univ., Center for Wetland Resources, Baton Rouge, LA. p. 141.

Somerville, C.C., Butler, L.C., Lee, T.J., Bourquin, A.W., and Spain, J.C. 1983. Degradation of jet fuel hydrocarbons by aquatic microbial communities. Abstr. Ann. Mtg. Amer. Soc. Microbiol. p. 284.

Somerville, H.J., Mason, J.R., and Ruffnell, R.N. 1977. Benzene degradation by bacterial cells immobilized in polyacrylamide gel. Europ. J. Appl. Microbiol. 4:75-85.

Sommers, L.E., Gilmore, C.M., Wildung, R.E., and Beck, S.M. 1981. The effect of water potential on decomposition processes in soils. In Water Potential Relations in Soil Microbiology. SSSA Special Publication No. 9. Soil Sci. Soc. Amer., Madison, WI. pp. 97-117.

Song, H.-G. and Bartha, R. 1986. Bacterial and fungal contributions to hydrocarbon mineralization in soil. Abstr. Ann. Mtg. Amer. Soc. Microbiol. p. 302.

Spain, J.C., Milhous, T., and Bourquin., A.W. 1981. Effects of adaptation on degradation of organic compounds by natural microbial populations. Abstr. Ann. Mtg. Amer. Soc. Microbiol. p. 207.

Spain, J.C., Pritchard, P.H., and Bourquin, A.W. 1980. Effects of adaptation on biodegradation rates in sediment/water cores from estuarine and freshwater environments. Appl. Environ. Microbiol. 40:726-734.

Spain, J.C. and Van Veld, P.A. 1983. Adaptation of natural microbial communities to degradation of xenobiotic compounds: effects of concentration, exposure, time, inoculum, and chemical structure. Appl. Environ. Microbiol. 45:428-435.

Spencer, W.F., Adam, J.D., Shoup, T.D., and Spear, R.C. 1980. Conversion of parathion to paraoxon on soil dusts and clay minerals as affected by ozone and UV light. J. Agr. Food Chem. 28:369.

Steen, W.C., Paris, D.F., and Latimer, B.E. 1981. Microbial adaptation and transformation of toxic substances in aquatic systems. Abstr. Ann. Mtg. Amer. Soc. Microbiol. p. 206.

Standard Methods for the Examination of Water and Wastewater, 1975.

Stanier, R.Y., Doudoroff, M., and Adelberg, E.A. (Eds.). 1970. The Microbial World. 3rd ed. Prentice-Hall, Inc., Englewood Cliffs, NJ. pp. 566-570.

Stanlake, G.J. and Finn, R.K. 1982. Isolation and characterization of a pentachlorophenol degrading bacterium. Appl. Environ. Microbiol. 44:1421-1427.

State of California, Leaking Underground Fuel Tank Task Force. 1987. Leaking Underground Fuel Tank Field Manual: Guidelines for Site Assessment, Cleanup, and Underground Stroage Tank Closure.

Steelman, B.L. and Ecker, R.M. 1984. Organics contamination of groundwater--an open literature review. 5th DOE Environmental Protection Meeting, Nov. 5, 1984. Albuquerque, NM.

Sterling, L.A., Watkinson, R.J., and Higgins, I.J. 1977. Microbial metabolism of alicyclic hydrocarbons: isolation and properties of a cyclohexane-degrading bacterium. J. Gen. Microbiol. 99:119-125.

Sterritt, R.M. and Lester, J.N. 1980. Interactions of heavy metals with bacteria. Sci. Total Environ. 14:5-17.

Stetzenbach, L.D., Kelley, L.M., and Sinclair, N.A. 1984. Isolation, identification and characterization of ground water bacteria. Abst. Ann. Mtg. Amer. Soc. Microbiol. p. 218.

Stetzenbach, L.D., Kelley, L.M., Stetzenbach, K.J., and Sinclair, N.A. 1985. Decreases in hydrocarbons by soil bacteria. Proc. Symp. on Groundwater Contam. and Reclamation. Amer. Water Res. Assoc. pp. 55-60.

Stetzenbach, L.D. and Sinclair, N.A. 1986. Degradation of anthracene and pyrene by soil bacteria. Abstr. Ann. Mtg. Amer. Soc. Microbiol. p. 302.

Stetzenbach, L.D., Sinclair, N.A., and Kelley, L.M. 1983. Isolation and growth of "naturally occuring" bacteria from ground water. In Parr, J.F., Marsh, P.B., and Kla, J.M. (Eds.), Land Treatment of Hazardous Wastes. Noyes Pub., Park Ridge, NJ. p. 28.

Stotzky, G. and Norman, A.G. 1961a. Factors limiting microbial activities in soil. I. The level of substrate, nitrogen, and phosphorus. Arch. Mikrobiol. 40:341-369.

Stotzky, G. and Norman, A.G. 1961b. Factors limiting microbial activities in soil. II. The effect of sulfur. Arch. Mikrobiol. 40:370-382.

Stotsky, G. and Krasovsky, V.N. 1981. Ecological factors that affect the survival, establishment, growth, and genetic recombination of microbes in natural habitats. In Levy, S.B., Clowes, R.C., and Koenig, E.L. (Eds.), Molecular Biology, Pathogenicity, and Ecology of Bacterial Plasmids. Plenum Press, New York, NY. pp. 31-42.

Stover, E.L. and Kincannon, D.F. 1983. Contaminated groundwater treatability--a case study. Research and Technology. pp. 292-298.

Stratton, R.G., Jr. 1981. M.S. thesis, Dept. of Civil Eng., Univ. of Illinois at U-C, Urbana, IL.

Stucki, G. and Alexander, M. 1987. Role of dissolution rate and solubility in biodegradation of aromatic compounds. Appl. Environ. Microbiol. 53:292-297.

Suflita, J.M. and Miller, G.D. 1985. The microbial metabolism of chlorophenolic compounds in ground water aquifers. Environ. Toxicol. Chem. 4:751-858.

Suidan, M.T., Cross, W.H., Fong, M., and Calvert, J.W. 1982. Proc., Amer. Soc. Civil Eng. Vol. 107, No. EE3.

Suntech Group. 1978. Petroleum Marketer. July-Aug. 1978. Reprint.

Suntech, Inc. 1977. Environmental Bioreclamation Brochure.

Sutton, P.M. 1987. Biological treatment of surface and groundwater. Poll. Eng. 19:86-89.

Swindoll, C.M., Aelion, C.M., Dobbins, D.C., Jiang, O.U., Long, S.C., and Pfaender, F.K. 1988. Aerobic biodegradation of natural and xenobiotic organic compounds by subsurface microbial communities. Environ. Toxicol. Chem. 7:291-299.

Swindoll, C.M., Aelion, C.M., and Pfaender, F.K. Draft. Influence of Mineral and Organic Nutrients on the Aerobic Biodegradation and Adaptation Response of Subsurface Microbial Communities.

Switzenbaum, M.S., Jewell, W.J. 1980. Anaerobic attached-film expanded-bed reactor treatment. J. Water Poll. Control Fed. 52:1953-1965.

Tabak, H.H., Quave, S.A., Mashni, C.I., and Barth, E.F. 1981. Biodegradability studies with organic priority pollutant pompounds. J. Water Poll. Control Fed. 53:1503-1518.

Tabor, P.S., Ohwada, K., and Colwell, R.R. 1981. Filterable marine bacteria found in the deep sea: distribution, taxonomy, and response to starvation. Microb. Ecol. 7:67-83.

Takai, Y. and Kamura, T. 1966. The mechanism of reduction in waterlogged paddy soil. Folia Microbiol. (Prague). 11:304-313.

Taylor, B.F., Campbell, W.L., and Chinoy, I. 1970. Anaerobic degradation of the benzene nucleus by a facultatively anaerobic microorganism. J. Bacteriol. 102:430-437.

Taylor, J.M., Parr, J.F., Sikora, L.J., and Willson, G.B. 1980. Considerations in the land treatment of hazardous wastes: Principles and practices. Proc. 2nd Oil and Hazardous Material Spills Conf. and Exhibition. Hazardous Materials Control Res. Inst., Silver Spring, MD.

Telegina, Z.P. 1963. A study of the capacity of individual saprophytic microflora species to adapt themselves to gaseous hydrocarbon oxidation. Mikrobiologiya. 32:398-402.

Texas Research Institute. 1982. Enhancing the microbial degradation of underground gasoline by increasing available oxygen. Report to the American Petroleum Institute, Washington, D.C.

Texas Research Institute, Inc. Feb. 1983. Progress Report: Biostimulation Study.

Thibault, G.T. and Elliott, N.W. 1979. Accelerating the biological clean-up of hazardous materials spills. Proceedings Oil and Hazardous Material Spills: Prevention--Control--Cleanup--Recovery--Disposal. Dec. 3-5, 1979. Sponsored by: Hazardous Materials Control Research Inst. and Information Transfer, Inc. pp. 115-120.

Thibault, G.T. and Elliot, N.W. 1980. Biological detoxification of hazardous organic chemical spills. Proc. 1980 Conf. Control of Hazardous Material Spills, Vanderbilt Univ., Nashville, TN. Sponsored by Environmental Protection Agency. pp. 398-402.

Thomas, J.M. and Alexander, M. 1983. Solubilization and mineralization rates of water-insoluble compounds. Abstr. Ann. Mtg. Amer. Soc. Microbiol. p. 274.

Thomas, J.M., Lee, M.D., Scott, M.J., and Ward, C.H. 1986. Microbial adaptation to a jet fuel spill. Abstr. Ann. Mtg. Amer. Soc. Microbiol. p. 303.

Thomas, J.M., Lee, M.D., and Ward, C.H. 1985. Microbial numbers and activity in the subsurface at a creosote waste site. Abstr. Ann. Mtg. Amer. Soc. Microbiol. p. 258.

Thomas, J.M., Yordy, J.R., Amador, J.A., and Alexander, M. 1986. Rates of dissolution and biodegradation of water-insoluble organic compounds. Appl. Environ. Microbiol. 52:290-296.

Thompson, G.A. and Watling, R.J. 1983. A simple method for the determination of bacterial resistance to metals. Bull. Environ. Contam. Toxicol. 31:705-711.

Thompson, L.M., Black, C.A., and Zoellner, J.A. 1954. Occurrence and mineralization of organic phophorus in soils, with particular reference to associations with nitrogen, carbon, and pH. Soil Sci. 77:185-196.

Thorn, P.M. and Ventullo, R.M. 1986. Growth of bacteria in aquifer soil as measured by ^{3}H thymidine incorporation. Abstr. Ann. Mtg. Amer. Soc. Microbiol. p. 300.

Thornton-Manning, J.R., Jones, D.D., and Federle, T.W. 1987. Effects of experimental manipulation of environmental factors on phenol mineralization in soil. Envir. Toxicol. Chem. 6:615-621.

Tiedje, J.M., Sexstone, A.J., Parkin, T.B., Revsbech, N.P., and Shelton, D.R. 1984. Anaerobic processes in soil. Plant Soil. 76:197-212.

Timoney, J.F., Port, J., Giles, J., and Spanier, J. 1978. Heavy-metal and antibiotic resistance in the bacterial flora of sediments of New York Bight. Appl. Environ. Microbiol. 36:465-472.

Tisdale, S.L. and Nelson, W.L. 1975. Soil Fertility and Fertilizers. Macmillan Publishing Co., Inc., New York, NY.

Tissot, B.P. and Welte, D.H. 1978. In Petroleum Formation and Occurence. Springer Verlag, Berlin. p. 361.

Tjessem, K. and Aaberg, A. 1983. Photochemical transformation and degradation of petroleum residues in the marine environment. Chemosphere. 12:1373-1394.

Tonomura, K. and Kanzaki, F. 1969. The reductive decomposition of organic mercurials by cell-free extract of a mercury-resistant pseudomonad. Biochim. Biophys. Acta. 184:227-229.

Traxler, R.W. 1972. Bacterial degradation of petroleum materials in low temperature marine environments. In Ahearn, D.G. and Meyers, S.P. (Eds.), The Microbial Degradation of Oil Pollutants, Workshop, Atlanta, GA, Dec. 4-6, 1972. Publ. No. LSU-SG-73-01. Louisiana State Univ., Center for Wetland Resources, Baton Rouge, LA. pp. 163-170.

Tripp, S., Barkay, T., and Olson, B.H. 1983. The effect of cadmium on the community structure of soil bacteria. Abstr. Ann. Mtg. Amer. Soc. Microbiol. p. 260.

Tsezos, M. and Benedek, A. 1980. Removal of organic substances by biologically activated carbon in a fluidized-bed reactor. J. Water Poll. Control Fed. 52:578-586.

Tuckett, J.D. and Moore, W.E.C. 1959. Production of filterable particles by Celvibrio gilvus. J. Bacteriol. 77:227-229.

Van Dam, J. 1967. The Migration of Hydrocarbons in a Water-Bearing Stratum. The Joint Problems of the Oil and Water Industries. Elsevier, New York, NY. pp. 55-96.

Vandenbergh, P.A., Olsen, R.H., and Coloruotolo, J.F. 1981. Isolation and genetic characterization of bacteria that degrade chloroaromatic compounds. J. Appl. Environ. Microbiol. 42:737-739.

Van Eyk, J. and Bartels, T.J. 1968. Paraffin oxidation in Pseudomonas aeruginosa. J. Bacteriol. 96:706-712.

Vanloocke, R., Verlinde, A.M., Verstraete, W., and DeBurger, R. 1979. Microbial release of oil from soil columns. Environ. Sci. Tech. 13:346-348.

Varga, G.M., Jr., Lieberman, M., and Avella, A.J., Jr. 1985. The effects of crude oil and processing on JP-5 composition and properties. Report to Naval Air Propulsion Center under Contract NOO140-81-C-9601. NAPC-PE-121C.

Vaughn, J.M., Landry, E.F., Beckwith, C.A., and Thomas, M.Z. 1981. Virus removal during groundwater recharge: effects of infiltration rate on adsorption of poliovirus to soil. Appl. Environ. Microbiol. 41:139-147.

Verma, L., Martin, J.P., and Haider, 1975. Decomposition of carbon-14-labeled proteins, peptides, and amino acids; free and complexed with humic polymers. Soil Sci. Soc. Amer. Proc. 39:279-284.

Visser, S.A. 1982. Surface active phenomena by humic substances of aquatic origin. Revue Francaise des Sciences de l' Eau. 1:285-295.

Visser, S.A. 1985. Physiological action of humic substances on microbial cells. Soil Biol. Biochem. 17:457-462.

Voerman, S. and Tammes, P.M. 1969. Adsorption and desorption of lindane and dieldrin by yeast. Bull. Environ. Contam. Toxicol. 4:271-277.

Vonk, J.W. and Sjipesteijn, A.K. 1973. Studies on the methylation of mercuric chloride by pure cultures of bacteria and fungi. Antonie van Leeuwenhoek. 39:505-513.

Waksman, S.A. 1924. Influence of microorganisms upon the carbon: nitrogen ratio in the soil. J. Agric. Sci. 14:555-562.

Walker, A. and Crawford, D.V. 1968. The role of organic matter in adsorption of the triazine herbicides by soil. In Isotopes and Radiation in Soil Organic Matter Studies. Proc. 2nd Symp. Intl. Atomic Energy Agency. Vienna, Austria.

Walker, J.D., Austin, H.F., and Colwell, R.R. 1975. Utilization of mixed hydrocarbon substrate by petroleum-dgrading microorganisms. J. Gen. Appl. Microbiol. 21:27-39.

Walker, J.D., Cofone, L., Jr., and Cooney, J.J. 1973. Microbial petroleum degradation: the role of Cladosporium resinae. In Proc. Joint Conf. on Prevention and Control of Oil Spills. American Petroleum Institute, Washington, D.C. pp. 821-825.

Walker, J.D. and Colwell, R.R. 1974a. Mercury-resistant bacteria and petroleum degradation. Appl. Microbiol. 27:285-287.

Walker, J.D. and Colwell, R.R. 1974b. Microbial degradation of model petroleum at low temperatures. Microb. Ecol. 1:63-95.

Walker, J.D. and Colwell, R.R. 1975. Factors affecting enumeration and isolation of actinomycetes from Chesapeake Bay and Southeastern Atlantic Ocean sediments. Mar. Biol. 30:193-201.

Walker, J.D. and Colwell, R.R. 1976a. Oil, mercury, and bacterial interactions. Environ. Sci. Technol. 10:1145-1147.

Walker, J.D. and Colwell, R.R. 1976b. Measuring the potential activity of hydrocarbon-degrading bacteria. Appl. Environ. Microbiol. 31:189-197.

Walker, J.D. and Colwell, R.R. 1976c. Enumeration of petroleum-degrading microorganisms. Appl. Environ. Microbiol. 31:198-207.

Walker, N., Janes, N.F., Spokes, J.R., and van Berkum, P. 1975. Degradation of 1-naphthol by a soil pseudomonad. J. Appl. Bact. 39:281-286.

Walker, J.D., Colwell, R.R., and Petrakis, L. 1975. Degredation of petroleum by an alga, Prototheca zopfii. Appl. Microbiol. 30:79.

Walker, J.D., Colwell, R.R., and Petrakis, L. 1976. Biodegradation of petroleum by Chesapeake Bay sediment bacteria. Can. J. Microbiol. 22:423-428.

Walton, G.C. and Dobbs, D. 1980. Biodegradation of hazardous materials in spill situations. Proc. 1980 Conf. on Control of Hazardous Material Spills, Louisville, KY. pp. 23-29.

Ward, C.H. and Lee, M.D. 1984. In Canter, L.W. and Knox, R.C. (Eds.), Ground Water Pollution Control. Lewis Publishers, Chelsea, MI. 526 pp.

Ward, C.H., Tomson, M.B., Bedient, P.B., and Lee, M.D. 1986. Transport and fate processes in the subsurface. In Loehr, R.C. and Malina, J., Jr. (Eds.), Land Treatment: A Hazardous Waste Management Alternative. Water Resources Symposium No. 13, Univ. of Texas, Austin, TX, 1986. pp. 19-39.

Ward, D.M. and Brock, T.D. 1974. Temperature and nutrient limitation of oil biodegradation in Lake Mendota, Wisconsin. Abstr. Ann. Meet. Amer. Soc. Microbiol. G 268.

Ward, D.M. and Brock, T.D. 1976. Environmental factors influencing the rate of hydrocarbon oxidation in temperate lakes. Appl. Environ. Microbiol. 31:764-772.

Webster, J.J., Hampton, G.J., Wilson, J.T., Ghiorse, W.C., and Leach, F.R. 1985. Determination of microbial cell numbers in subsurface samples. Ground Water. 23:17-25.

Wentsel, R.S., et al. 1981. EPA Report No. EPA-600/2-7-81-1-208. Environmental Protection Agency, Cincinnati, OH.

Werner, P. 1982. Vom Wasser. p. 157 (in German).

Westlake, D.W.S., Jobson, A.M., and Cook, F.D. 1978. In situ degradation of oil in a soil of the boreal region of the Northwest Territories. Can. J. Microbiol. 24:254-260.

Westlake, D.W.S., Jobson, A., Phillippe, R., and Cook, F.D. 1974. Biodegradability and crude oil composition. Can. J. Microbiol. 20:915-928.

Wetzel, D.M. and Reible, D.D. 1982. Report to LSU Hazardous Waste Research Center, Louisiana State Univ., Baton Rouge, LA.

Wetzel, R.S. 1986. Proc. 12th Ann. EPA Res. Symp. on Land Disposal, Remedial Action, Incineration, and Treatment of Hazardous Waste, Cincinnati, OH, 1986.

Wetzel, R.S., Davidson, D.H., Durst, C.M., and Sarno, D.J. 1986. Field demonstration of in situ biological treatment of contaminated groundwater and soils. In Proc. 12th Ann. EPA Reseach Symp. on Land Disposal, Remedial Action, Incineration, and Treatment of Hazardous Waste, Apr. 21-23, 1986. Sponsored by Environmental Protection Agency, Hazardous Waste Engineering Research Lab, Cincinnati, OH.

Wetzel, R.S., Henry, S.M., Spooner, P.A., James, S.C., and Heyse, E. 1985. In situ treatment of contaminated groundwater and soils, Kelly Air Force Base, TX. Proc. 11th Ann. EPA Research Symp. on Land Disposal, Remedial Action, Incineration, and Treatment of Hazardous Waste, Cincinnati, OH.

White, D.C. 1983. Analysis of microorganisms in terms of quantity and activity in natural environments. In Slater, J.H., Whittenbury, R., and Wimpenny, J.W.T. (Eds.), Microbes in Their Natural Environment. Cambridge Univ. Press, New York, NY. pp. 37-66.

White, D.C., Smith, G.A., Gehron, M.J., Parker, J.H., Findlay, R.H., Matz, R.F., and Fredrickson, H.L. 1982. The groundwater aquifer microbiota: biomass, community structure and nutritional status. Dev. Ind. Microbiol. 24:201-211.

Wickerham, L.J. 1951. Taxonomy of yeasts. U.S. Dept. Agr. Tech. Bull. 1029:1-55.

Widdle, F. and Nobert, P. 1981. Studies on a dissimilatory sulfate-reducing bacterium that decomposes fatty acids. 1. Isolation of new sulfate-reducing bacteria enriched with acetate from saline environments. Description of Desulfobacter postgatei, gen. nov., sp. nov. Arch. Microbiol. 129:395-400.

Wiggins, B.A. and Alexander, M. 1986. Role of protozoa in microbial acclimation to low concentrations of organic chemicals. Abst. Ann. Mtg. Amer. Soc. Microbiol. p. 300.

Wilkinson, R.R., Kelso, G.L., and Hopkins, F.C. 1978. State of the art report: pesticide disposal research. EPA Report No. 600/2-78-183. Environmental Protection Agency, Cincinnati, OH.

Williams, D.E. and Wilder, D.G. 1971. Gasoline pollution of a ground-water reservoir--a case history. Ground Water. 9(6):50-56.

Willians, G.R., Cumins, E., Gardener, A.C., Palmier, M., and Rubidge, T. 1981. The growth of Psuedomonas putida in AVTUR aviation turbine fuel. J. Appl. Bacteriol. 50:551-557.

Williams, P.A. 1981a. Genetics of biodegradation. In Leisinger, T., Hutter, R., Cook, A.M., and Nuesch, J. FEMS Symposium No. 12. Microbial degradation of xenobiotics and recalcitrant compounds. 12:97.

Williams, R.J.P. 1981b. Phil. Trans. Roy. Soc. Lond. Ser. B. 57:294.

Williams, R.J.P. 1983. Dahlem Konferenzen on Changing Biogeochemical Cycles of Metals and Human Health, Berlin, F.R.G., March 20-25, 1983.

Williams, R.J. and Evans, W.C. 1973. Anaerobic metabolism of aerobic substrates by certain microorganisms. Transactions Biochem. Soc. (London). 1:186-187.

Wilson, B.H., Bledsoe, B.E., Armstrong, J.M., and Sammons, J.H. 1986. Biological fate of hydrocarbons at an aviation gasoline spill site. Petroleum Hydrocarbons and Organic Chemicals in Ground Water: Prevention, Detection and Restoration Conf. and Expo., Nov. 12-14, 1986, Houston, TX. Presented by NWWA and API.

Wilson, B.H. and Rees, J.F. 1985. Proc. biotransformation of gasoline hydrocarbons in methanogenic aquifer material. NWWA/API Conf. on Petroleum Hydrocarbons and Organic Chemicals in Ground Water, Nov. 13-15, Houston, TX. pp. 128-141.

Wilson, J.T., Leach, L.E., Henson, M., and Jones, J.N. 1986. In situ biorestoration as a ground water remediation technique. Ground Water Mon. Rev. 56-64.

Wilson, J.T. and McNabb, J.F. 1983. Biological transformation of organic pollutants in groundwater. EOS. 64:505-507.

Wilson, J.T., McNabb, J.F, Balkwill, D.L., and Ghoirse, W.C. 1983. Enumeration and characterization of bacteria indigenous to a shallow water table aquifer. Ground Water. 21:134-142.

Wilson, J.T., McNabb, J.F., Cochran, J.W., Wang, T.H., Tomson, M.B., and Bedient, P.B. 1985. Influence of microbial adaptation on the fate of organic pollutants in ground water. Environ. Toxicol. Chem. 4:721-726.

Wilson, J.T., McNabb, J.F., Cochran, J.W., Wang, T.H., Tomson, M.B., and Bedient, P.B. 1986. Influence of microbial adaptation on the fate of organic pollutants in ground water. Environ. Toxic. Chem. 4:721-726.

Wilson, J.T., McNabb, J.F., Wilson, B.H., and Noonan, M.J. 1983. Biotransformation of selected organic pollutants in ground water. Dev. Ind. Microbiol. 24:225-233.

Wilson, J.T., Noonan, M.J., and McNabb, J.F. 1983. In Ward, C.H. (Ed.), Proc., 1st Intl. Conf. on Ground Water Quality Research. John Wiley and Sons, Inc., New York, NY.

Wilson, J.T. and Ward, C.H. 1986. Opportunities for bioreclamation of aquifers contaminated with petroleum hydrocarbons. Dev. in Ind. Microbiol. 27:109-116.

Wilson, J.T. and Wilson, B.H. 1985. Biotransformation of trichloroethylene in soil. Appl. Environ. Microbiol. 49:242-243.

Winograd, I.J. and Robertson, F.N. 1982. Deep oxygenated ground water: anomaly or common occurrence? Science. 216:1227-1228.

Wiseman, A., Lim, T.-K., and Woods, L.F.J. 1978. Regulation of the biosynthesis of cytochrome P-450 in Brewer's yeast. Role of cyclic AMP. Biochim. Biophys. Acta. 544:615-623.

Wodzinski, R.S. and Coyle, J.E. 1974. Physical state of phenanthrene for utilization by bacteria. Appl. Microbiol. 27:1081-1084.

Wodzinski, R.S. and Johnson, M.J. 1968. Yields of bacterial cells from hydracarbons [sic]. Appl. Microbiol. 16:1886-1891.

Wood, J.M. 1983. Chem. Scr. 21:155.

Wood, J.M., Cheh, A., Dizikes, L.J., Ridley, W.P., Rackow, S., and Lakowicz, J.R. 1978. Mechanisms for the biomethylation of metals and metalloids. Fed. Proc. 37:16-21.

Wood, J.M. and Wang, Hong-Kang. 1983. Microbial resistance to heavy metals. Environ. Sci. Technol. 17:582A-590A.

Woodward, R.L. 1963. Review of the bactericidal effectiveness of silver. J. Amer. Water Works Assn. 55:881-888.

Woolson, E.A. 1977. Fate of arsenicals in different environmental substrates. Environ. Health Perspect. 19:73-81.

Yagi, O. and Sudo, R. 1980. Degradation of polychlorinated biphenyls by microorganisms. J. Water Poll. Control Fed. 52:1035-1043.

Yalkowshy, S.H., Valvani, S.C., and MacKay, D. 1983. Estimation of the aqueous solubility of some aromatic compounds. Residue Rev. 85:43-55.

Yang, J.T. and Bye, W.E. 1979. EPA Report No. EPA-570/9-79-017. Environmental Protection Agency, Washington, D.C.

Yaniga, P.M. and Smith, W. 1984. Aquifer restoration via accelerated in situ biodegradation of organic contaminants. Proc. NWWA/API Conf. on Petroleum Hydrocarbons and Organic Chemicals in Ground Water--Prevention, Detection and Restoration. Houston, TX. Nat. Water Works Assn., Worthington, OH. pp. 451-472.

Young, L.Y. 1984. Anaerobic degradation of aromatic compounds. In Gibson, D.T. (Ed.), Microbial Degradation of Organic Compounds. Marcel Dekker, Inc., New York, NY. pp. 487-523.

Young, L.Y. and Bossert, I. 1984. Degradation of chlorinated and non-chlorinated phenols and aromatic acids by anaerobic microbial communities. Abstr. Ann. Mtg. Amer. Soc. Microbiol. p. 217.

Zaidi, B.R., Stucki, G., and Alexander, M. 1986. Inoculation of lake water to promote biodegradation. Abstr. Ann. Mtg. Amer. Soc. Microbiol. p. 286.

Zajic, J.E. 1964. Chapter 2. Biochemical Reactions in hydrocarbon metabolism. Dev. Ind. Microbiol. 6:16-27.

Zajic, J.E. 1969. Microbial Biogeochemistry. Academic Press, New York, NY. 345 p.

Zajic, J.E. and Daugulis, A.J. 1975. Selective enrichment processes in resolving hydrocarbon pollution problems. In Bourquin, A.W., Ahearn, D.G., and Meyers, S. P. (Eds.), Proc. Impact of the Use of Microorganisms on the Aquatic Environment. EPA Report No. EPA 660-3-75-001. Environmental Protection Agency, Corvallis, OR. p. 169.

Zajic, J.E. and Gerson, D.F. 1977. Amer. Chem. Soc. Symposium on Oil Sands and Oil Shale. 22:195.

Zajic, J.E. and Panchal, C.J. 1977. Bioemulsifiers. CRC Crit. Rev. Microbiol. 5:39-66.

Zajic, J.E. and Supplisson, B. 1972. Emulsification and degradation of "Bunker C" fuel oil by microorganisms. Biotech. Bioeng. 14:331-343.

Zajic, J.E., Suplisson, B., and Volesky, B. 1974. Bacteria degradation and emulsification of no. 6 fuel oil. Environ. Sci. Technol. 8:664-668.

Zeikus, J.G. 1977. The biology of methanogenic bacteria. Bact. Rev. 41:514-541.

Zeikus, J.G. 1980. Chemical and fuel production by anaerobic bacteria. Ann. Rev. Microbiol. 34:423-464.

Zeyer, J., Kuhn, E.P., and Schwarzenback, R.P. 1986. Rapid microbial mineralization of toluene and 1,3-dimethylbenzene in the absence of molecular oxygen. Appl. Environ. Microbiol. 52:944-947.

Zilber, I.K., Rosenberg, E., and Gutnick, D. 1980. Incorporation of ^{32}P and growth of pseudomonad UP-2 on n-tetracosane. Appl. Environ. Microbiol. 40:1086-1093.

Zitrides, T.G. 1978. Pesticide Disposal Research and Development Symp., Sept. 6-7, 1978, Reston, VA. 23:1082-1089.

Zitrides, T. 1983. Biodecontamination of spill sites. Poll. Eng. 15:25-27.

ZoBell, C.E. 1946. Action of microorganisms on hydrocarbons. Bacteriol. Rev. 10:1-49.

ZoBell, C.E. 1969. Microbial modification of crude oil in the sea. In Proc. Joint Conf. on Prevention and Control of Oil Spills. American Petroleum Institute, Washington, D.C. pp. 317-326.

ZoBell, C.E. 1973. Microbial-facilitated degradation of oil: a prospectus. In Ahearn, D.G. and Meyers, S.P. (Eds.), The Microbial Degradation of Oil Pollutants, Workshop, Atlanta, GA, Dec. 4-6, 1972. Publ. No. LSU-SG-73-01. Louisiana State Univ., Center for Wetland Resources, Baton Rouge, LA. p. 3.

ZoBell, C.E. and Prokop, J.F. 1966. A. Alg. Mikrobiol. 6:143-162.

BIOREMEDIATION

OF PETROLEUM

CONTAMINATED SITES

Appendix

Appendix Table of Contents

Appendix List of Tables

Appendix List of Illustrations

APPENDIX SECTION 1

Introduction

1.1 BACKGROUND

1.1.1 Environmental Contamination

1.1.2 Groundwater Contamination

1.1.3 Pollution Legislation

Table A.1-1 lists the classes of compounds established by the U.S. Environmental Protection Agency as organic priority pollutants (Leisinger, 1983)

The organic priority pollutants have been detected in the groundwater of all 10 EPA regions (Plumb, 1985). Table A.1-2 presents some of the most common contaminants.

1.2 BIODEGRADATION AS A TREATMENT ALTERNATIVE

1.2.1 On-Site Biological Treatment Techniques

The biological process reactors available for water and wastewater treatment can be classified according to the nature of their biological growth (Sutton, 1987). Those in which the active biomass is suspended

Table A.1-1.   Organic Priority Pollutants According to the U.S. Environmental Protection Agency (Leisinger, 1983)

Chemical Class	Number of Compounds
Aliphatics	3
Halogenated aliphatics	31
Nitrosamines	3
Aromatics	14
Chloroaromatics (including TCDD)	16
Polychlorinated biphenyls (PCBs)	7
Nitroaromatics	7
Polynuclear aromatic hydrocarbons	16
Pesticides and metabolites (including DDT)	17

Table A.1-2.   Some of the Most Frequently Detected Groundwater Contaminants within Three Classes of Organic Priority Pollutants (Plumb, 1985)

	Detection Frequency (%)
Volatile Compounds	
Acetone	12.4
Acid Compounds	
Phenol	13.6
4-Methyl phenol	5.8
2-Methyl phenol	4.2
2,4-Dimethylphenol	1.9
Benzoic acid	1.3
Base/Neutral Compounds	
Naphthalene	4.1

Based upon a compilation of data from 183 disposal sites located across the U.S. All frequencies expressed in percent.

as free organisms or microbial aggregates can be regarded as suspended growth reactors, whereas, those in which growth occurs on or within a solid medium can be termed supported growth or fixed-film reactors.

1.2.1.1 Aerobic Biological Systems

Aerobic treatment systems include conventional activated sludge processes, as well as modifications, such as sequencing batch reactors, and aerobic attached growth biological processes, such as rotating biological contactors (RBCs) and trickling filters (Roberts, Koff, and

Karr, 1988). Aerobic processes are capable of reducing significantly a wide range of organic toxic and hazardous compounds; however, treatment is limited to dilute aqueous wastes (usually not exceeding 1 percent). Genetically engineered bacteria have been recently developed for effective biological treatment of specific hazardous wastes that are relatively uniform in composition. Such systems are typically used to treat aqueous wastes contaminated with low levels of nonhalogenated organic or certain halogenated organics. This treatment requires consistent, stable operating conditions.

1.2.1.1.1 Activated Sludge. There are many variations of the conventional activated sludge process, all of which use the same principles of unit operation (Roberts, Koff, and Karr, 1988). The first step in the process involves aeration in open tanks, in which the organic biodegradable matter in the waste is degraded by microorganisms in the presence of oxygen. The hydraulic detention time of this process is usually from 6 to 24 hr, depending upon the process mode. This is followed by a sludge-liquid separation step in a clarifier. The organisms multiply during the process. A zoogleal sludge is settled out and a portion of the organisms (Return Activated Sludge, or RAS) is recycled to the aeration basin, which allows growth of an acclimated population. The remaining sludge is wasted, while the clarified water is discharged in a manner appropriate to its quality. Organic loading rates can vary from 10 to 180 lb of BOD applied per 1000 cf, depending upon the mixed liquor suspended solids (MLSS) concentration, the food-to-microorganism (F/M) ratio, and oxygen supply. Variations of the conventional activated sludge system that utilize pure oxygen or oxygen-enriched air, instead of air, produce a more rapid breakdown of chemical solutes. Extended aeration involves longer detention times than conventional activated sludge and relies on a higher population of microorganisms. Contact stabilization involves only short contact of the aqueous wastes and suspended microbial solids, with subsequent settling of sludge and treatment of the sludge to eliminate the sorbed organics. Use of powdered activated carbon is also reported to have excellent pollutant removal capabilities for wastes that are difficult to treat.

Activated sludge has a great potential for treatment of hazardous waste, since it can be easily controlled (Shuckrow, Pajak, and Touhill, 1982). However, nonbiodegradable organics and metals can be adsorbed by the biomass, interfere with metabolism, and cause the generated sludge to be hazardous. Activated sludge processes can handle organic loadings as high as 10,000 ppm BOD, but are sensitive to shock loads (JRB and Associates, Inc., 1982).

1.2.1.1.2 Trickling Filters or Fixed-Film Systems. Trickling filters or fixed-film systems involve contact of the aqueous waste stream with microorganisms attached to some inert medium, such as rock or specially designed plastic material (Roberts, Koff, and Karr, 1988). The original trickling filter consisted of a bed of rocks over which the contaminated water was sprayed. Microbial deposits form slime layers on the rocks where metabolism of the solute organics occurs. Oxygen is provided with the air being introduced countercurrently to the wastewater flow. Present technology suggests, however, that gas-suspended biomass systems are better applicable to treating oily sludges than are fixed-film systems. Trickling filters are not as effective as activated sludge, but are less sensitive to shock loads and have lower energy costs (Lee and Ward, 1986). Their principal use is for secondary treatment or as a roughing filter to even out loading (JRB and Associates, Inc., 1982).

In fixed-film processes, the cell retention time is long compared with suspended growth processes (Stratton, 1981; Switzenbaum and Jewell, 1980). Fixed-film processes foster long cell retention and enhance growth of slow-growing microorganisms. Such populations are particularly advantageous when sorption is the main mechanism for the removal of a compound. One advantage of these processes is that they can provide cell concentrations of an order of magnitude higher than those found in suspended growth systems. A study of the partitioning of organic compounds into biomass indicates that efficient removal is possible only when the biomass concentration is large (Stratton, 1981; Matter-Muller, Gujer, Giger, and Strumm, 1980).

The subsurface environment is generally characterized by low substrate and nutrient concentrations and high specific surface area, which favor predominance of bacteria attached to solid surfaces in the form of biofilms (Bouwer and McCarty, 1984). Attached bacteria have an advantage over suspended bacteria, as they can remain near the source of fresh substrate and nutrients contained in groundwater that flows by them.

1.2.1.1.3 Biological Towers. Biological towers are a modification of the trickling filter (Roberts, Koff, and Karr, 1988). The medium (e.g., of polyvinyl chloride, polyethylene, polystyrene, or redwood) is stacked into towers, which typically reach 16 to 20 ft. The wastewater is sprayed across the top, and as it moves downward, air is drawn

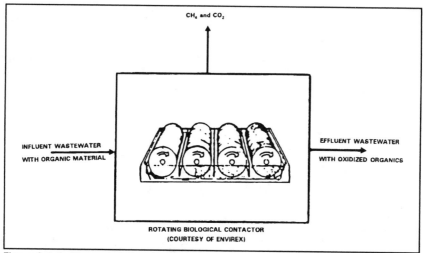

Figure A.1-1 Diagram of Rotating Biological Contactor (Roberts, Koff, and Karr, 1988)

upward through the tower. A slime layer of microorganisms forms on the medium and removes the organic contaminants as the water flows over the slime layer.

1.2.1.1.4 Rotating Biological Contactor (RBC). A rotating biological contactor consists of a series of rotating discs, connected by a shaft set in a basin or trough, as in Figure A.1-1 (Roberts, Koff, and Karr, 1988). The contaminated water passes through the basin where the microorganisms, attached to the discs, metabolize the organics present in the water. Approximately, 40 percent of the disc's surface area is submerged. This allows the slime layer to come alternately into contact with the contaminated water and the air where the oxygen is provided to the microorganisms. These units are compact, and they can handle large flow variations and high organic shock loads. They do not require the use of aeration equipment. Due to the varied composition of oily sludges and high concentrations of solids, oils, and heavy metals, the applicability of the RBC to this problem material is questionable.

1.2.1.1.5 Lagoons and Waste Stabilization Ponds. Lagoons and waste stabilization ponds are similar to activated sludge processes, except for the biomass recycle (Wilkinson, Kelso, and Hopkins, 1978). Lagoons provide dilution and buffering of load fluctuations; but they require more land, and the operational controls are less flexible

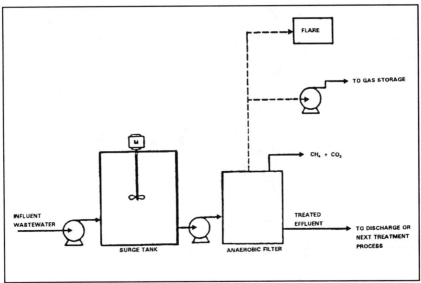

Figure A.1-2 Diagram of Typical Anaerobic Filter System (Lee and Ward, 1986)

(Shuckrow, Pajak, and Touhill, 1982). Oxygen can be supplied by aerating the lagoons to speed up the degradation (Johnson, 1978). Anaerobic or facultative lagoons are also available, with their ease of operation and low costs, but these will produce a lower quality effluent (JRB and Associates, Inc., 1982; Johnson, 1978). They utilize anaerobic degradation pathways, which may be more efficient for the removal of some compounds, and sludge production is minimized.

Waste stabilization ponds are principally a polishing technique useful for low organic wastewaters (Johnson, 1978). Since natural biodegradation processes are employed, requirements for energy and chemical additions are low; however, large land areas are needed.

1.2.1.1.6 Fluidized-Bed Reactors. Particles, such as sand or coal, are fluidized by the action of the aeration gas stream and the wastewater stream and are colonized by a dense growth of microorganisms, which gives rapid treatment (McCarty, Rittmann, and Bouwer, 1984).

1.2.1.2 Anaerobic Digesters

Anaerobic biological digester systems promote the reduction of organic matter to methane and carbon dioxide in an oxygen-free environment (Lee and Ward, 1986).   The most common anaerobic attached growth treatment process is the anaerobic filter. This process consists of a column filled with solid media. A number of proprietary anaerobic biotechnology processes are on the market, each with distinct features, but all the processes involve the fundamental anaerobic bacterial conversion to methane.   The digester gas can be flared or fired in boilers, gas turbines, or reciprocating engines with or without the prior removal of sulfurous gases. A typical anaerobic filter system is depicted in Figure A.1-2 (Lee and Ward, 1986).

These systems are used to treat aqueous wastes with low to moderate levels of organics.   Anaerobic digestion can dechlorinate certain halogenated organics better than aerobic treatment. Stable, consistent operating conditions must be maintained.   Anaerobic degradation can take place in native soils; however, when used as a controlled treatment process, an air-tight reactor is required. Hazardous organic substances that have been found to be amenable to anaerobic treatment include acetaldehyde, acetic anhydride, acetone, acrylic acid, aniline, benzoic acid, butanol, creosol, ethyl acrylate, MEK, phenol, and vinyl acetate.

APPENDIX SECTION 2

Organic Compounds
in Refined Fuels and Fuel Oils

2.1 CHEMICAL COMPOSITION OF FUEL OILS

2.1.1 Naptha

2.1.2 Kerosene

2.1.3 Fuel Oil and Diesel #2

Table A.2-1 lists organic compounds found in diesel fuel #2, as reported by Clewell, 1981.

Table A.2-1. Composition of Diesel Fuel #2 (Clewell, 1981)

Component	Concentration (% Volume)	Component	Concentration (% Volume)
C_{10} paraffins	0.9	C_{15} paraffins	7.4
C_{10} cycloparaffins	0.6	C_{15} cycloparaffins	5.5
C_{10} aromatics	0.4	C_{15} aromatics	3.2
C_{11} paraffins	2.3	C_{16} paraffins	5.8
C_{11} cycloparaffins	1.7	C_{16} cycloparaffins	4.4
C_{11} aromatics	1.0	C_{16} aromatics	2.5
C_{12} paraffins	3.8	C_{17} paraffins	5.5
C_{12} cycloparaffins	2.8	C_{17} cycloparaffins	4.1
C_{12} aromatics	1.6	C_{17} aromatics	2.4
C_{13} paraffins	6.4	C_{18} paraffins	4.3

Table A.2-1. continued

Component	Concentration (% Volume)	Component	Concentration (% Volume)
C_{13} cycloparaffins	4.8	C_{18} cycloparaffins	3.2
C_{13} aromatics	2.8	C_{18} aromatics	1.8
C_{14} paraffins	8.8	C_{19} paraffins	0.7
C_{14} cycloparaffins	6.6	C_{19} cycloparaffins	0.6
C_{14} aromatics	3.8	C_{19} aromatics	0.3

2.1.4 Gasoline

Tables A.2-2, A.2-3, and A.2-4 list the organic compounds and trace elements found in gasoline, as reported by different references.

Table A.2-2. Composition of Various Gasolines (Ghassemi, Panahloo, and Quinlivan, 1984b)

Paraffins

Propane	Dimethyl pentanes
Isobutane ($1C_4$)	Methyl hexanes
n-Butane (nC_4)	Trimethyl pentanes
Isopentane ($1C_5$)	Normal heptane
n-Pentane ($1C_5$)	Dimethyl hexanes
Dimethyl butanes (C_6)	Methylethyl pentanes
Methyl pentanes (C_6)	Dimethyl hexanes
n-Hexane	Trimethyl hexanes
n-Octane	

Naphthenes

Olefins

Methylcyclopentane	Methylbutene
Cyclohexane	Pentene
Methylcyclohexane	Methylpentene
Other cyclic saturates	Other olefins

Aromatics

Benzene	Propylbenzene
Toluene	Methylethylbenzenes
Ethylbenzene	Trimethylbenzene
Xylenes	Other aromatics

Table A.2-3.   Components of Gasoline (Jamison, Raymond, and Hudson, 1976)

Component	Component
n-Propane	2,5-Dimethylhexane
n-Butane	2,4-Dimethylhexane
n-Pentane	2,3-Dimethylhexane
n-Hexane	3,4-Dimethylhexane
n-Heptane	2,2-Dimethylhexane
n-Octane	2,2-Dimethylheptane
n-cis-Butene-2	1,1-Dimethylcyclopentane
n-Pentane-2	1,2- and 1,3-Dimethylcyclopentane
2,3-Dimethylbutene-1	1,3- and 1,4-Dimethylcyclohexane
Olefins C_4	1,2-Dimethylcyclohexane
Olefins C_5	2,2,3-Trimethylbutane
Olefins C_6	2,2,4-Trimethylpentane
Isobutane	2,2,3-Trimethylpentane
Cyclopentane	2,3,4-Trimethylpentane
Cyclohexane	2,3,3-Trimethylpentane
Methylcyclopentane	2,2,5-Trimethylpentane
Methylcyclohexane	1,2,4-Trimethylcyclopentane
2-Methylbutane	Ethylpentane
2-Methylpentane	Ethylcyclopentane
3-Methylpentane	Ethylcyclohexane
2-Methylhexane	Benzene
3-Methylhexane	Ethylbenzene
2-Methylheptane	Toluene
3-Methylheptane	o-Xylene
4-Methylheptane	m-Xylene
2,2-Dimethylbutane	p-Xylene
2,3-Dimethylbutane	
2,2-Dimethylpentane	
2,4-Dimethylpentane	
3,3-Dimethylpentane	
2,3-Dimethylpentane	

Table A.2-4.   Composition of Gasoline (State of California, 1987)

Compound	Number of Carbons	Concentration (Weight Percent)
Straight Chain Alkanes		
Propane	3	0.01 to 0.14
n-Butane	4	3.93 to 4.70
n-Pentane	5	5.75 to 10.92
n-Hexane	6	0.24 to 3.50
n-Heptane	7	0.31 to 1.96
n-Octane	8	0.36 to 1.43
n-Nonane	9	0.07 to 0.83

Table A.2-4.   continued

Compound	Number of Carbons	Concentration (Weight Percent)
n-Decane	10	0.04 to 0.50
n-Undecane	11	0.05 to 0.22
n-Dodecane	12	0.04 to 0.09
Branched Alkanes		
Isobutane	4	0.12 to 0.37
2,2-Dimethylbutane	6	0.17 to 0.84
2,3-Dimethylbutane	6	0.59 to 1.55
2,2,3-Trimethylbutane	7	0.01 to 0.04
Neopentane	5	0.02 to 0.05
Isopentane	5	6.07 to 10.17
2-Methylpentane	6	2.91 to 3.85
3-Methylpentane	6	2.4 (vol)
2,4-Dimethylpentane	7	0.23 to 1.71
2,3-Dimethylpentane	7	0.32 to 4.17
3,3-Dimethylpentane	7	0.02 to 0.03
2,2,3-Trimethylpentane	8	0.09 to 0.23
2,2,4-Trimethylpentane	8	0.32 to 4.58
2,3,3-Trimethylpentane	8	0.05 to 2.28
2,3,4-Trimethylpentane	8	0.11 to 2.80
2,4-Dimethyl-3-ethyl-pentane	9	0.03 to 0.07
2-Methylhexane	7	0.36 to 1.48
3-Methylhexane	7	0.30 to 1.77
2,4-Dimethylhexane	8	0.34 to 0.82
2,5-Dimethylhexane	8	0.24 to 0.52
3,4-Dimethylhexane	8	0.16 to 0.37
3-Ethylhexane	8	0.01
2-Methyl-3-ethylhexane	9	0.04 to 0.13
2,2,4-Trimethylhexane	9	0.11 to 0.18
2,2,5-Trimethylhexane	9	0.17 to 5.89
2,3,3-Trimethylhexane	9	0.05 to 0.12
2,3,5-Trimethylhexane	9	0.05 to 1.09
2,4,4-Trimethylhexane	9	0.02 to 0.16
2-Methylheptane	8	0.48 to 1.05
3-Methylheptane	8	0.63 to 1.54
4-Methylheptane	8	0.22 to 0.52
2,2-Dimethylheptane	9	0.01 to 0.08
2,3-Dimethylheptane	9	0.13 to 0.51
2,6-Dimethylheptane	9	0.07 to 0.23
3,3-Dimethylheptane	9	0.01 to 0.08
3,4-Dimethylheptane	9	0.07 to 0.33
2,2,4-Trimethylheptane	10	0.12 to 1.70
3,3,5-Trimethylheptane	10	0.02 to 0.06
3-Ethylheptane	10	0.02 to 0.16

Table A.2-4.   continued

Compound	Number of Carbons	Concentration (Weight Percent)
2-Methyloctane	9	0.14 to 0.62
3-Methyloctane	9	0.34 to 0.85
4-Methyloctane	9	0.11 to 0.55
2,6-Dimethyloctane	10	0.06 to 0.12
2-Methylnonane	10	0.06 to 0.41
3-Methylnonane	10	0.06 to 0.32
4-Methylnonane	10	0.04 to 0.26

Cycloalkanes

Compound	Number of Carbons	Concentration (Weight Percent)
Cyclopentane	5	0.19 to 0.58
Methylcyclopentane	6	Not quantified
1-Methyl-cis-2-ethyl-cyclopentane	8	0.06 to 0.11
1-Methyl-trans-3-ethyl-cyclopentane	8	0.06 to 0.12
1-cis-2-dimethylcyclo-pentane	7	0.07 to 0.13
1-trans-2-dimethylcyclo-pentane	7	0.06 to 0.20
1,1,2-trimethylcyclo-pentane	8	0.06 to 0.11
1-trans-2-cis-3-tri-methylcyclopentane	8	0.01 to 0.25
1-trans-2-cis-4-tri-methylcyclopentane	8	0.03 to 0.16
Ethylcyclopentane	7	0.14 to 0.21
n-Propylcyclopentane	8	0.01 to 0.06
Isopropylcyclopentane	8	0.01 to 0.02
1-trans-3-dimethyl-cyclohexane	8	0.05 to 0.12
Ethylcyclohexane	8	0.17 to 0.42

Straight Chain Alkenes

Compound	Number of Carbons	Concentration (Weight Percent)
cis-2-Butene	4	0.13 to 0.17
trans-2-Butene	4	0.16 to 0.20
Pentene-1	5	0.33 to 0.45
cis-2-Pentene	5	0.43 to 0.67
trans-2-Pentene	5	0.52 to 0.90
cis-2-Hexene	6	0.15 to 0.24
trans-2-Hexene	6	0.18 to 0.36
cis-3-Hexene	6	0.11 to 0.13
trans-3-Hexene	6	0.12 to 0.15
cis-3-Heptene	7	0.14 to 0.17
trans-2-Heptene	7	0.06 to 0.10

Branched Alkenes

Compound	Number of Carbons	Concentration (Weight Percent)
2-Methyl-1-butene	5	0.22 to 0.66
3-Methyl-1-butene	5	0.08 to 0.12
2-Methyl-2-butene	5	0.96 to 1.28
2,3-Dimethyl-1-butene	6	0.08 to 0.10

Table A.2-4.   continued

Compound	Number of Carbons	Concentration (Weight Percent)
2-Methyl-1-pentene	6	0.20 to 0.22
2,3-Dimethyl-1-pentene	7	0.01 to 0.02
2,4-Dimethyl-1-pentene	7	0.02 to 0.03
4,4-Dimethyl-1-pentene	7	0.6 (vol)
2-Methyl-2-pentene	6	0.27 to 0.32
3-Methyl-cis-2-pentene	6	0.35 to 0.45
3-Methyl-trans-2-pentene	6	0.32 to 0.44
4-Methyl-cis-2-pentene	6	0.04 to 0.05
4-Methyl-trans-2-pentene	6	0.08 to 0.30
4,4-Dimethyl-cis-2-pentene	7	0.02
4,4-Dimethyl-trans-2-pentene	7	Not quantified
3-Ethyl-2-pentene	7	0.03 to 0.04
Cycloalkenes		
Cyclopentene	5	0.12 to 0.18
3-Methylcyclopentene	6	0.03 to 0.08
Cyclohexene	6	0.03
Alkyl Benzenes		
Benzene	6	0.12 to 3.50
Toluene	7	2.73 to 21.80
o-Xylene	8	0.68 to 2.86
m-Xylene	8	1.77 to 3.87
p-Xylene	8	0.77 to 1.58
1-Methyl-4-ethylbenzene	9	0.18 to 1.00
1-Methyl-2-ethylbenzene	9	0.19 to 0.56
1-Methyl-3-ethylbenzene	9	0.31 to 2.86
1-Methyl-2-n-propyl-benzene	10	0.01 to 0.17
1-Methyl-3-n-propyl-benzene	10	0.08 to 0.56
1-Methyl-3-isopropyl-benzene	10	0.01 to 0.12
1-Methyl-3-t-butyl-benzene	11	0.03 to 0.11
1-Methyl-4-t-butyl-benzene	11	0.04 to 0.13
1,2-Dimethyl-3-ethyl-benzene	10	0.02 to 0.19
1,2-Dimethyl-4-ethyl-benzene	10	0.50 to 0.73
1,3-Dimethyl-2-ethyl-benzene	10	0.21 to 0.59
1,3-Dimethyl-4-ethyl-benzene	10	0.03 to 0.44
1,3-Dimethyl-5-ethyl-benzene	10	0.11 to 0.42
1,3-Dimethyl-5-t-butyl-benzene	12	0.02 to 0.16
1,4-Dimethyl-2-ethyl-benzene	10	0.05 to 0.36
1,2,3-Trimethylbenzene	9	0.21 to 0.48
1,2,4-Trimethylbenzene	9	0.66 to 3.30
1,3,5-Trimethylbenzene	9	0.13 to 1.15

Table A.2-4.   continued

Compound	Number of Carbons	Concentration (Weight Percent)
1,2,3,4-Tetramethyl-benzene	10	0.02 to 0.19
1,2,3,5-Tetramethyl-benzene	10	0.14 to 1.06
1,2,4,5-Tetramethyl-benzene	10	0.05 to 0.67
Ethylbenzene	8	0.36 to 2.86
1,2-Diethylbenzene	10	0.57
1,3-Diethylbenzene	10	0.05 to 0.38
n-Propylbenzene	9	0.08 to 0.72
Isopropylbenzene	9	<0.01 to 0.23
n-Butylbenzene	10	0.04 to 0.44
Isobutylbenzene	10	0.01 to 0.08
sec-Butylbenzene	10	0.01 to 0.13
t-Butylbenzene	10	0.12
n-Pentylbenzene	11	0.01 to 0.14
Isopentylbenzene	11	0.07 to 0.17
Indan	9	0.25 to 0.34
1-Methylindan	10	0.04 to 0.17
2-Methylindan	10	0.02 to 0.10
4-Methylindan	10	0.01 to 0.16
5-Methylindan	10	0.09 to 0.30
Tetralin	10	0.01 to 0.14

Polynuclear Aromatic Hydrocarbons

Naphthalene	10	0.09 to 0.49
Pyrene	16	Not quantified
Benz(a)anthracene	18	Not quantified
Benz(a)pyrene	20	0.19 to 2.8 mg/kg
Benzo(e)pyrene	20	Not quantified
Benzo(g,h,i)perylene	21	Not quantified

Elements

Bromine	80 to 345 ug/g
Cadmium	0.01 to 0.07 ug/g
Chlorine	80 to 300 ug/g
Lead	530 to 1120 ug/g
Sodium	<0.6 to 1.4 ug/g
Sulfur	0.10 to 0.15(ASTM)
Vanadium	<0.02 to 0.001 ug/g

Additives

Ethylene dibromide	0.7 to 177.2 ppm
Ethylene dichloride	150 to 300 ppm
Tetramethyl lead	
Tetraethyl lead	

2.1.5 JP-5

Tables A.2-5, A.2-6, and A.2-7 list the organic compounds and trace elements found in JP-5, as reported by different references.

Table A.2-5.   Selected Compound Types Occurring in JP-5 (Varga, Lieberman, and Avella, 1985)

Aromatic	Partial Saturation	Saturated
Benzene		Cyclohexane
Indene	Indane (Indian)	Hydrindane (Hydroindane)
Naphthalene	Tetralin (Tetrahydronaphthalene)	Decalin (Decahydronaphthalene)
Acenaphthalene	Acenaphthene	Perhydroacenaphthalene
Phenanthrene	Tetrahydrophenanthrene	Perhydrophenanthrene

Table A.2-6.   Major Components of JP-5 (Smith, Harper, and Jaber, 1981)

Fuel Component	Concentration (Weight Percent)
n-Octane	0.12
1,3,5-Trimethylcyclohexane	0.09
1,1,3-Trimethylcyclohexane	0.05
m-Xylene	0.13
3-Methyloctane	0.07
2,4,6-Trimethylheptane	0.09
o-Xylene	0.09
n-Nonane	0.38
1,2,4-Trimethylbenzene	0.37
n-Decane	1.79
n-Butylcyclohexane	0.90
1,3-Diethylbenzene	0.61
1,4-Diethylbenzene	0.77
4-Methyldecane	0.78
2-Methyldecane	0.61
1-Ethylpropylbenzene	1.16
n-Undecane	3.95
2,6-Dimethyldecane	0.72
1,2,3,4-Tetramethylbenzene	1.48
Naphthalene	0.57
2-Methylundecane	1.39

Table A.2-6.   continued

Fuel Component	Concentration (Weight Percent)
n-Dodecane	3.94
2,6-Dimethylundecane	2.00
1,2,4-Triethylbenzene	0.72
2-Methylnaphthalene	0.90
1-Methylnaphthalene	1.44
1-Tridecene	0.45
Phenylcyclohexane	0.82
n-Tridecane	3.45
1-t-Butyl-3,4,5-trimethylbenzene	0.24
n-Heptylcyclohexane	0.99
n-Heptylbenzene	0.27
Biphenyl	0.70
1-Ethylnaphthalene	0.32
2,6-Dimethylnaphthalene	1.12
n-Tetradecane	2.72
2,3-Dimethylnaphthalene	0.46
n-Octylbenzene	0.78
n-Pentadecane	1.67
n-Hexadecane	1.07
n-Heptadecane	0.12

Table A.2-7.   Trace Elements in Shale-Derived JP-5 (Ghassemi, Panahloo, and Quinlivan, 1984a)

Element	ppm	Element	ppm	Element	ppm
Al	0.048	Cu	<0.02	Si	< =10
Sb	<3	Fe	<0.01	Ag	<0.02
As	<0.5	Pb	<0.06	Na	0.14
Be	<0.01	Mg	<5.3	Sr	< =0.94
Cd	<0.02	Mn	<0.02	Tl	<6
Ca	< =0.6	Hg	<2	Sn	< =0.93
Cl	<2	Mo	< =0.03	Ti	<0.4
Cr	< =0.094	Ni	<3.9	V	< =0.0008
Co	< =0.04	Se	<0.5	Zn	< =0.02

2.1.6 JP-4

Tables A.2-8, A.2-9, and A.2-10 list the organic compounds and trace elements found in JP-4, as reported by different references.

Table A.2-8.   Composition of JP-4 (Clewell, 1981)

Component	Concentration (% Volume)	Component	Concentration (% Volume)
C_5 hydrocarbons	3.9	Napthalene	0.2
C_6 paraffins	8.1	C_{11} paraffins	4.8
C_6 cycloparaffins	2.1	C_{11} cycloparaffins	2.5
Benzene	0.3	Dicycloparaffins	3.4
C_7 paraffins	9.4	C_{11} aromatics	1.1
C_7 cycloparaffins	7.1	C_{11} napthalenes	0.2
Toluene	0.7	C_{12} paraffins	2.8
C_8 paraffins	10.1	C_{12} cycloparaffins	1.2
C_8 cycloparaffins	7.4	C_{12} aromatics	0.5
C_8 aromatics	1.6	C_{12} napthalenes	0.2
C_9 paraffins	9.1	C_{13} paraffins	1.1
C_9 cycloparaffins	4.3	C_{13} cycloparaffins	0.4
C_9 aromatics	2.4	C_{13} aromatics	0.1
C_{10} paraffins	7.3	C_{14} hydrocarbons	0.2
C_{10} cycloparaffins	3.7	C_{15} hydrocarbons	0.1
C_{10} aromatics	1.8	Tricycloparaffins	1.8
		Residual hydrocarbons	0.1

Table A.2-9.   Major Components of JP-4 (Smith, Harper, and Jaber, 1981)

Fuel Component	Concentration (Weight Percent)
n-Butane	0.12
Isobutane	0.66
n-Pentane	1.06
2,2-Dimethylbutane	0.10
2-Methylpentane	1.28
3-Methylpentane	0.89
n-Hexane	2.21
Methylcyclopentane	1.16
2,2-Dimethylpentane	0.25
Benzene	0.50
Cyclohexane	1.24
2-Methylhexane	2.35
3-Methylhexane	1.97

Table A.2-9. continued

Fuel Component	Concentration (Weight Percent)
trans-1,3-Dimethylcyclopentane	0.36
cis-1,3-Dimethylcyclopentane	0.34
cis-1,2-Dimethylcyclopentane	0.54
n-Heptane	3.67
Methylcyclohexane	2.27
2,2,3,3-Tetramethylbutane	0.24
Ethylcyclopentane	0.26
2,5-Dimethylhexane	0.37
2,4-Dimethylhexane	0.58
1,2,4-Trimethylcyclopentane	0.25
3,3-Dimethylhexane	0.26
1,2,3-Trimethylcyclopentane	0.25
Toluene	1.33
2,2-Dimethylhexane	0.71
2-Methylheptane	2.70
4-Methylheptane	0.92
cis-1,3-Dimethylcyclohexane	0.42
3-Methylheptane	3.04
1-Methyl-3-ethylcyclohexane	0.17
1-Methyl-2-ethylcyclohexane	0.39
Dimethylcyclohexane	0.43
n-Octane	3.80
1,3,5-Trimethylcyclohexane	0.99
1,1,3-Trimethylcyclohexane	0.48
2,5-Dimethylheptane	0.52
Ethylbenzene	0.37
m-Xylene	0.96
p-Xylene	0.35
3,4-Dimethylheptane	0.43
4-Ethylheptane	0.18
4-Methyloctane	0.86
2-Methyloctane	0.88
3-Methyloctane	0.79
o-Xylene	1.01
1-Methyl-4-ethylcyclohexane	0.48
n-Nonane	2.25
Isopropylbenzene	0.30
n-Propylbenzene	0.71
1-Methyl-3-ethylbenzene	0.49
1-Methyl-4-ethylbenzene	0.43
1,3,5-Trimethylbenzene	0.42
1-Methyl-2-ethylbenzene	0.23

Table A.2-9.   continued

Fuel Component	Concentration (Weight Percent)
1,2,4-Trimethylbenzene	1.01
n-Decane	2.16
n-Butylcyclohexane	0.70
1,3-Diethylbenzene	0.46
1-Methyl-4-propylbenzene	0.40
1,3-Dimethyl-5-ethylbenzene	0.61
1-Methyl-2-i-propylbenzene	0.29
1,4-Dimethyl-2-ethylbenzene	0.70
1,2-Dimethyl-4-ethylbenzene	0.77
n-Undecane	2.32
1,2,3,4-Tetramethylbenzene	0.75
Naphthalene	0.50
2-Methylundecane	0.64
n-Dodecane	2.00
2,6-Dimethylundecane	0.71
2-Methylnaphthalene	0.56
1-Methylnaphthalene	0.78
n-Tridecane	1.52
2,6-Dimethylnaphthalene	0.25
n-Tetradecane	0.73

Table A.2-10. Trace Elements in Petroleum-Based JP-4 (Ghassemi, Panahloo, and Quinlivan, 1984a)

Element	ppm	Element	ppm	Element	ppm
Al	NA	Cu	<0.05	Si	NA
Sb	<0.5	Fe	<0.05	Ag	NA
As	0.5	Pb	0.09	Na	NA
Be	NA	Mg	NA	Sr	NA
Cd	<0.03	Mn	NA	Th	NA
Ca	NA	Hg	<1	Sn	NA
Cl	NA	Mo	NA	Ti	NA
Cr	<0.05	Ni	<0.05	V	<0.05
Co	NA	Se	<0.03	Zn	<0.05

2.2 FACTORS AFFECTING CONTAMINATION AND BIODEGRADATION

The fate of toxic pollutants in the environment is determined by a variety of chemical, physical, biological, and environmental processes that interact in a complex manner (Pfaender, Shimp, Palumbo, and Bartholomew, 1985). The factors that appear to affect the overall chemical movement and availability for biodegradation in soil are 1) adsorption, 2) physical properties of the soil, and 3) climatic conditions (Panchal and Zajic, 1978). The four principal means for chemical transport within soils are 1) downward flowing water, 2) upward moving water, 3) diffusion in soil water, and 4) diffusion in the air space of the soil.

A common technique for measuring the biodegradability of an organic compound is calculation of the BOD/COD ratio. This ratio is an indication of the amount of degradation that occurs (BOD), relative to the amount of material available to be degraded (COD). In Table A.2-11 there are relative biodegradabilities by adapted sludge cultures of various substances in terms of a BOD/COD ratio, after five days of incubation (Environmental Protection Agency, 1985b). A higher ratio represents a higher relative biodegradability. As the ratio approaches zero, the compound becomes less degradable.

2.2.1 Physical, Chemical, Biological, and Environmental Factors

2.2.1.1 Solubility

Organic compounds differ widely in their solubility, from infinitely miscible polar compounds, such as methanol, to extremely low solubility nonpolar compounds, such as high molecular weight polynuclear aromatic hydrocarbons (Horvath, 1982). Many synthetic chemicals have low water solubilities (Stucki and Alexander, 1987). The availability of a compound to an organism will dictate its biodegradability (Environmental Protection Agency, 1985b). Compounds with greater aqueous solubilities are generally more available to degradative enzymes. An example is cis-1,2-dichloroethylene, which is preferentially degraded relative to trans-1,2-dichloroethylene. This is probably due to "cis" being more polar than "trans" and, therefore, more water soluble. However, the rate of solubilization may not be the sole factor determining the degradation of lipophilic compounds (Thomas and Alexander, 1983). Surfactants can increase the solubility and, thus, the degradability of compounds.   Table A.2-12 lists a number of common organic contaminants and relates their biodegradability to their solubility in water (Brubaker and O'Neill, 1982).

The main solute components of seawater are known to be inorganic salts (mainly NaCl), which are 35 percent of the total concentration. Thus, the seawater solubility of organic chemicals will be lower than distilled water solubility due to salting-out effects (Hashimoto, Tokura, Kishi, and Strachan, 1984).   See Table A.2-13 for the solubilities of chemicals in seawater.

Table A.2-11. BOD$_5$/COD Ratios for Various Organic Compounds (Environmental Protection Agency, 1985b)

Compound	Ratio
Relatively Undegradable	
Heptane	0
Hexane	0
o-Xylene	<0.008
m-Xylene	<0.008
Ethylbenzene	<0.009
Moderately Degradable	
Gasolines (various)	0.02
Nonanol	>0.033
Undecanol	<0.04
Dodecanol	0.097
p-Xylene	<0.11
Toluene	<0.12
Jet Fuels (various)	0.15
Kerosene	0.15
Relatively Degradable	
Naphthalene (molten)	<0.20
Hexanol	0.20
Benzene	<0.39
p-Xylene	<0.11
Toluene	<0.12
Jet Fuels (various)	0.15
Kerosene	0.15
Benzaldehyde	0.62
Phenol	0.81
Benzoic acid	0.84

Table A.2-12. Solubility and Biodegradability of Some Common Organic Contaminants (Brubaker and O'Neill, 1982)

Contaminant	Solubility in Water[a]	Biodegradability
Acetone	Miscible	+ +
Aniline	35 g/l	+ +
Anthracene	1 mg/l	+
Benzene	320 mg/l	+ +
o-Cresol	31 g/l	+ +
Isopropanol	Miscible	+ +
Methanol	Miscible	+ +
Methylene Chloride	20 g/l	(?)
Methylethylketone	370 g/l	+ +
Naphthalene	29 ug/ml[b]	nr
Phenol	82 g/l	+
Pyrene	0.2 mg/l	+
Toluene	470 mg/l	+ +
Benzo(a)pyrene[c]	3.8 g/l	nr

+ + = readily biodegradable
+ = slow biodegradability
? = materials of uncertain biodegradability
nr = not reported
a = The actual solubility may be influenced greatly by other chemicals in the water.
b = (Thomas, Yordy, Amador, and Alexander, 1986)
c = (Mackay and Shiu, 1977)

Table A.2-13. Solubilities of Chemicals in Pacific Seawater at 20°C (Hashimoto, Tokura, Kishi, and Strachan, 1984)

Chemical	Solubility
Anthracene	1.18×10^{-7}
Pyrene	3.22×10^{-7}
Phenanthrene	4.15×10^{-6}
Biphenyl	2.39×10^{-5}
Naphthalene	1.34×10^{-4}
p-Nitrotoluene	1.83×10^{-3}
p-Toluidine	4.89×10^{-2}
o-Nitrophenol	8.34×10^{-3}
m-Nitrophenol	6.49×10^{-2}
p-Nitrophenol	7.76×10^{-2}
Phenol	1.35

The interaction between the groundwater and an organic contaminant depends upon the solubility of the substance and the organic carbon content already present in the aquifer (Mackay, Roberts, and Cherry, 1985). Hydrophobic compounds that do not dissolve well in water and are attracted to solids will have a more retarded flow velocity than the water itself. The density and viscosity of the material will determine if it will float on the water table, mix with groundwater, or sink. Slightly soluble compounds, such as benzene, will develop a plume in the saturated zone. Plumes containing different compounds may overlap. More information is needed about the rates and products of biotransformation in such superimposed plumes and the effects of complex contaminant distributions on the activities of microorganisms.

Organic compounds are only rarely found in groundwater at concentrations approaching their solubility limits (Mackay, Roberts, and Cherry, 1985). The observed concentrations are usually more than a factor of 10 lower than the solubility, presumably because of the diffusional limitations of dissolution and the dilution of the dissolved organic contaminants by dispersion. Therefore, the volume of groundwater contaminated by an organic liquid phase could be much larger than that calculated by assuming dissolution to the solubility limit.

Microorganisms use various mechanisms to metabolize organic substrates present at concentrations that exceed their water solubility (Thomas, Yordy, Amador, and Alexander, 1986). The physical state of the insoluble phase of a compound, whether it is a liquid or a solid, may affect its degradation. Liquid hydrocarbons can be taken up and incorporated into the cell membrane (Johnson, 1964); whereas, the mechanism of utilization of solid substrates is not fully understood. Polycyclic aromatic hydrocarbons might be used only in the dissolved state. Growth of pure cultures of bacteria on naphthalene, phenanthrene, and anthracene are faster on the solid substrates having the highest water solubilities (Wodzinski and Johnson, 1968). Rates of dissolution of naphthalene have been found to be directly related to its surface area (Thomas, Yordy, Amador, and Alexander, 1986).

Many microorganisms may excrete emulsifiers that increase the surface area of the substrate (Thomas, Yordy, Amador, and Alexander, 1986), or they may modify their cell surface to increase its affinity for hydrophobic substrates and, thus, facilitate their absorption (Kappeli, Walther, Mueller, and Fiechter, 1984). It has been suggested that microbial degradation of the insoluble phase of crystalline hydrocarbons is difficult because of the large amount of energy needed to disperse the solid (Zilber, Rosenberg, and Gutnick, 1980). If an organism cannot use the insoluble form of a chemical, it may be expected that the organism will first metabolize that portion of a chemical that is in solution and that the subsequent rate of transformation of the compound will be limited by the rate of dissolution. It has been found that increasing the surface area of hexadecane increases the microbial destruction of the alkane (Fogel, Lancione, Sewall, and Boethling, 1985). Emulsification can be employed for this purpose (Liu, 1980), and some organisms produce their own emulsifiers (Thomas, Yordy, Amador, and Alexander, 1986). Some microbes may even be able to utilize an insoluble substrate. Such an example is a marine pseudomonad, which was found to grow on bound and free n-tetracosane (Zilber, Rosenberg, and Gutnick, 1980).

While growth of bacteria appears to be limited by the rate of dissolution of a hydrocarbon, exponential growth does not always continue in parallel with the available material (Stucki and Alexander, 1987). Growth of strains of Flavobacterium and Beijerinckia in media containing 84 uM phenanthrene began to decline at densities of about 4 x 10^6/ml. The dissolution rate should have allowed exponential growth to a fivefold higher cell density. It is not clear why exponential growth ended so soon.

Many organic solutes with a low water solubility preferentially leave dilute water solutions and concentrate primarily in lipids of animals and in organic material of soil and sediments (Karickhoff, Brown, and Scott, 1979). This action is proportional to the partitioning between octanol-1 and water. The octanol-water partition coefficient (P) is defined as

$$P = C_o/C_w$$

where C_o and C_w are the concentration of the solute in n-octanol and water. P values measured in the laboratory can be used to predict the environmental behavior of organic pollutants (Mallon and Harrison, 1984). The hydrophobic nature of a pollutant, as measured by the octanol/water partition coefficient or by the dielectric constant, is important in predicting its flow through clay soils (Green, Lee, and Jones, 1981).

The subsurface transport of immiscible organic liquids is governed by a different set of factors than those for dissolved contaminants (Mackay, Roberts, and Cherry, 1985). The migration of an immiscible organic liquid phase is governed largely by its density and viscosity. Commonly encountered groundwater contaminants, including halogenated aliphatics with one or two carbon atoms, tend to have moderately low solubilities. These may migrate as discrete nonaqueous phases, with some components dissolving into the surrounding groundwater. The concentration of many organic compounds in groundwater is often limited by their very low solubilities; however, they may be toxic at very low concentrations (Freeze and Cherry, 1979). The presence of large quantities of high-density, low-solubility contaminants can provide a "hidden" source for long-term contamination of the groundwater.

The rate at which dissolved organic contaminants migrate in groundwaters is determined by the processes of advection, dispersion, sorption, chemical and biological transformations, and perhaps by volatilization at the water table (Barker and Patrick, 1985). All processes except advection can bring about lower aqueous concentrations and, therefore, can contribute to natural attenuation. Attenuation of organic chemicals by dilution in groundwater is not as great as in surface water because the flow velocities of groundwaters are generally low (Steelman and Ecker, 1984). Therefore, these contaminants tend to maintain much of their integrity as they move through an aquifer.

2.2.1.2 Advection

In sand and gravel aquifers, the dominant factor in the migration of a dissolved contaminant is advection, the process by which solutes are transported by the bulk motion of flowing groundwater (Mackay, Roberts, and Cherry, 1985). Groundwater generally flows from regions of the subsurface where water level is high to regions where water level is low, a process called hydraulic gradient. In most cases, the flow velocities under natural gradient conditions can be between 10 and 100 m/year or even lower. In the zone of influence of a high-capacity well or wellfield, however, the artificially increased gradient substantially increases the local velocity.

2.2.1.3 Dispersion and Diffusion

The rate of movement of organic chemicals through air, water, and organic matter is directly proportional to the concentration of the toxicant and its diffusion coefficient (Kaufman, 1983).

Dissolved contaminants spread as they move with the groundwater (Mackay, Roberts, and Cherry, 1985). This dispersion results from molecular diffusion and mechanical mixing, and causes a net flux of the solutes from a zone of high concentration to a zone of lower concentration. This movement causes the concentrations to diminish with increasing distance from the source and the plume to become more uniform. Dispersion in the direction of flow is often much greater than dispersion in the directions transverse to the flow.

Percolating water is the principal means of movement of relatively nonvolatile chemicals, and diffusion in soil water is important only for transport over very small distances (Kaufman, 1983).

2.2.1.4 Sorption

Sorption is perhaps the most important single factor affecting the behavior of organic chemicals in the soil environment (Kaufman, 1983). Adsorption to soil constituents will affect the rate of volatilization, diffusion, or leaching, as well as the availability of chemicals to microbial or chemical degradation. Some dissolved contaminants may interact with the aquifer solids encountered along the flow path through physical adsorption, partitioning, ion exchange, (Freeze and Cherry, 1979), oppositely charged surfaces, or formation of a bond (Roberts and Valocchi, 1982).

Compounds that sorb strongly onto solids are retarded in their movement through an aquifer (Roberts and Valocchi, 1982). Sorption takes place because the compound has either a low affinity for water (hydrophobic) or has a high affinity for the solid. The advancing front of sorbing contaminants moves at a linear velocity that is slower than that of the carrier groundwater. The main subsurface solids responsible for adsorption of organic chemicals are solid organic matter (strong associations with hydrophobic organic compounds), clay minerals, and amorphous minerals (e.g., iron hydroxides) (Pettyjohn and Hounslow, 1983).

These interactions distribute the contaminants between the aqueous phase and the aquifer solids, diminish concentrations in the aqueous phase, and retard movement of the contaminant relative to groundwater flow (Rubin, 1983). The higher the fraction of contaminant sorbed, the more retarded is its transport. Also, the more hydrophobic a compound is, the more it should be retarded. Sorption equilibrium may require weeks or months and, thus, may not always be reached in the field (Karickhoff, 1984). The interaction of these variable factors make it difficult to predict accurately groundwater transport of the contaminants (Mackay, Roberts, and Cherry, 1985).

A large amount of organic material is necessary to adsorb low molecular weight hydrocarbons, such as chloroform. Highly water-soluble organic substances, such as acetone and methanol, are only slightly retarded by sorption, but fortunately are easily degraded (Pettyjohn and Hounslow, 1983).

The pKa, or dissociation constant, of a compound indicates the degree of acidity or basicity that a compound will exhibit and, therefore, should be very important in determining both the extent of adsorption and the ease of desorption (Kaufman, 1983). Functional groups can affect the degree of sorption of a chemical (Brindley and Thompson, 1966). Chain molecules terminating in -OH, -COOH, and $-NH_2$ readily form complexes with montmorillonite, whereas, similar molecules terminating in -Cl and -Br do not. All compounds are adsorbed strongly at low pH; anionic substances are adsorbed negatively at slightly basic conditions; and nonionic compounds are moderately adsorbed (Frissel, 1961). Adsorption processes are exothermic and desorption processes are endothermic, and an increase in temperature should reduce adsorption and favor the desorption process (Kaufman, 1983).

2.2.1.5 Volatility

The surface sheen residue emanating from an emulsified oil slick after evaporation of the light boiling components gradually becomes enriched in high molecular constituents concomitant with a change in chemical composition. Such constituents exhibit efficient absorption over a broad spectral range extending into the IR region. A film of crude oil will, therefore, absorb a considerable amount of solar energy, leading to transformation and degradation of the film through photoinduced oxidation. Petroleum resins have been shown to be particularly unstable in air and sunlight, and they may aromatize and evolve into structures more like asphaltenes (Tissot and Welte, 1978).

2.2.1.6 Viscosity

Viscosity and surface-wetting properties affect the transport of an organic liquid phase (Mackay, Roberts, and Cherry, 1985). Viscosity of polluting oils is an important property that determines, in part, the spreading and dispersion of the hydrocarbon mixture and, thus, the surface area available for microbial attack (Atlas, 1978c). It affects a chemical's migration in groundwater (Noel, Benson, and Beam, 1983). For example, about four times the volume of a light fuel oil in the high viscosity range would be retained by the average soil, compared with gasoline, a distillate with a lower viscosity. Gasoline would also spread over a wider area of an aquifer than a light fuel oil. Contaminants that are highly water soluble must be handled differently than those that float on the water table, like gasoline (Nielsen, 1983).

Large quantities of immiscible liquid organic contaminants could be stored as droplets dispersed within the pores of aquifer media, even if the bulk of the migrating mass of liquid is removed (Mackay, Roberts, and Cherry, 1985). These droplets may then dissolve over time into the groundwater flowing past them.

2.2.1.7 Density

Density differences of only 1 percent can influence fluid movement in the subsurface (Mackay, Roberts, and Cherry, 1985). The specific gravities of hydrocarbons (gasoline and other petroleum distillates) may be as low as 0.7, and halogenated hydrocarbons, are almost without exception, significantly more dense than water.

Density determines where in the aquifer the contaminant will most likely be concentrated. Low-density hydrocarbons have a tendency to float on water and may be found in the upper portions of an aquifer. High-density hydrocarbons would sink to the lower portions of the aquifer. It is important to recognize that the migration of dense organic liquids is largely uncoupled from the hydraulic gradient that drives advective transport and that the movement may have a dominant vertical component, even in horizontally flowing aquifers.

An organic liquid contaminant, such as gasoline, which is immiscible with and less dense than groundwater, would migrate vertically to and then float on the water table, spreading out in the downgradient direction (Mackay, Roberts, and Cherry, 1985). If the organic liquid contains a contaminant slightly soluble in water, e.g., benzene, a plume would form in the saturated zone. A complex pattern of overlapping plumes can develop when many contaminants are involved.

2.2.1.8 Chemical Structure

The structure, concentration, and toxicity of a chemical are important in determining whether it is accumulated in the environment and the environmental impact of the accumulation (Leisinger, 1983). The chemical will accumulate if its structure prevents mineralization or biodegradation by organisms. This may be due to its insolubility or to a novel chemical structure to which microorganisms have not been exposed during evolutionary history. Such compounds are termed xenobiotic. This name refers to compounds of anthropogenic (man-made) origin, as well as to compounds that may occur naturally but exceed normal levels in the environment. Various laboratory culture techniques applied to samples from nature can be used to select or develop bacteria with the ability to biodegrade many of these chemicals.

The chemical structure of a contaminant will affect its biodegradation in two ways (Hutzinger and Veerkamp, 1981). First, the molecule may contain groups or substituents that cannot react with available or inducible enzymes (i.e., these chemical bonds cannot be broken). Second, the structure may determine the compound to be in a physical state (adsorbed, gas-phase) where microbial degradation does not easily occur. This seems to be a problem with many of the lipophilic compounds, which have very low solubilities in water.

Generally, the larger and more complex the structure of a hydrocarbon, the more slowly it is oxidized. This may depend upon the type of organism involved and the medium in which it was developed (Texas Research Institute, Inc., 1982). Conflicting conclusions have been reported on this subject. Some authors have proposed that aliphatic, long-chain molecules are attacked more readily than short chains, with hydrocarbons in the range of $C_{10}H_{22}$ to $C_{16}H_{34}$ being oxidized by soil bacteria more readily than those of lower weight. Saturated compounds, e.g., n-alkanes, are highly degraded, while asphaltenes and aromatics are often resistant to microbial attack (Jobson, Cook, and Westlake, 1972). Other workers have suggested that although n-alkanes are probably metabolized more rapidly than naphthenes or other aromatics, the reactions appear to be slower with increasing chain length, possibly because of differences in water solubility, and because the n-alkanes with shorter chains (from C_5 to C_9) are more easily used as a source of carbon and energy by microorganisms than those with longer chain lengths (from C_{10} to C_{14}) (Williams, Cumins, Gardener, Palmier, and Rubidge, 1981). These undergo oxidation to form alcohols, aldehydes, and acids. The most degradable alkanes are those with molecular weights in the C_6 to C_{28} range (Perry, 1968). Aromatic compounds are the least degradable by microbes, and straight-chain paraffins are the easiest to degrade (Evans, Deuel, and Brown, 1980).

The degree of substitution affects biodegradation. Compounds that possess amine, methoxy, and sulfonate groups, ether linkages, halogens, branched carbon chains, and substitutions at the meta position of benzene rings are generally persistent (Knox, Canter, Kincannon, Stover, and Ward, 1968). Addition of aliphatic sidechains increases the susceptibility of cyclic hydrocarbons to microbial attack (Atlas, 1978c). Linear nonbranched compounds are more easily biodegraded than branched forms and rings (Pettyjohn and Hounslow, 1983). The side chains of the latter are generally attacked first. Changes in n-alkane to isoprenoid hydrocarbon ratios occur in oil spills as biodegradation proceeds (Haines Pesek, Roubal, Bronner, and Atlas, 1981). Phenanthrenes and dibenzothiophenes with C_2 and greater substitution are relatively resistant to biodegradation, while unsubstituted and C_1 substituted 2- and 3-ring condensed aromatics are subject to abiotic and biotic losses. The number and locations of fused rings in polynuclear aromatics are important in determining the rates of their decomposition (Sims and Overcash, 1983). Hydrocarbons that are strong fat solvents may be less readily tolerated or assimilated than those that are less likely to dissolve cell lipids (ZoBell, 1946).

Table A.2-14 summarizes organic groups subject to microbial metabolism by aerobic respiration, anaerobic respiration, and fermentation (Environmental Protection Agency, 1985b). Oxidation indicates that the compound is used as a primary substrate, and cooxidation indicates that the compound is cometabolized. These tables provide only a general indication of degradability of compounds, and treatability studies will usually be required to determine the degradability of specific waste components.

Several linkages may be readily susceptible to biodegradation (Kearney and Plimmer, 1970):

1. $R-NH-CO_2R'$

2. $R-NH-COR'$

3. $\begin{matrix} -O \\ \\ -O \end{matrix} \Big\rangle P-S-R$

4. $\begin{matrix} -O \\ \\ -O \end{matrix} \Big\rangle P-O-R$

5. $R-CHCl-COO-$

6. $R-CCl_2-COO-$

Table A.2-14. Summary of Organic Groups Subject to Biodegradation (Environmental Protection Agency, 1985b)

Substrate Compounds	Respiration		Fermentation	Oxidation	Cooxidation
	Aerobic	Anaerobic			
Straight Chain Alkanes	+	+	+	+	+
Branched Alkanes	+	+	+	+	+
Alcohols	+	+		+	
Aldehydes, Ketones	+	+		+	
Carboxylic Acids	+	+		+	
Cyclic Alkanes	+		+	+	+
Unhalogenated Aromatics	+	+		+	+
Phenols	+	+	+	+	+
Fused Ring Hydroxy Compounds	+				
Phenols - Dihydrides, Polyhydrides	+			+	+
Two- and Three-ring Fused Polycyclic Hydrocarbons	+			+	

Aliphatic acids, anilides, carbamates, and phosphates are generally degraded within a short time in the soil (Kaufman, 1983). The rate at which linkages are hydrolyzed will depend upon the nature of R and R'. Chemicals that degrade with an initial ester hydrolysis reaction are relatively short-lived in soil, whereas those that initially undergo dealkylation tend to be somewhat more persistent (Kaufman and Plimmer, 1972). Chemicals that are initially dehalogenated are variable in their persistence. Halogenated aliphatic acids are readily degraded, whereas, halogenated benzoic acids and s-triazines are intermediate in their persistence. Table A.2-15 lists some of the major chemical classes and their relative persistence and initial degradation reactions (Kaufman, 1983).

An inverse relationship and a high correlation between microbial transformation rates and van der Waal's radius of eight phenols have been found, suggesting the latter as useful for predicting degradability of xenobiotics (Paris, Wofe, Steen, and Baughman, 1983).

2.2.1.9 Environmental Factors

Metabolism by indigenous microflora is influenced by environmental factors, such as light, temperature, pH, presence of cometabolites, reactive radicals, other organic and inorganic compounds, and available oxygen, nitrogen, and phosphorus, as well as the physical state of the oil (Cooney, Silver, and Beck, 1985). The environment influences biodegradation by regulating both the bioavailability of the compound and the activity of the degraders. Salinity, temperature, chlorophyll, nitrogen, and phosphorus concentrations have been correlated with rates of biodegradation in surface water environments. The numbers of hydrocarbon-using organisms may also be enhanced by prior pollution of a site.

2.2.1.10 Toxicity

Organic compounds may not be readily degraded in groundwater when the microbial population is low, the nutrient balance is inadequate, or because of toxicity from contaminant overloading (Pettyjohn and Hounslow, 1983). Crude oils are mixtures of tremendous complexity, containing hundreds of hydrocarbon and nonhydrocarbon components, many of them still unidentified and some of these very toxic toward microorganisms (Atlas and Bartha, 1973b). Some compounds may be more toxic to microbes than others (Scholze, Wu, Smith, Bandy, and Basilico, 1986), and the presence of inhibitory substances in oil can delay or prevent the biodegradation of otherwise suitable hydrocarbon substrates (Bartha and Atlas, 1977).

When the structural features necessary for toxicity are compared with those features permitting degradation in the environment for target organisms, differences are found among the various chemical classes (Kaufman and Plimmer, 1972). In some classes, those structural features contributing to toxicity are coincident with those necessary for degradability; in other chemical classes they are diametrical. In all classes, however, the relationships are mediated by substituent type, number, and position.

The structure-toxicity and structure-degradability relationships of certain halogenated aliphatic acids are quite similar; i.e., the most phytotoxic structures were also the most readily degradable (Kaufman, 1983). Meta-substitution (para to a free ortho position) confers resistance to biodegradation and eliminates phytotoxic activity. Halogenation in the para position increases both phytotoxicity and biodegradability of phenoxyacetates. Increasing the length of the side chain affects both phytotoxicity and biodegradability.

Table A.2-15. Relative Persistence and Initial Degradative Reactions of Nine Major Organic Chemical Classes (Panchal and Zajic, 1978)

Chemical Class	Persistence	Initial Degradative Process
Carbamates	2 to 8 weeks	Ester hydrolysis
Aliphatic acids	3 to 10 weeks	Dehalogenation
Nitriles	4 months	Reduction
Phenoxyalkanoates	1 to 5 months	Dealkylation, ring hydroxylation or oxidation
Toluidine	6 months	Dealkylation (aerobic) or reduction (anaerobic)
Amides	2 to 10 months	Dealkylation
Benzoic acids	3 to 12 months	Dehalogenation or decarboxylation
Ureas	4 to 10 months	Dealkylation
Triazines	3 to 18 months	Dealkylation or dehalogenation

Methyl-, dimethyl-, and trimethyl-naphthalenes are more toxic than naphthalene to the freshwater alga, Selenastrum capricornatum, while dibenzofuran, fluorene, phenanthrene, and dibenzothiophene are even more toxic to this organism (Hsieh, Tomson, and Ward, 1980). In general, compounds with higher boiling points are more toxic.

C_2 to C_6 alkanes are inhibitory to some microorganisms possibly because their size allows them to penetrate into cell membranes (Hornick, Fisher, and Paolini, 1983). This is also seen with cycloalkanes of similar size and could be the reason for the "toxicity" of short chain alkanes seen with a few microorganisms.

The toxicity of PAHs to microorganisms is also related to their water solubility (Sims and Overcash, 1983). Aromatic hydrocarbons in water-soluble fractions of petroleum products are toxic to aquatic organisms, but rapid volatilization of low molecular weight hydrocarbons limits the exposure time (Coffey, Ward, and King, 1977). The vapor phase of short chain alkanes is less toxic than the liquid phase. The toxicity of the short chain alkanes is also related to temperature, since a higher temperature will increase the amount of alkane in the vapor phase and decrease the concentration of the liquid alkane. In cold water, however, these compounds may delay the onset of biodegradation for several weeks (Bartha and Atlas, 1977).

While low molecular weight aromatic hydrocarbons are quite toxic to microorganisms, they can be metabolized when present in low concentrations. Condensed polyaromatic hydrocarbons are less toxic to microorganisms but are metabolized only rarely and at slow rates. Cycloalkanes are highly toxic and serve as growth substrates for isolated organisms only in exceptional cases. Some are readily degraded, however, by the cometabolic attack of mixed microbial communities. In general, the biotransformation process for PAHs with more than three rings appears to be cometabolism (Sims and Overcash, 1983).

Low molecular weight hydrocarbons solvate and, hence, destroy the lipid-containing pericellular and intracellular membrane structures (Bartha and Atlas, 1977). Liquid hydrocarbons of the n-alkane, iso-alkane, cycloalkane, and aromatic type with carbon numbers under 10 all share this property to varying degrees.

Floating oil is able to concentrate hydrophobic pollutants (Bartha and Atlas, 1977). This makes the material more toxic and interferes with microbial degradation.

In certain situations, it may be possible to modify the chemical composition of oil, rendering it more susceptible to biodegradation (Atlas, 1977). The toxic components in the low molecular weight and low boiling range can be removed by temporary heating, ignition, and burning of the oil, and artificially increasing air movement over the oil (Atlas and Bartha, 1972a; Atlas, 1975). Oil is particularly difficult to ignite in many aquatic environments (Fay, 1969). Burning would probably remove toxic components and many other hydrocarbons. However, burning of the substance is not without problems, and, depending upon the chemical composition of the oil, may create a residual that is more resistant to biodegradation.

2.2.1.11 Concentration

The chemical concentration may also affect the level of tolerance (see Section 2.2.2) (Scholze, Wu, Smith, Bandy, and Basilico, 1986). The term "xenobiotic compound" refers not only to compounds with structural features foreign to life but also to those compounds that are released in the environment by the action of man and, thereby, occur in a concentration that is higher than natural (Leisinger, 19830). The concentration of hydrocarbons in water has two effects (Texas Research Institute, Inc., 1982). At low concentrations, all fractions are likely to be attacked, but at high concentrations, only those fractions most susceptible to degradation will be attacked. Also, if the hydrocarbon mixture contains water-soluble toxic substances, their effect is intensified at high concentrations (see also Section 2.2.3).

2.2.1.12 Naturally Occurring Organic Materials

Naturally occurring organic materials can influence the ability of microorganisms to degrade pollutants (Shimp and Pfaender, 1984). After adaptation of a microbial community to four types of compounds, it was found that amino acids, fatty acids and carbohydrates stimulated biodegradation of monosubstituted phenols, while humics decreased biodegradation rates. Many aromatic compounds bind to particulate material and, thus, eventually reside in sediments of natural aquatic ecosystems (Horowitz and Tiedje, 1980).

2.2.1.13 Abiotic Hydrolysis and Oxidation

The effects of these processes on particular contaminants in the groundwater zone are unknown (Mackay, Roberts, and Cherry, 1985). It is believed, however, that most chemical reactions in the groundwater are likely to be slow in comparison with transformations mediated by microorganisms (Cherry, Gillham, and Barker, 1984).

2.2.1.14 Biological Factors

Increased persistence of chemicals may result from several types of biological interactions: 1) the biocidal properties of the chemicals to soil microorganisms may preclude their biodegradation, 2) direct inhibition of the adaptive enzymes of effective soil microorganisms, and 3) inhibition of the proliferation processes of effective microorganisms (Kaufman, 1983). Inhibition of microbial degradation may ultimately affect mobility of a chemical in soil.

It should be realized that biodegradability of a petroleum compound is a result of the action of the mixed flora present at the location, and that the material being degraded is actually a complex mixture of hydrocarbons, some of which contribute to the breakdown of others (Cooney, Silver, and Beck, 1985). It is also possible that problems could arise involving the degradation, persistence, or toxicity of organic chemicals when several wastes or their residues are present in the soil together (Kaufman, 1983). These factors should be taken into account when assessing the microbiological potential for petroleum degradation.

It has been proposed that the observed recalcitrance of many compounds in vitro may be due to the lack of properly designed experiments under the appropriate conditions that are conducive to degradation (Hegeman, 1972). Recalcitrance could also be due to insufficient time to evolve enzymatic pathways to degrade certain chemicals. The acclimation or induction of enzymes that catalyze the necessary reactions in the microbial population is an important factor in determining biodegradability (Paris, Wolfe, Steen, and Baughman, 1983).

2.2.2 Rate of Biodegradation

When the biodegradability or composition of a waste constituent is unknown, it is prudent to undertake a laboratory investigation of the kinetics of biodegradation of the material (Thibault and Elliott, 1979). Respirometric techniques have been used to establish biodegradation rates (kinetics), the potential for inhibition of these rates at various waste concentrations, oxygen and nutrient requirements, and temperature effects.

In many instances, it is possible that the minimum set of factors or variables (at least for substrates that are mineralized) that need to be considered in assessing the rate of degradation of a compound are the concentration of the compound and the abundance and activity of organisms capable of mineralizing the compound (Alexander, 1986). For instance, studies indicate that the rate of mineralization of naphthalene is determined primarily by the presence of elevated hydrocarbon-degrading microbial populations and may not be directly related to elevated populations of heterotrophic bacteria or sediment organic carbon content (Heitkamp, Freeman, and Cerniglia, 1987). At 1 mg/l tertiary butyl alcohol, it appears the microbial population receives insufficient energy to cause a population increase and utilization rates remain slow (Novak, Goldsmith, Benoit, and O'Brien, 1985). Rates are faster at higher concentrations, where growth can be better supported. Environmental factors can also affect the rate of biodegradation. For example, the action of a Nocardia sp. on hexadecane suggested that the rate of natural biodegradation of oil in marine environments was limited by low temperatures and phosphorus concentration, but not by the concentrations of naturally occurring nitrogen (Mulkins-Phillips and Stewart, 1974b).

An approach has been developed to measure rates of degradation based upon the examination of metabolic kinetics using radiolabeled substrates (Pfaender and Klump, 1981). The measurements require short incubations (8 to 10 hr), which should yield rates close to those occurring in nature. The method has been applied to fresh water, estuarine, and oceanic environments. There is a fivefold decrease in rates of metabolism from fresh to estuarine water and a tenfold further decrease from estuarine to ocean water.

Very rapid mineralization rates (e.g., for naphthalene) have been reported for some sediments that are chronically exposed to very high concentrations of degradable hydrocarbons (Heitkamp, Freeman, and Cerniglia, 1987). The mineralization rate and half-life calculated for naphthalene was about 2.9 percent/day and 2.4 weeks (17 days), respectively, for such a source; while the half-life was 4.4 weeks with sediment from a pristine environment.

Mineralization of glucose by Salmonella at ng/ml levels stopped, and cell death occurred before the substrate was exhausted, suggesting a significant role for maintenance energy in determining the kinetics of mineralization of organic chemicals at low concentrations (Simkins, Schmidt, and Alexander, 1984).

Microorganisms have been found to be able to degrade one- and two-ringed aromatic hydrocarbons with high reaction rates down to extremely low concentration levels (i.e., < 1 ug/l), given sufficient oxygen and nutrients (Gray, 1978). It is estimated that polluting oil in the sea might be biodegraded at rates as high as 100 to 960 mg/m^3/day (Bartha and Atlas, 1977). Degradation rates for hexadecane have been measured to be 0.050 g (Knetting and Zajic, 1972) and 0.015 g/m^3/day (Knetting and Zajic, 1972; Seki, 1976) at summer temperatures, 0.001 g/m^3/day in the colder waters of Alaska, and a rate even lower than this in the open waters of the Arctic Ocean (Robertson, Arhelger, Kinney, and Button, 1973).

While microbes are important in degrading many petroleum hydrocarbons in ocean waters, the low degradation rates for fluorene and benzopyrene suggest that this does not hold for high molecular weight aromatics (Lee and Ryan, 1976). Biodegradation of polynuclear aromatics appears to be inversely correlated to molecular size and ring condensation (Bossert, 1983). PAHs are degraded at much lower rates than mono- and dinuclear aromatics and n-paraffins in ocean waters (Bartha and Atlas, 1977). Turnover times of the order of months are found for paraffins and light aromatics and of the order of several years for polynuclear aromatics. Degradation rates are high in spring and low in winter.

Naphthalene had the highest degradation rate of the relatively nonvolatile hydrocarbons tested, followed in decreasing order by methylnaphthalene, heptadecane, hexadecane, octadecane, fluorene, and benzopyrene (Thomas and Alexander, 1983). The degradation rate of all hydrocarbons was higher in the estuarine area than in water from offshore. The calculated turnover times (time required to convert all hydrocarbon to carbon dioxide) are presented in Table A.2-16 (Science Applications International Corporation, 1985a). Relatively little is known of the kinetics of degradation of mixed substrates at low concentrations, or the possible interactions among primary and secondary substrates and bacteria (McCarty, Reinhard, and Rittmann, 1981). The presence of additional substrates in the soil or lake water may also alter the kinetics of mineralization of low concentrations of organic pollutants (Schmidt, Scow, and Alexander, 1985). With a pure culture of Pseudomonas acidovorans, acetate and phenol disappeared at an equal rate, when they were at low concentrations. However, phenol mineralization was repressed at high acetate concentrations.

2.2.3 Effect of Hydrocarbon Concentrations on Degradation

Table A.2-17 lists concentrations at which certain compounds have been found to be toxic in industrial waste treatment (Environmental Protection Agency, 1985b). Microorganisms present in the subsurface, however, may be more tolerant to high concentrations of these compounds. This must be determined on a case-by-case basis.

The concentration of a contaminant will affect the number of organisms present. It has been noted that higher concentrations of gasoline in contaminated water were related to higher counts of microorganisms (McKee, Laverty, and Hertel, 1972; Litchfield and Clark, 1973). Waters containing less than 10 ppm of hydrocarbon had populations of bacteria less than 10^3, while concentrations of hydrocarbon in excess of 10 ppm sometimes supported growth of 10^6 bacteria/ml.

It has been shown that the persistence of a compound increases as the initial concentration increases (Hamaker, 1972). The reduced rate is explained either by the limited active sites available (Hance and McKone, 1971), or by a toxic effect on microorganisms or enzyme inhibition (Hurle and Walker, 1980). With PAHs, there is an increasing trend of initial rate of degradation as the initial concentration increases (Sims and Overcash, 1983). Table A.2-18 summarizes rates of degradation from the literature.

Table A.2-16. Turnover Times for Microbial Hydrocarbon Degradation in Coastal Waters (Lee and Ryan, 1976)

Compound	Concentration (ppb)	Date Locality	Turnover Time (days)
Benzpyrene	5	Feb--Sk	0
	5	May--Sk	3,500
	5	May--O	0
Fluorene	30	Feb--Sk	0
	30	Feb--O	0
	30	May--Sk	0
	30	June--Sk	1,000
Heptadecane	8	May--Sk	80
	15	May--Sk	60
	30	May--Sk	54
	30	May--Sk	170
Hexadecane	25	Feb--Sk	500
	25	April--Sk	210
	25	April--O	1,000
Naphthalene	40	Feb--Sk	500
	40	May--Sk	46
	40	May--Sk	79
	130	May--Sk	30
	130	May--O	330
Methyl-naphthalene	40	Feb--Sk	500
	40	May--Sk	50
Octadecane	16	May--Sk	100
Toluene	20	May--Sk	17
	20	May--Sk	17
	20	May--O	40

Sk = Skidaway River (3 m)
O = offshore water (10 m)

Table A.2-17. Problem Concentrations of Selected Chemicals (Environmental Protection Agency, 1985b)

| Chemical | Problem Concentration (mg/l) | |
	Substrate Limiting[a]	Nonsubstrate Limiting[b]
Formaldehyde	--	50 - 100
Acetone	--	>1000
Phenol	>1000	300 - 1000
Ethyl benzene	>1000	--
Dodecane	>1000	--

a = Substrate limiting represents the condition in which the subject compound is the sole carbon and energy source.

b = Nonsubstrate limiting represents the condition in which other carbon and energy sources are present.

Table A.2-18. Kinetic Parameters Describing Rates of Degradation of Aromatic Compounds in Soil Systems (Sims and Overcash, 1983)

PAH	Initial Concentration (ug/g soil)	k (day^{-1})	Rate of Transformation (ug/g-day)	$t_{1/2a}$ (days)	
Pyrocatechol	500	3.47	1,735	0.2	m
Phenol	500	0.693	364.5	1.0	m
Phenol	500	0.315	157.5	2.2	l
Fluorene	0.9	0.018	0.016	39	m
Fluorene	500	0.347	173.3	2	m
Indole	500	0.693	364.5	1.0	m
Indole	500	0.315	157.5	2.2	l
Naphthol	500	0.770	385	0.9	m
Naphthalene	7.0	5.78	40.4	0.12	m
Naphthalene	7.0	0.005	0.035	125	l
Naphthalene	25,000	0.173	4,331	4	h
1,4-Naphthoquinone	500	0.578	288.8	1.2	m
Acenaphthene	500	0.173	86.6	4	m
Acenaphthene	5	2.81	22.6	0.3	m
Anthracene	3.4	0.21	0.714	3.3	l
Anthracene	13.7	0.004	0.054	175	m
Anthracene	10.3	0.005	0.050	143	m
Anthracene	11.4	0.006	0.073	108	m
Anthracene	40.0	0.005	0.208	138	m
Anthracene	36.4	0.005	0.196	129	m
Anthracene	25,000	0.198	4,950	3.5	h

Table A.2-18. continued

PAH	Initial Concentration (ug/g soil)	k (day⁻¹)	Rate of Transformation (ug/g-day)	t₁/₂ₐ (days)	
Phenanthrene	2.1	0.027	0.056	26	m
Phenanthrene	25,000	0.277	6,930	2.5	h
Carbazole	500	0.067	33	10.5	m
Carbazole	5	0.231	1.16	3	m
Benz(a)anthracene	0.12	0.046	0.005	15.2	l
Benz(a)anthracene	0.12	0.0001	0.00001	6,250	m
Benz(a)anthracene	3.5	0.007	0.024	102	m
Benz(a)anthracene	20.8	0.003	0.062	231	m
Benz(a)anthracene	25.8	0.005	0.134	133	m
Benz(a)anthracene	17.2	0.008	0.060	199	m
Benz(a)anthracene	22.1	0.006	0.130	118	m
Benz(a)anthracene	42.6	0.003	0.118	252	m
Benz(a)anthracene	72.8	0.004	0.257	196	m
Benz(a)anthracene	25,000	0.173	4,331	4	h
Fluoranthene	3.9	0.016	0.061	44	m
Fluoranthene	18.8	0.004	0.072	182	m
Fluoranthene	23.0	0.007	0.152	105	m
Fluoranthene	16.5	0.005	0.080	143	m
Fluoranthene	20.9	0.006	0.125	109	m
Fluoranthene	44.5	0.004	0.176	175	m
Fluoranthene	72.8	0.005	0.379	133	m
Pyrene	3.1	0.020	0.061	35	m
Pyrene	500	0.067	33	10.5	m
Pyrene	5	0.231	1.16	3	m
Chrysene	4.4	0	0	-	
Chrysene	500	0.067	33	10.5	m
Chrysene	5	0.126	0.63	5.5	m
Benz(a)pyrene	0.048	0.014	0.007	50	l
Benz(a)pyrene	0.01	0.001	0.00001	694	l
Benz(a)pyrene	3.4	0.012	0.041	57	m
Benz(a)pyrene	9.5	0.002	0.022	294	m
Benz(a)pyrene	12.3	0.005	0.058	147	m
Benz(a)pyrene	7.6	0.003	0.020	264	m
Benz(a)pyrene	18.5	0.023	0.312	30	m
Benz(a)pyrene	17.0	0.002	0.028	420	m
Benz(a)pyrene	32.6	0.004	0.129	175	m
Benz(a)pyrene	1.0	0.347	0.347	2	h
Benz(a)pyrene	0.515	0.347	0.179	2	h
Benz(a)pyrene	0.00135	0.139	0.0002	5	h
Benz(a)pyrene	0.0094	0.002	0.00002	406	l
Benz(a)pyrene	0.545	0.011	0.006	66	l
Benz(a)pyrene	28.5	0.019	0.533	37	l
Benz(a)pyrene	29.2	0	0	-	
Benz(a)pyrene	9,100	0.018	161.7	39	h
Benz(a)pyrene	19.5	0.099	1.93	7	h

Table A.2-18. continued

PAH	Initial Concentration (ug/g soil)	k (day^{-1})	Rate of Transformation (ug/g-day)	$t_{1/2a}$ (days)	
Benz(a)pyrene	19.5	0.139	2.70	5	h
Benz(a)pyrene	19.5	0.231	4.50	3	h
Benz(a)pyrene	130.6	0.173	22.63	4	h
Benz(a)pyrene	130.6	0.116	15.08	6	h
Dibenz(a,h)anthracene	9,700	0.033	320.1	21	h
Dibenz(a,h)anthracene	25,000	0.039	962.5	18	h

^{a}l = low temperature range (<15°C)

m = medium temperature range (15 to 25°C)

h = high temperature range (>25°C)

2.2.3.1 Examples of Degradable Levels of Organic Compounds

Examples of concentrations of various compounds that have been shown to be tolerated and degraded are discussed below and summarized in Table A.2-19.

- Biological degradation is the generally recognized treatment method for phenolic wastewater containing phenol concentrations up to 500 ppm (Roberts, Koff, and Karr, 1988). At phenol concentrations of 10,000 ppm, biodegradation of 99 to 100 percent have been reported for phenol, methylphenols, nitrophenols, and chlorophenols.
- Aeromonas, Alcaligenes, Pseudomonas, and Vibrio were found to degrade South Louisiana Crude Oil and motor oil at concentrations of 1.0 and 5.0 percent, but growth on 0.01 percent was very limited (Frieze and Oujesky, 1983).
- Approximately 30,000 gal of gasoline contaminated over 75,000 ft^2 (Brown, Norris, and Brubaker, 1985). Five pumping wells recovered 18,500 gal, and bioreclamation was initiated to remediate the soil and groundwater. Nutrients were periodically injected and air was continuously sparged over 10 months through 14 wells, and the gasoline level in the soil was reduced to less than 50 ppm.
- From 700 to 1400 gal of mixed fuels/solvents (45 percent aromatics/55 percent alkanes) were confined to a tank vault (Brown, Loper, and McGarvey, 1985). Free product recovery of about 700 gal was followed by enhanced bioreclamation and ended with carbon treatment. Groundwater levels during bioreclamation dropped from 22 to 45 ppm to <550 ppb in 2 1/2 months.

Table A.2-19. Initial and Final Concentrations of Compounds Susceptible to Biodegradation

Compound	Time for Degradation	Initial Concentration	Final Concentration	Organism/ Source
South Louisiana Crude Oil and Motor Oil[a]		1.0%, 5.0%		Aeromonas, Alcaligenes, Pseudomonas, Vibrio
Mixed Fuels/ Solvents[b]	2 1/2 mo	22 to 45 ppm (Groundwater)	<550 ppb	
Gasoline[c]		100 to 500 ppm	2 to 5 ppm	
Methylene chloride[d]	1 yr	91 ppm (Groundwater)	<1 ppm	
Acetone[d]	1 yr	54 ppm	<1 ppm	
Acrylonitrile[e]	3 mo	1000 ppm (Groundwater)	1 ppm	Mutant bacteria
Acrylonitrile[f]	1 mo	1000 ppm	Iod	Mutant bacteria
Phenol[f]	40 d	31 ppm	30 ppm	Mutant bacteria
Organic chemicals[g]		<1000 ppm (Soil)	<1 ppm	Indigenous and hydro-carbon degrading bacteria
Methylene chloride[h]	2 1/2 mo	2500 mg/l	<100 mg/l	Commercial hydrocarbon degrading bacteria
Dichlorobenzene[h]	2 1/2 mo	800 mg/l	<50 mg/l	"
p-Cresol[i]		8 ppm		
Hydrocarbon[j]		10 ppm		
Gasoline[k]	10 mo	11,500 gal/ 75,000 ft^2	<50 ppm	
Gasoline[k]	18 mo another 6 mo	5 to 8 ppm 2.4 ppm (Groundwater)	2.4 ppm <500 ppb	Indigenous organisms

Table A.2-19. continued

Compound	Time for Degradation	Initial Concentration	Final Concentration	Organism/ Source
Gasoline[k]	25 mo (air sparging) 10 mo (microbial nutrient)	15 ppm (Ground- water)	2.5 ppm 200 to 1200 ppb	Indigenous organisms
Petroleum distillate[l]	21 d	12,000 ppm	>1 ppm	BI-CHEM-SUS-8
Formaldehyde[l]	22 d	1400 ppm	>1 ppm	PHENOBAC
Phenols[m]	7 hr	1500 ppm	>1 ppm	Azotobacter
Phenols[m]		10,000 ppm	0 to 100 ppm	
Phenol[n]		32 ng/g soil		
Solvent/fuel mixture (aliphatic and aromatic hydrocarbons)[o]	2-1/2 mo	23 ppm	0.5 ppm	Landfarming
Gasoline[o]	10 mo	30 to 40 ppm (groundwater)	>1 ppm	
	10 mo	2,000 to 3,000 ppm (soil)	>50 ppm	
Phenol[p]	7 days 7 days	5 mg/l 10 mg/l	0 mg/l 0 mg/l	Domestic wastewater
Naphthalene[p]	7 days 7 days	5 mg/l 10 mg/l	0 mg/l 0 mg/l	Domestic wastewater
Benzenepe[p]	7 days	5 mg/l	0 mg/l	Domestic
Benzene[p]	14 days	10 mg/l	0 mg/l	wastewater
Toluene[p]	7 days 7 days	5 mg/l 10 mg/l	0 mg/l 0 mg/l	Domestic wastewater
Anthracene[p]	21 days 21 days	5 mg/l 10 mg/l	0.4 mg/l 5 mg/l	Domestic wastewater
Phenanthrene[p]	7 days 7 days	5 mg/l 10 mg/l	0 mg/l 0 mg/l	Domestic wastewater
1,2-Benzanthracene[p]	7 days 7 days	5 mg/l 10 mg/l	3 mg/l 6 mg/l	Domestic wastewater

Table A.2-19. continued

Compound	Time for Degradation	Initial Concentration	Final Concentration	Organism/ Source
Pyrene[p]	7 days 21 days	5 mg/l 10 mg/l	0 mg/l 10 mg/l	Domestic wastewater
Methanol[q]	>30 days >200 days	100 mg/l 1000 mg/l	<lod <lod	Soil (aerobic and anaerobic)
Tertiary butyl alcohol[q]	>1 mo >1 yr	10 mg/l 70 mg/l	<lod <lod	Soil
m-Xylene[r]		0.4 mM		Denitrifying bacteria
Aliphatic and aromatic hydro-carbons (fuels/solvents)[s]	2 1/2 mo	23 ppm	0.05 ppm	
Gasoline[s]	10 mo	30 to 40 ppm	<1 ppm	
Formaldehyde[t]	24 d	>700 ppm	1 ppm	Hydrobac[tm]
Toluene[u]	100 d	10,329 ppm (Groundwater)	>10 ppb	

lod = limits of detection = 50 ppb
a = (Frieze and Oujesky, 1983)
b = (Brown, Loper, and McGarvey, 1985)
c = (Minugh, Patry, Keech, and Leek, 1983)
d = (Jhaveria and Mazzacca, 1982)
e = (Polybac Corporation, 1983)
f = (Walton and Dobbs, 1980)
g = (Ohneck and Gardner, 1982)
h = (Quince and Gardner, 1982)
i = (Pritchard, Van Veld, and Cooper, 1981)
j = (Ehrlich, Schroeder, and Martin, 1985)
k = (Brown, Norris, and Brubaker, 1985)
l = (Environmental Protection Agency, 1985b)
m = (Roberts, Koff, and Karr, 1988)
n = (Scow, Simkins, and Alexander, 1986)
o = (Niaki, Pollock, Medlin, Shealy, and Broscious, Draft)
p = (Tabak, Quave, Mashni, and Barth, 1981)
q = (Novak, Goldsmith, Benoit, and O'Brien, 1985)
r = (Zeyer, Kuhn, and Schwarzenbach, 1986)
s = (Brown, Longfield, Norris, and Wolfe, 1985)
t = (Sikes, 1984)

- After nine months of treatment of soil contaminated with gasoline, soils in a highly contaminated tank storage area still showed signs of gasoline contamination at levels of 500 to 100 ppm and the average concentrations of dissolved carbon (DOC) was 20 ppm (Minugh, Patry, Keech, and Leek, 1983). After the biostimulation program ended, gasoline odors and a cloudy sheen were detected in some of the pits. However, samples showed continued improvement. The DOC in the water had fallen to the point where 71 percent of the measurements fell below 5 ppm and 50 percent were under 2 ppm. Since the DOC levels initially were not given, it is difficult to evaluate the efficiency of the cleanup.
- The contaminant levels in a pumping wells declined from an average of 91 ppm methylene chloride and 54 ppm acetone to less than 1 ppm within a year of treatment (Jhaveria and Mazzacca, 1982). The COD of the groundwater from the pumping well was significantly reduced, also. The area near the pumping wells was the last to be treated and still showed fairly high levels of contamination (up to 40 mg/l of methylene chloride), but the monitoring wells closer to the reinjection system had very low levels of methylene chloride; a reduction from about 70 ppm to less than 0.02 ppm was noted for two of the monitoring wells.
- An acrylonitrile spill contaminated the soil and groundwater of a site in Indiana (Polybac Corporation, 1983). Treatment was by pumping groundwater from several wells to a biotreator seeded with mutant bacteria from Polybac and then injection into the groundwater table. The concentrations of acrylonitrile fell from 1000 to 1 ppm within three months. The report did not contain sufficient details to judge the importance of microbial activity in the removal of the acrylonitrile.
- Groundwater contaminated by 7,000 gal of acrylonitrile was treated by air stripping the recovered groundwater and, after concentrations had been reduced enough to allow microbial growth, by adding mutant bacteria (Walton and Dobbs, 1980). The degradation of acrylonitrile occurred rapidly, as the levels fell from 1,000 ppm to the limits of detection within a month. No data were presented that conclusively linked the drop in the concentration of acrylonitrile to activity by the mutant bacteria.
- An accidental spill of 130,000 gal of organic chemicals entered a 15-ft-thick shallow unconfined aquifer and produced contaminant levels as high as 10,000 ppm (Ohneck and Gardner, 1982). An investigation into stimulating microbial activity showed that the contaminants were biodegradable at concentrations below 1,000 ppm. Both the indigenous microflora and a specific facultative hydrocarbon degrader were able to biodegrade the materials in a soil/water matrix rapidly when supplied with additional nutrients. The biological treatment process accelerated the removal of the compounds, as shown by a series of soil borings during the treatment process, and reduced the levels of contaminants in the groundwater to less than 1 ppm. The specific hydrocarbon degraders did not increase

degradation in laboratory tests beyond that of the native microbes and may not have significantly increased in situ biodegradation. No data were presented that showed that the number of the specific hydrocarbon degraders increased in the subsurface or that they were able to outcompete the indigenous microflora in degrading contaminants.

- Dichlorobenzene, methylene chloride, and trichloroethane contaminated groundwater at levels up to 2,500 mg/l (Quince and Gardner, 1982). The treated water was inoculated with commercial hydrocarbon-degrading bacteria and nutrients injected into the subsurface. The levels of bacteria increased until optimum conditions were established in the reactor and then were injected into the soil. In 2 1/2 months, the levels of methylene chloride fell from 2,500 mg/l to less than 100 mg/l, and the levels of dichlorobenzene fell from 800 mg/l to less than 50 mg/l in a monitoring well. The inoculated bacteria were expected to continue to degrade the contaminants beyond the 95 percent reduction reached before the treatment was terminated. The importance of microbial activity could not be determined from the data presented.

- Soil and rail ballast contaminated with 20,000 gal of 40 percent formaldehyde initially showed a concentration of leachate with greater than 700 ppm formaldehyde (Sikes, 1984). Microorganisms and nutrients were sprayed over the contaminated area. Within 24 days of operation, the concentration had dropped to 1 ppm.

- An artificial outdoor stream channel was dosed with 8 ppm p-cresol for a period of 48 hr, after which time the concentration of the compound decreased rapidly (Pritchard, Van Veld, and Cooper, 1981).

- As toluene levels fell from 10,329 ppb to less than 10 ppb in about 100 days, benzene levels rose, possibly due to demethoxylation of the toluene to benzene (Wilson, Bledsoe, Armstrong, and Sammons, 1986). Then levels of both compounds fell.

2.2.3.2 Low Concentrations

There appears to be a minimum concentration to which a single organic material can be decomposed under steady-state conditions (McCarty, Reinhard, and Rittmann, 1981). Trace organic materials may be biodegradable, but are often below this minimum concentration. In general, biodegradation of such materials can occur only if they are used as secondary substrates; i.e., if there is an abundant primary organic substrate available and bacteria capable of decomposing both. An alternative is decomposition within a nonsteady-state system that has a sufficiently large population of bacteria previously grown on a primary substrate. Biodecomposition is also possible, if several organic substrates are present in a sufficiently large total concentration.

Low levels, or trace concentrations, of some organic substrates would be less than 100 ng/ml, which is characteristic of the concentrations of pollutants in many fresh, estuarine, and marine waters (Alexander, 1986). This level of contamination is also important, since criteria and standards for water quality refer to maximum acceptable levels of many organic pollutants that are below 100 ng/ml (Patrick and Mahapatra, 1968) and since numerous toxicants are harmful at levels in the parts-per-billion range (Batterton, Winters, and van Baalen, 1978).

Microorganisms might not assimilate carbon from chemicals present in trace amounts in natural environments (Alexander, 1985). They might not grow or produce the large, acclimated populations needed for enhanced biodegradation. It is possible to predict the minimum concentration of a chemical necessary to support microbial growth. However, erroneous conclusions may be reached, if data from laboratory studies of chemicals at high concentrations are extrapolated to environments in which the chemicals exist at low concentrations (Alexander, 1986). It is important to use concentrations characteristic of those in nature for laboratory investigations. It is also possible for a chemical to be mineralized at one concentration and cometabolized at another.

Several anomalies have been detected in biodegradation of low concentrations of organic compounds (Alexander, 1986).

1. The rate of mineralization may be less than anticipated if it is assumed that the rates are linearly related to concentration.
2. Chemicals mineralized at one concentration may not be converted to carbon dioxide at lower levels.
3. Organic compounds may not be mineralized at low and presumably nontoxic levels in water, but they may be metabolized to carbon dioxide at still lower concentrations.
4. Mineralization may not follow the commonly described kinetics but may proceed in a biphasic manner.
5. The extent of mineralization in samples from a single body of water may vary markedly.
6. Microbial communities may acclimate to mineralize a substrate even though the substrate concentration is below the threshold level to sustain growth.
7. Compounds may be mineralized in some but not all waters.

Indigenous Microorganisms in Biodegradation

3.1 MICROORGANISMS IN SOIL

Often, microbes capable of degrading a compound may remain in the soil after the last treatment, and pure cultures capable of degrading many of these chemicals can still be isolated several years later (Kaufman, 1983). Hydrocarbonoclastic organisms can be recovered from the deep subsurface. Many bacteria and yeasts that could catabolize methanol under aerobic or anaerobic conditions and at low and high concentrations are isolated from deep soil (to 100 ft) (Benoit, Novak, Goldsmith, and Chadduck, 1985).

Table A.3-1 lists a number of organisms, other than true bacteria, and the compounds they are able to degrade (Kobayashi and Rittmann, 1982).

Table A.3-2 shows the species or groups of microorganisms that might be useful for treating contamination by specific types of organic compounds and the conditions that would be most favorable for their development (Kobayashi and Rittmann, 1982).

3.1.1 Bacteria

Of 41 cultures isolated from the soil and groundwater at a site contaminated with gasoline from a leaking gas station tank, 17 were considered to be Pseudomonas, 4 Flavobacterium, 11 Nocardia, and 9 unidentified (Raymond, Jamison, and Hudson, 1978). Many of the cultures were composed of very small cells.

Small cells have been produced experimentally as a result of starvation (Novitsky and Morita, 1978). These "ultramicrobacteria" are particularly suited to survival at low nutrient concentrations and low population densities. The cells are less than 0.3 um in diameter, demonstrate slow growth, and do not significantly increase in size when inoculated on only a nutrient-rich agar medium. Biochemical characterization might not be performed on them due to lack of sufficient growth of the organisms. The need for minimal nutrient concentration and a prolonged incubation time appears to be extremely important for recovering ultramicrobacteria (Tabor, Ohwada, and Colwell, 1981).

Cellovibrio spp. occur in the natural environment as small filterable particles that cannot be subcultured or reverted to normal cells (Tuckett and Moore, 1959). Aerobacter spp. have also been found to form starvation-resistant, small cells (Harrison and Lawrence, 1963). Dwarf cells, some less than 0.08 um in diameter, have been observed in soil samples with electron microscopy (Bae, Costa-Robles, and Casida, 1972). In the last case, many of the cells are too small to be observed by light microscopy and possess fine structural differences from normal-sized soil bacteria.

The propensity of organisms to be attached also affects the type of compounds they will attack. For instance, free-living bacteria are responsible for a large proportion of the metabolism of m-cresol and chlorobenzene, while particle-associated organisms are more important in the degradation of NTA (Palumbo, Pfaender, Paerl, Bland, Boyd, and Cooper, 1983).

3.1.1.1 Aerobes

Selected aerobic hydrocarbon-degrading bacterial strains are discussed below.

1. Pseudomonas

Pseudomonas appears to be the most ubiquitous, and able to adapt to many different man-made compounds (Kobayashi and Rittmann, 1982). Pseudomonas spp. are responsible for the degradation of most of the aromatics in gasoline (Jamison, Raymond, and Hudson, 1976), although efficiency in degrading aromatic hydrocarbons can vary among strains (Kobayashi and Rittmann, 1982). At optimal conditions of 20°C and after 20 days, one isolate had degraded 70 percent of the aromatic fraction while the other had utilized only 40 percent (Solanas and Pares,

1984).   Degradation of alkylnaphthalenes by this genus was more extensive with the less substituted aromatics. The site of attack appeared to be the ring system rather than the alkyl chains.

P. putida was isolated from aviation fuel and grew rapidly using the jet fuel as the sole source of carbon and energy (Williams, Cumins, Gardener, Palmier, and Rubidge, 1981). It grew on and oxidized a wide range of alkanes, with optimum growth occurring on octane. It was able to oxidize straight chain fatty acids ranging from acetate (C_2) to palmitic acid (C_{16}).

Table A.3-1. Degradation of Anthropogenic Compounds by Different Groups of Microorganisms (Kobayashi and Rittmann, 1982)

Microorganism	Compound
Cyanobacteria (blue-green algae)	
Microcystis aeruginosa	Benzene, toluene, naphthalene, phenanthrene, pyrene
Algae	
Selanastrum capricornatum	Benzene, toluene, naphthalene, phenanthrene, pyrene
Actinomycetes	
Nocardia spp.	*n*-paraffins: pentane, hexane, heptane, octane, 2-methylbutane, 2-methylpentane, 3-methylheptane, 2,2,4-trimethylpentane, ethylbenzene[a], hexadecane and kerosene (at 2% but not 4%)[b]
Yeasts	
Trichosporon, Pichia Rhodosporidium, Rhodotorula, Debaryomyces, Endomycopsis,	hexadecane and kerosene (at 2% but not 4%)[b]
Candida parapsilosis, C. tropicalis, C. guilliermondii, C. lipolytica, C. maltosa, Debaryomyces hansenii, Trichosporon sp., *Rhodosporidium toruloides*	(naphthalene, biphenyl, benzo(a)pyrene)[c]

a = (Jamison, Raymond, and Hudson, 1976)
b = (Ahearn, Meyers, and Standard, 1971)
c = (Cerniglia and Crow, 1981)

Table A.3-2.   Selective Use of Microorganisms for Removal of Different Anthropogenic Compounds (Kobayashi and Rittmann, 1982)

Microorganism	Selective[a] Characteristics	Significance
Fungi		
Yeast	pH < 5, ae-mae; high O_2 tension, pH < 5 moisture about 50%	Attacks and partially degrades compounds not readily metabolized by other organisms. Wide range of nonspecific enzymes
Mold		
Algae	ae-mae; light: 600 to 700 nm; low carbon flux	Self-sustaining population, light is primary energy source, partially degrades certain complex compounds, photochemical reactions, oxygenates medium during daylight, supports growth of other microbes, effective in bioaccumulation of hydrophobic substances
Cyanobacteria (blue-green algae)	ae-mae, an; light: 600 to 700 nm; low carbon flux	See algae
Bacteria		
Heterotrophs (aerobic)	ae; proper organic substrate, growth factors as required; Eh: 0.45 to 0.2V	For many compounds degradation is more complete and faster than under anaerobic conditions; high sludge production
Anaerobic (fastidious)	an; Eh: < -0.2 to -0.4 V	Conditions for abiotic or biological reductive dechlorination, certain detoxification reactions not possible under aerobic conditions; no aeration, little sludge produced
Facultative anaerobes	ae, mae-an; Eh: < -0.2V	No aeration, reductive dechlorination possible

Table A.3-2.   continued

Microorganism	Selective[a] Characteristics	Significance
Photosynthetic bacteria		
Purple sulfur	an (light), mae (dark); Eh: 0 to -0.2 V; S^{-2}: 2 to 8mM, 0.4 to 1 mM; light: 800 to 890 nm at 1000 to 2000 lux, high intensities near limit; low C flux	Self-sustaining population able to use light energy, conditions right for reductive dechlorination, no aeration
Purple nonsulfur	an; Eh: 0 to -0.2 V; light: 800 to 890 nm; low C flux	See purple sulfur bacteria, also nonspecific enzymes
Actinomycetes	ae, moisture: 80 to 87%, temp.: 23 to 28°C, urea as nitrogen source	Universal scavengers with range of complex organic substrates often not used by other microbes
Oligotrophs (from almost any group above)	ae; carbon flux of <1mg/l/d; favorable attachment sites	Removal of organic contaminants intrace concentrations, many inducible enzymes for multiple substrates

ae = aerobic; mae = Microaerophilic (<0.2 atm oxygen); an = anaerobic
a = Possible characteristics for selection, not growth range

A mutant of P. putida CB-173 (ATCC 31800), which is active at temperatures as low as 1° to 4°C, has been developed by Sybron Corp. for degrading phenolic wastewaters during the winter months (Roberts, Koff, and Karr, 1988).

Pseudomonas spp. were found to grow on aromatic constituents of gasoline as a sole source of carbon (Jamison, Raymond, and Hudson, 1976). Moderate to heavy growth was obtained with some isolates on n-pentane, n-hexane, n-heptane, n-octane, 2-methylbutane, toluene, and m- and p-xylene. Light or questionable growth was observed on 3-methylheptane and 2,2,4-trimethylpentane.

P. oleovorans is being used to convert C_6 to C_{12} alkanes to 1,2-epoxides and n-alkanes first to primary alcohols, then to aldehydes, then to carboxylic acids (Roberts, Koff, and Karr, 1988).

2. Beijerinckia

Strains of this genus are very active in hydrocarbon degradation. After growth on succinate in the presence of biphenyl, a mutant strain oxidized benzo(a)pyrene and benzo(a)anthracene to cis-9, 10-dihydroxy-9, 10-dihydro-benzo(a)pyrene and cis-1, 2-dihydroxy-1, 2-dihydrobenzo(a)anthracene, respectively (Gibson, 1976).

3. Acinetobacter

These organisms have been frequently encountered as hydrocarbon degraders. They grow rapidly with cyclohexanol as the sole source of carbon, with attack of cyclohexane by other organisms being initiated by hydroxylation (Donoghue, Griffin, Norris, and Trudgill, 1976).

4. Actinomycetes

Morphologically intermediate between bacteria and fungi, these organisms (which are considered bacteria) are found in environments in which unusual compounds are encountered (Kobayashi and Rittmann, 1982). They attack a wide variety of complex organic compounds, including phenols, pyridines, glycerides, steroids, chlorinated and nonchlorinated aromatic compounds, paraffins, other long-chain carbon compounds, and lignocellulose.

A common actinomycete in aquatic systems is Nocardia; e.g., N. aramae, which grows under low nutrient conditions (oligotrophically), such as in distilled water. N. globerula grows rapidly with cyclohexanol as the sole source of carbon, with attack of cyclohexane by other organisms being initiated by hydroxylation (Donoghue, Griffin, Norris, and Trudgill, 1976).

Advantages to using these organisms in surface treatment systems are that they produce less sludge than bacteria and fungi, grow over a wide temperature range (from psychrophilic to thermophilic), are resistant to desiccation, and have a wide pH range (Kobayashi and Rittmann, 1982). Nocardia spp. are able to utilize n-paraffins from gasoline as a sole source of carbon (Jamison, Raymond, and hudson, 1976) (see Table A.3-1). Actinomycetes might be especially useful in treatment of contaminated soil where a composting technique would be practical (Kobayashi and Rittmann, 1982).

Organic decomposition by actinomycetes results in metabolites that can be mineralized by other organisms. Therefore, mixed cultures are necessary, if actinomycetes are to be used.

5. Pseudomonas methanica

This organism has the ability to use only methane and methanol for growth (Zajic, 1964). It can also cometabolize ethane, propane, and n-butane.

3.1.1.2 Anaerobes

3.1.1.3 Oligotrophs

3.1.1.4 Counts in Uncontaminated Soil

Direct count in uncontaminated soil ranged from 10^6 to 10^7 organisms/g in the literature, while viable counts were reported from zero to 10^8 CFU/g. Table A.3-3 compares counts in different soil types at different depths. These counts assumed that there are 50 umoles phospholipid/g dry weight of bacteria and that there are 10^{12} bacteria/g (Gehron and White, 1983). However, these estimates may be low since subsurface bacteria are smaller than surface bacteria, due to severe nutrient limitation (Webster, Hampton, Wilson, Ghiorse, and Leach, 1985), and a gram of bacteria may contain many more than 10^{12} organisms/g of soil. Apparently, many of the bacteria in soil environments exist as cells smaller than 1 um in size (Ghiorse and Balkwill, 1985b).

Table A.3-3. Plate Counts _versus_ Soil Depth and Type (Federle, Dobbins, Thornton-Manning, and Jones, 1986)

Soil Type/Horizon	Approximate Depth (cm)	Plate Count (CFU/g)
A	0 to 20	10^5 to 10^8
Ap	0 to 12	7.1×10^6
Sterrett/B23t	100 to 125	1.3×10^6
Dewey/B23t	50 to 120	2.5×10^5
Whitwell/C	95 to 200	9.9×10^4
Nauvoo/B3t	80 to 105	4.7×10^4

Other investigations used two methods for measuring bacterial populations in soil at different depths (Novak, Goldsmith, Benoit, and O'Brien, 1985). These detected considerable differences among plate counts but little variation in direct counts with depth. These counts are presented in Table A.3-4 for samples taken from different sites.

Table A.3-4. Bacterial Populations in Subsurface Soils (Novak, Goldsmith, Benoit, and O'Brien, 1985)

Depth (m)	Soil Extract (CFU/g)	AO Direct Count (Organisms/g)
Site 1		
0	9.7×10^6 +- 5.7×10^5	7.6×10^6 +- 3.0×10^6
3	$< 10^3$	5.4×10^6 +- 2.7×10^6
4.5	3.3×10^6 +- 4.0×10^5	2.3×10^6 +- 1.6×10^6
9	5.6×10^5 +- 7.1×10^3	2.9×10^6 +- 2.5×10^6
Site 2		
0	3.0 +- 0.3×10^7	5.6 +- 1.9×10^7
3-4	3.5 +- 2.1×10^3	3.9 +- 1.4×10^7

A number of studies were conducted to determine bacterial counts in uncontaminated soil. These are summarized in Table A.3-5 and described below.

Table A.3-5. Summary of Viable and Direct Counts in Uncontaminated Soils from Several Studies

Viable Counts (CFU/g)	Direct Counts (Organisms/g)
[a]8×10^3 to 3.4×10^6	[a]3.4×10^6 to 9.8×10^6
[b]1.5×10^5 to 8×10^5	[b]4×10^6 to 9×10^6
	[b]1.2×10^7 to 1.6×10^7
[c]5×10^5	[c]10^6
	[d]2.1×10^7

a = One study determined the number of bacteria in the B and C horizons of an undefined soil series by use of fluorescent microscopy (Wilson, McNabb, Balkwill, and Ghiorse, 1983). The number of bacteria did not decline with depth and ranged from 3.4 to 9.8×10^6 bacteria/g dry wt soil. The viable cell counts were more variable, ranging from 8×10^3 to 3.4×10^6.

b = Another investigation used fluorescent microscopy on uncontaminated soil (Webster, Hampton, Wilson, Ghiorse, and Leach, 1985). Total counts ranged from 4×10^6 to 9×10^6 bacteria/g at one site and 1.2×10^7 to 1.6×10^7 bacteria/g at another. Only a small percentage (<5 percent) of these cells were actively respiring, as measured by their ability to reduce INT. Based on this, the active bacteria ranged from 1.5×10^5 to 8×10^5 bacteria/g.

c = The microflora of saturated and unsaturated subsurface samples (depths of 4 to 16 ft) were examined (Balkwill and Ghiorse, 1982). Total cells, determined by epifluorescence (EF) light microscopy counts of acridine orange-stained preparations, numbered 10^6/g dry weight in all samples. The population appeared to be entirely bacterial. The predominant cell types were small, coccoid rods, mainly gram positive. Plating on soil extract agar showed that at least 50 percent of the cells counted by EF were viable. Counts on a nutritionally rich medium were three to five orders of magnitude lower.

d = High permeability subsurface soils in a pristine area contained 2.1×10^7 cells/g dry soil using acridine orange direct counts (Thomas, Lee, Scott, and Ward, 1986).

3.1.1.5 Counts in Contaminated Soil

Direct counts in contaminated soil were found to range from 10^3 to 10^8 organisms/g, while viable counts were reported from less than 100 to 10^6 CFU/g. These results were derived from the studies below and are summarized in Table A.3-6.

Table A.3-6. Summary of Viable and Direct Counts in Contaminated Soils from Several Studies

Viable Counts (CFU/g)	Direct Counts (Organisms/g)	Hydrocarbon-degraders (Organisms/g)
[a]10^3 to 10^5	[a]7.8×10^6	[a]8.5×10^5 [a]1.2×10^5
	[b]10^6	
[c]<100 to 7×10^6	[c]7.6×10^6 to 1.7×10^8	
		[d]10^3

a = High permeability subsurface soils in an area contaminated with jet fuel contained 7.8×10^6 cells/g dry soil using acridine orange direct counts (Thomas, Lee, Scott, and Ward, 1986). Viable counts were one to three orders of magnitude lower; they were higher in contaminated than uncontaminated soil. Additions of 1000 ppb of benzene and 1000, 100, 10, and 1 ppb toluene could not be detected after 4 weeks. The MPN of benzene and toluene degraders in contaminated soil was 8.5 $\times 10^5$ and 1.2×10^5 cells/g dry soil, while none of these organisms were detected in uncontaminated soil. This indicates the microflora exposed to jet fuel have adapted and multiplied to degrade these compounds.

b = Core samples collected from petroleum-contaminated and uncontaminated soil revealed an even distribution of bacteria for both soil conditions from 0.3 to 2.0 m (10^6 bacteria/g dry weight of soil) (Stetzenbach, Kelley, Stetzenbach, and Sinclair, 1985). However, bacteria isolated from the contaminated soil were able to degrade naphthalene more quickly in the laboratory than the isolates from the uncontaminated soil. Some PAHs (fluorene, anthracene, pyrene, and naphthalene) were used as sole carbon sources, indicating utilization by the indigenous population.

c = In contaminated soil from Kelly Air Force Base, direct counts of organisms from subsurface samples ranged from 7.6×10^6 to 1.7×10^8 cells/g; viable cells counts ranged from less than 100 to 7×10^6 cells/g (Wetzel, Davidson, Durst, and Sarno, 1986). Similar yields of cells for seven different substrate media indicated the presence of highly adaptive bacteria.

d = A natural flora of gasoline-utilizing organisms were present at levels of 10^3/ml (Jamison, Raymond, and Hudson, 1975) in an area contaminated with over 3000 barrels of high-octane gasoline. This population was increased a thousand-fold by supplementing the groundwater with air, inorganic nitrogen, and phosphate salts.

3.1.2 Fungi

Naphthalene oxidation predominates in the order Mucorales, which includes species of Cunninghamella, Syncephalastrum, and Mucor (Cerniglia, Hebert, Szaniszlo, and Gibson, 1978). Cunninghamella elegans oxidizes naphthalene and benzo(a)pyrene (Zajic and Gerson, 1977). Other degradative fungi are Neurospora crassa, Claviceps paspale, and Psilocybe strains.

Phanerochaete chrysosporium is a white-rot fungus with a demonstrated ability to degrade chlorinated organics to carbon dioxide in pure liquid culture (Lamar, Larsen, Kirk, and Glaser, 1987). It has also been shown to degrade slowly benzo(a)pyrene in nutrient nitrogen-deficient cultures (Bumpus, Tien, Wright, and Aust, 1985). This ability suggests that the fungus may have potential as an in situ hazardous waste degrader (Lamar, Larsen Kirk, and Glaser, 1987). Degradation of mixtures of complex hydrocarbons by P. chrysosporium often proceeds faster than the rate of degradation of the pure chemicals (Bumpus, Fernando, Mileski, and Aust, 1987). Toxicity of chemicals to this fungus is rare but can be circumvented by using mature mycelia instead of fungal spores.

Soil type has a significant effect on growth and growth habit (Lamar, Larsen, Kirk, and Glaser, 1987). Nitrogen content appears to play a major role in mediating growth of this fungus in the soils. Increasing soil water potential from -1.5 MPa to -0.03 MPa results in greatly increased growth of this organism. It might benefit from soil water potentials above -0.03 MPa. The fungus is mesophilic and has a temperature optimum of about 30°C on 2 percent malt agar (Kirk, Schultz, Connors, Lorenz, and Zeikus, 1978). Growth of the fungus is significantly greater at 30° and 39°C than at 25°C (Lamar, Larsen, Kirk, and Glaser, 1987). Organic matter content of the soil does not affect growth.

The ability of supplemental glucose to increase the rate and extent of biodegradation of DDT suggests a dependency upon the availability of a carbon source that can serve as a growth substrate (Bumpus, Fernando, Mileski, and Aust, 1987). The glucose may simply allow an increase in the overall rate of fungal metabolism or it may provide the substrate for fungal production of hydrogen peroxide, a required cosubstrate for ligninases that are partly responsible for oxidation of many organopollutants. Bulking agents, such as wood chips or corn cobs, can also serve as a carbon source for the fungus. Several Candida spp. are able to metabolize naphthalene, biphenyl, and benzo(a)pyrene (see Table A.3-1).

A strain of soil yeast, Trichosporon cutaneum, which uses phenol in preference to glucose, has been reported (Shoda and Udaka, 1980). This unusual metabolic response could be valuable in transforming xenobiotics. T. cutaneum has a broad specificity of enzyme induction and is actually as versatile in aromatic catabolism as any pseudomonad (Dagley, 1981).

3.1.3 Photosynthetic Microorganisms

Some cyanobacteria, e.g., an Oscillatoria sp., can oxidize hydrocarbons, such as naphthalene and biphenyl to a limited extent when growing photoautotrophically (Fewson, 1981). Some green, brown, and red algae and diatoms can partially oxidize hydrocarbons, such as naphthalene.

Species of Oscillatoria, Microcoleus, Anabaena, Cocochloris, Nostoc, Chlorella, Dunaliella, Chlamydomonas, Ulva, Cylindretheca, Ampora, and Porphyridium have been found to be capable of oxidizing naphthalene (Cerniglia, Gibson, and Baalen, 1980b).

A few species, such as Rhodopseudomonas palustris (Gottschalk, 1979) or Rhodocyclus purpureus (Pfennig, 1978b), can use thiosulfate or sulfide as an electron donor in addition to organic compounds. Generally, however, the sulfide concentration must be low (Schnitzer, 1982). R. palustris can use benzoate as the sole substrate under aerobic conditions via respiration or anaerobically by photometabolism (Proctor and Scher, 1960). The enzyme system used by this organism to photometabolize aromatic substrates is inducible and lacks substrate specificity (Dutton and Evans, 1969).

An organic compound may be nontoxic in the amounts that exist free in the water or outside the microbial cell in soil, but if the chemical is subject to bioconcentration, species at higher trophic levels may be harmed (Alexander, 1986). Phototrophic organisms can bioaccumulate hydrophobic compounds (Kobayashi and Rittmann, 1982). The organic compounds that can be accumulated by phototrophs as they break down other organic compounds are shown in Table A.3-7. Bioaccumulation of pollutants is also observed in other organisms (Finnerty, Kennedy, Lockwood, Spurlock, and Young, 1973). For example, Acinetobacter sp., yeasts, and filamentous fungi can accumulate hydrocarbons in their cytoplasm.

A scheme has been devised to separate and treat refractory compounds (Kobayashi and Rittmann, 1982). Degradable compounds are treated directly by the organisms. Refractory compounds are removed by sorption to microorganisms, concentrated, and may be disposed of by other means, such as burial or incineration. Relatively recalcitrant compounds can be degraded if they are first sorbed. This could be accomplished with algae. Other microorganisms may then further degrade intermediate products.

3.1.4 Higher Life Forms and Predation

3.1.5 Cometabolism

Cometabolism is an important process in the breakdown of polycyclic aromatic hydrocarbons (Texas Research Institute, Inc., 1982). It may play a major role in PAH degradation, such as phenanthrene (PHE), in polluted sediments (Shiaris and Cooney, 1981). In sediments, PHE utilizers ranged from <5 x 10^2 to 5.54 x 10^5 CFU/g dry sediment, which corresponded to <0.1 to 1.8 percent of the total viable count. PHE cooxidizers ranged from 1 x 10^2 to 1.12 x 10^6 CFU/g dry sediment, which corresponded to <0.1 to 1.6 percent of the total viable count.

Table A.3-7. Bioaccumulation of Anthropogenic Compounds by Phototrophs (Kobayashi and Rittmann, 1982)

Microorganism	Compound
Cyanobacteria	
Microcystis aeruginosa	Benzene, toluene, naphthalene, phenanthrene, pyrene
Algae	
Selanastrum capricornatum	Benzene, toluene, naphthalene, phenanthrene, pyrene

Soil microorganisms capable of cometabolically degrading a given xenobiotic may not be enriched; i.e., there may not be proliferation at the expense of the xenobiotic substrate (Kaufman, 1983). Therefore, unique techniques may be necessary to isolate such organisms. Such techniques could possibly utilize, as a primary substrate, a chemical analog that would permit enrichment of microbial populations having the necessary cometabolic requirements to degrade the xenobiotic.

3.1.6 Microbial Interactions

Mutualistic relationships (all members derive some benefit) are based not only upon growth-factor interdependence, but may also encompass removal of a (toxic) product of metabolism produced by one component of the mixed population and used by another (483), combined metabolic attack (Johanides and Hrsak, 1976), or relief of substrate inhibition (Osman, Bull, and Slater, 1976). Some microorganisms thrive on metabolic products or products from lysis of other organisms, as a result of a commensalistic relationship (Harder, 1981). For example, a Nocardia sp. has been identified as the primary cyclohexane utilizer, but growth occurred only in the presence of an unidentified pseudomonad that provided biotin and possibly other growth factors (Sterling, Watkinson, and Higgins, 1977).

3.2 MICROORGANISMS IN GROUNDWATER

Table A.3-8 lists some of the growth requirements for microorganisms present in groundwater (Bitton and Gerba, 1985).

3.2.1 Bacteria

3.2.1.1 Aerobes

3.2.1.2 Anaerobes

3.2.1.3 Counts in Uncontaminated Groundwater

Bacterial populations, measured by acridine orange direct microscopy and by plate count, ranged from nondetectable to as high as 10^7 cells or CFU/g of sediment (Fredrickson and Hicks, 1987). Some of the highest counts were from the Tuscaloosa aquifer that attains depths to 400 m below the surface. The greatest diversity and activity of microorganisms and the highest population densities were observed in the sandy water-bearing strata; whereas the dense, dry-clay layer zones had the least microbiological activity.

Using direct counting methods (acridine orange), bacterial populations were not shown to decline appreciably with depths to 800 ft in some wells (Federle, Dobbins, Thornton-Manning, and Jones, 1986).   In contrast, indirect plating methods showed declines in counts to zero with depth, indicating that bacteria persist and even grow at relatively deep depths but that they will not grow on plating media (e.g., are probably fastidius and oligotrophic).

Table A.3-8.   Microbial Growth Requirements in the Groundwater Environment (Bitton and Gerba, 1985)

Parameter	Comments
Bacterial Concentrations	Almost 0 to 10^6/ml
Environmental Factors:	
Temperature:	Increases with depth ($3°C/100$ m)
	May preclude growth at great depths
Osmotic pressure:	May have an effect only in saline aquifers
Hydrostatic pressure:	Not likely to exclude microbial growth
pH:	Not likely to exclude microbial growth
Redox potential:	Certain microorganisms may be excluded in very reduced environments
Role of surfaces:	Biodegradation probably occurs on surfaces in the aquifer material
Growth Requirements:	
Carbon sources:	Inorganic C: Carbonates and bicarbonates
	Organic C: "Humic" substances and wastewater organics, some of which are recalcitrant
Mode of utilization of organics:	At very low concentrations, organics may be utilized as secondary substrates
Other elements:	N,P,S,Na,Ca,Mg,...may be present in sufficient quantities to allow growth
Electron acceptors: (Respiration)	O_2: Absent in most deep aquifers, but other electron acceptors, such as NO_3^- and SO_4^{2-} may be available
Sampling for Groundwater Indigenous Microorganisms	Methodological difficulties due mostly to contamination during drilling

Table A.3-9 shows the similarity of counts recovered from different depths of shallow water-table aquifers and vadose zones (Ghiorse and Balkwill, 1983).

Table A.3-9.   Numbers of Organisms in the Subsurface Environment (Ghiorse and Balkwill, 1983)

Site	Depth to Water Table (m)	Subsoil[a]	Just Above Water Table[a]	Just Below Water Table[a]
Lula, OK				
February 1981	3.6	6.8	3.4	6.8
June 1981	6.0	9.8	3.7	3.4
Fort Polk, LA	6.0	3.4	1.3	3.0
"	5.0	7.0	1.3	9.8
Conroe, TX	6.0	0.5	0.3	0.6
LongIsland, NY	6.0	---	---	36
"	3.0	170	---	---
Pickett, OK	5.0	---	---	5.2

[a]In millions per gram dry material

Other studies also supported the findings that counts are similar at different depths (Novak, Goldsmith, Benoit, and O'Brien, 1985). Table A.3-10 shows the number of bacteria detected at various levels of the saturated zone in samples taken from different sites.

Table A.3-10. Bacterial Populations in Aerobic and Anaerobic Aquifers (Novak, Goldsmith, Benoit, and O'Brien, 1985)

Depth (m)	Plate Counts Soil Extract (CFU/g)	AO Direct Count (bacteria/g)
Site 1		
15 to 17	$5.2 \times 10^6 \pm 9.2 \times 10^5$	$8.0 \times 10^6 - 4.7 \times 10^6$
24 to 25	$9.8 \times 10^5 \pm 5.3 \times 10^4$	$5.4 \times 10^6 - 4.1 \times 10^6$
31	$1.1 \times 10^5 \pm 2.8 \times 10^4$	$3.5 \times 10^6 - 3.3 \times 10^6$
Site 2		
9	$1.4 \times 10^5 \pm 0.8 \times 10^5$	$4.6 \times 10^7 - 2.7 \times 10^7$
Site 3		
0	$1.0 \times 10^7 \pm 4.0 \times 10^6$	$1.0 \times 10^8 - 4.1 \times 10^7$
2	$9.3 \times 10^5 \pm 1.1 \times 10^5$	$7.6 \times 10^7 - 3.8 \times 10^7$
3 to 4	$1.1 \times 10^6 \pm 8.5 \times 10^4$	$8.0 \times 10^7 - 7.4 \times 10^7$

The viable and direct counts obtained in groundwater samples by other researchers are and summarized in Table A.3-11 and described below Table A.3-11. As was observed with soil samples and the previous two studies, direct counts showed greater consistency with depth than did plate counts. Overall, direct counts in the saturated zone averaged 10^6 to 10^7 organism/g, ranging from 10^4 to 10^8 organism/g. On the other hand, the viable counts ranged from 10^{-2} to 10^6 CFU/g.

Table A 3.11. Summary of Viable and Direct Counts in Uncontaminated Groundwater from Several Studies

Viable Count (CFU/ml or /g)	Direct Count (Organisms/ml or /g)
[a]10^3 to 10^5	
[b]3.4×10^6 (February, 5 m)	[b]6.8×10^6 (February, 5 m)
[b]3.4×10^6 (June, 5 m)	[b]4.1×10^5 (June, 5 m)
	[c]10^6 (shallow)
	[d]10^4 (20 ft)
	[d]10^5 (450 ft)
[e]1.4×10^5 (10 m)	[e]3.4×10^6 (10 m)
[f]10^{-2} to 10^2 (4.5 to 5.5 m)	[f]5.5×10^7 to 2.2×10^8 (4.5 to 5.5 m)
[g]10^4 to 10^6	[g]10^6 to 10^7
[h]7×10^3 to 9×10^3	1.21×10^6

a = Sixty percent of the water samples from 621 wells had a plate count of aerobic bacteria of 10^3 to 10^5, 7 percent exceeded a count of 10^6, and 17 percent had a count of zero (McCabe, Symons, Lee, and Robeck, 1970). This showed that groundwater is not sterile.

b = Similar numbers of organisms were found at levels of 1.2, 3.0, and 5.0 m, (where the water table was at 3.6 m and the bedrock at 6.0 m), with a total count ranging from 3.4×10^6 to 6.8×10^6/g dry material at 5 m, and a viable count ranging from 4.1×10^5 to 3.4×10^6/g dry material at 5 m (Wilson, McNabb, Balkwill, and Ghiorse, 1983). The indigenous bacteria recovered could rapidly degrade toluene.

c = In shallow wells, bacterial concentration was shown to be as high as 10^6 bacteria/ml, using epifluorescence microscopy (Ladd, Ventullo, Wallis, and Costerton, 1982).

d = In wells 20-ft and 450-ft deep, 10^4 and 10^5 bacteria/ml have been found, respectively (Bitton and Gerba, 1985).

e = Groundwater (saturated zone, 10 m deep) had a microscopic count of 3.4×10^6 bacteria/g and an agar plate count of 1.4×10^5 CFU/g (Benoit, Allen, and Novak, 1984). The unsaturated soil above the groundwater had a microscopic and plate count of 2.9×10^6 and 3.5×10^3, respectively (see Table A.3-5).

f = Acridine orange direct counts (AODC) were used to determine a population of 5.5×10^7 to 2.2×10^8 cells/g of soil from a pristine aquifer (Swindoll, Aelion, Dobbins, Jiang, Long, and Pfaender, 1988). The water table occurred at 3.6 m and the bedrock, at 6 m. A fine sand from a depth of 4.5 to 5.5 m was evaluated. Viable bacteria in this soil varied between 0.01 and 1.0×10^2 cells/g. The number of bacteria actually found to be degrading the test substrates in this study was a fraction of the total counts but higher than the viable plate count estimates.

g = Subsurface samples from Oklahoma and Texas contained 10^6 to 10^7 cells/g of subsurface material at depths of 2 to 9 m (Webster, Hampton, Wilson, Ghiorse, and Leach, 1985). Between 1 and 10 percent of the cells were metabolically active, a significant number for modifying pollutants.

h = A groundwater microbial community was found to have counts comparable to those in oligotrophic surface waters (Larson and Ventullo, 1983). AODC counts were 1.21×10^6/ml; MPN, 10^4 to 10^5/ml; and PC, 7 to 9×10^3/ml. The number of bacteria in Canadian groundwater is similar to that found in a pristine groundwater site in Germany and two- to ten-fold lower than those at a polluted site in the same aquifer. The variation in AODC observed from different geographic areas may be due to differences in sampling methods or inherent variation in the biomass in diverse groundwaters; however, too few data are currently available to generalize on this observation.

3.2.1.4 Counts in Contaminated Groundwater

3.3 MICROORGANISMS IN LAKE, ESTUARINE, AND MARINE ENVIRONMENTS

Organisms in eutrophic lakes and ponds are better able to degrade organic compounds than organisms in oligotrophic systems. A greater number of heterotrophic microorganisms can be supported in dense algal communities, which contributes to hydrocarbon degradation.

Exposure of water environments to natural oil seepages and accidental spillages can rapidly alter the composition of the biological communities (Atlas, Schofield, Morelli, and Cameron, 1976). Some of the changes in composition are beneficial in removing the oil, but they may also result in the disappearance of some members of the ecosystem. The effects may be temporary, but the Arctic biological communities recover slowly, if at all.

For example, water collected from Alaskan coastal waters and incubated with Prudhoe crude oil resulted in an increase in the bacterial populations by several orders of magnitude (Atlas, Schofield, Morelli, and Cameron, 1976). The amoeboid protozoa were replaced by flagellated protozoa, coccoid green algae completely disappeared, diatoms increased, while blue-green and green filamentous algae appeared to be unaffected. On the other hand, the organisms present in water associated with a natural oil seepage were very different. There was no vascular plant cover and, in some areas, no bacteria. Fungi (Rhodotorula sp., Candida sp., and a Mucor sp.) were abundant in the bacteria-free regions, with lichen flourishing in older sections. These populations appeared to be adapted to this environment. The fungi were probably tolerant to both the oil and the low pH.

The immediate impact of a spill of fuel oil #2 on a freshwater ecosystem was studied during a 7-day spill period and a 7-day recovery period (Johnson and Romanenko, 1983). An immediate complete collapse of phytoplanktonic an zoophytoplanktonic populations was found. There were dramatic changes in microbiota of sediments, both in functional and microbial diversity. The bacteria were all small rods and cocci; bacterial oil degraders increased 1000 times and saprophytic bacteria increased 100 times. The dissolved oxygen and redox potential decreased. There was no mortality in the fish population.

When contained oil slicks were floated in Prudhoe Bay in Alaska, the bacterial populations directly underlying the slicks increased, especially when the slick was supplemented with an oleophilic fertilizer (octyl phosphate and paraffinized urea) (Atlas, Schofield, Morelli, and Cameron, 1976). The increase was in the oil-degrading psychrophilic Pseudomonas and Staphyloccus epidermidis, a nonhydrocarbon-utilizing mesophile. These organisms were able to degrade the oil extensively and remove the pollutant. The rest of the normal population was unaffected by the oil. Fungi and algae were in concentrations of less than 10/ml and did not increase, even under the fertilized slicks.

Initial toxicity of fuel components can retard biodegradation. When a mixture of 15 hydrocarbons, representative of those in jet fuels, was added to water and sediment containing natural microbial communities collected from estuarine and freshwater sites, C_6 to C_9 compounds (hexane, cyclohexane, n-heptane, methylcyclohexane, toluene, n-octane, ethylcyclohexane, p-xylene, cumene, 1,3,5-trimethylbenzene, n-tetradecane, and 2,3-dimethylnaphthalene) volatilized quickly (Somerville, Butler, Lee, Bourquin, and Spain, 1983). The fuel mixture was initially toxic, causing a 24-hour lag period. However, after that time, the less volatile indan, naphthalene, and 2-methylnaphthalene were rapidly biodegraded.

Motorboat oils in river water could be degraded using microorganisms in activated sewage sludge as an inoculum (Ludzack and Kinkead, 1956). Reseeding was necessary to maintain the degradation. Rates of degradation increased when elevated temperatures, increased aeration, nutrient supplementation, and oil emulsification were coupled with seeding.

Table A.3-12 lists the most common hydrocarbon-degrading microorganisms isolated from aquatic environments (Bartha and Atlas, 1977).

Table A.3-12. Hydrocarbon-Degrading   Microorganisms   Isolated   from   Aquatic Environments (Bartha and Atlas, 1977)

Bacteria	Fungi	Algae
Achromobacter	Aspergillus	Cytophaga[d]
Acinetobacter	Aureobasidium	Prototheca
Actinomyces	Candida	
Aeromonas	C. guilliermondif[c]	
Alcaligenes	C. lipolytica[c]	
Arthrobacter	C. parapsilosis[c]	
Bacillus	C. tropicalis[c]	
Bacterium	Cephalosporium	
Beneckea	Cladosporium	
Brevibacterium	Cunninghamella	
Cladosporium resinae[a]	Debaryomyces[c]	
Corynebacterium	Endomycopsis[c]	
Flavobacterium	Hansenula	
Micrococcus	Penicillium	
Mycobacterium	Pichia[c]	
Nocardia	Rhodosporidium	
Proactinomyces	R. toruloides[c]	
Pseudobacterium	Rhodotorula	
Pseudomonas	Saccharomyces	
Sarcina	Sporobolomyces	
Spirillum	Torulopsis	
Staphylococcus epidermidis[b]	Trichosporon[c]	
Thermomicrobium[e]		
Vibrio		
Xanthomonas[d]		

a = (Colwell, Walker, and Nelson, 1973)
b = (Atlas, Schofield, Morelli, and Cameron, 1976)
c = (Ahearn, Meyers, and Standard, 1971)
d = (Atlas, 1981)
e = (Merkel, Stapleton, and Perry, 1978)

3.3.1 Counts

Hydrocarbon utilizers are widely distributed, being found in similar concentrations in temperate to Arctic marine environments (Atlas, 1978a). Multiplication of marine bacteria depends upon the composition of the medium and on temperature (Butkevich and Butkevich, 1936). A considerable portion of the bacteria in the sea are probably present in a resting stage.

Approximately 100 oil-degrading bacteria/l were found in most arctic coastal waters tested (Atlas and Busdosh, 1976). The largest number was

500/l from a saltwater pond, and the smallest was 2/l from a freshwater pond. Populations of hydrocarbonoclastic microorganisms were found to occur in concentrations 10 to 100 times greater in the surface ocean layer than at a 10-cm depth (Crow, Cook, Ahearn, and Bourquin, 1976).

The distribution of hydrocarbon-utilizing microorganisms reflects the historical exposure of the environment to hydrocarbons (Bartha and Atlas, 1977). In polluted sediments, hydrocarbon degraders reached 10^8 CFU/ml (ZoBell and Prokop, 1966). In polluted seawater, up to 10^6 CFU/ml were reported (Polyakova, 1962). In marine environments, hydrocarbon degraders have been found in higher concentrations in regions chronically polluted with petroleum (Environmental Protection Agency, 1985b). Increased rates of uptake and mineralization of [^{14}C]hexadecane were observed for bacteria in samples collected from an oil-polluted harbor compared with samples from a relatively unpolluted, shellfish-harvesting area, with turnover times of 15 and 60 min for these areas, respectively (Walker and Colwell, 1976b). Hydrocarbon-utilizing microbial populations were 5 to 12 times greater in lake sediment after chronic petrogenic chemical exposure than in sediment from an uncontaminated ecosystem (Heitkamp, Freeman, and Cerniglia, 1987). The fraction of the total heterotrophic bacteria represented by the hydrocarbon utilizers ranged up to 100 percent, depending upon the area's previous history of oil spillage; most values were less than 10 percent (Mulkins-Phillips and Stewart, 1974a). Comparison of data from various investigators is difficult because of the widely varying enumeration techniques employed (Bartha and Atlas, 1977).

Microbial Degradation and Transformation of Petroleum Constituents and Related Elements

Microorganisms require energy to maintain themselves (Alexander, 1980). They must carry out oxidations to obtain sufficient energy for their essential functions. Different microorganisms utilize different metabolic processes to derive their energy, and based on what they use for an energy source, these organisms can be classified as autotrophic or heterotrophic (Davis, Dulbecco, Eisen, Ginsberg, and Wood, 1970).

Autotrophic bacteria are able to live in a strictly inorganic environment, utilizing carbon dioxide or carbonates as a sole source of carbon (Zinsser, 1960). The chemosynthetic autotrophs (also called lithotrophs) obtain their energy by oxidation of an inorganic substrate (e.g., iron, sulfur, ammonia, nitrite). Included with the chemoautotrophs are the hydrogen bacteria, which derive energy from the oxidation of hydrogen, and the iron bacteria, which oxidize ferrous salts to ferric hydroxide. Those organisms that can oxidize hydrogen can often similarly utilize carbon compounds (Davis, Dulbecco, Eisen, Ginsberg, and Wood, 1970).

The photosynthetic autotrophs are pigmented anaerobic organisms that obtain energy by the utilization of radiant energy (Zinsser, 1960). There are three kinds of photosynthetic bacteria: purple suflur bacteria, which reduce CO_2 at the expense of H_2S; purple nonsulfur bacteria, which reduce CO_2 at the expense of organic compounds, require growth factors, and cannot use H_2S as a hydrogen donor; and green sulfur bacteria, which reduce CO_2 at the expense of H_2S. Bacterial photosynthesis differs from the process in green plants in that molecular oxygen is not produced. Also, in green plants, the hydrogen donor is H_2O, while in bacteria, a variety of oxidizable substances may be utilized.

Heterotrophic aerobes employ respiration to oxidize organic compounds as a source of carbon and energy. The heterotrophic bacteria require a more complex source of carbon than CO_2 and obtain their energy by the degradation of organic matter. These organisms are the most active in degradation of petroleum hydrocarbons.

Table A.4-1 presents examples of autotrophic modes of metabolism.

Table A.4-2 summarizes several microbial processes and shows an approximate relationship between the process and the environmental redox potential (Berry, Francis, and Bollag, 1987). The more negative the number, the stronger the reducing environment, as can be seen by the strict requirement for anaerobiosis.

4.1 ORGANIC COMPOUNDS

Table A.4-3 lists the aromatic hydrocarbons known to be oxidizable by microorganisms (Gibson, 1977). Individual components of petroleum and organisms capable of degrading them are given in Table A.4-4, originally published by Poglazova, Fedoseeva, Khesina, Meissel, and Shabad, 1967, and expanded in this report by inclusion of additional references. Thirty-two organisms were isolated from groundwater contaminated with high-octane gasoline (Jamison, Raymond, and Hudson, 1976). These were identified and used to study biodegradation of selected constituents of the gasoline (Table A.4-5). Table A.4-6 shows the biodegradation of gasoline components by mixed normal flora (Jamison, Raymond, and Hudson, 1976). The mixed population of natural flora in the groundwater biodegraded more constituents of the gasoline than the individual isolates. This suggests that a form of mutualism takes place in the degradation of petroleum, where a variety of organisms are necessary for complete degradation. The percent of each constituent biodegraded by the mixture is given in the table.

Table A.4-1.   Autotrophic Modes of Metabolism (Davis, Dulbecco, Eisen, Ginsberg, and Wood, 1970)

Organism or Group	Source of Energy	Remarks
Aerobic lithotrophs		Inorganic (litho-) electron donors
Hydrogen bacteria	$H_2 + 1/2\ O_2 -- H_2O$	
Sulfur bacteria (colorless)	$H_2S + 1.2\ O_2 -- H_2O + S$ $S + 1.5\ O_2 + H_2O -- H_2SO_4$	Can produce H_2SO_4 to pH as low as 0
Iron bacteria	$2\ Fe^{+2} + 1/2\ O_2 + H_2O --$ $2\ Fe^{+3} + 2\ OH^-$	
Nitrifying bacteria		
Nitrosomonas *Nitrobacter*	$NH_3 + 1.5\ O_2 -- HNO_2 + H_2O$ $HNO_2 + 1/2\ O_2 -- HNO_3$	Convert soil N to nonvolatile form Most can also use electron donors
Anaerobic respirers		
Denitrifiers[a]	$nH_2 + NO_3^- -- N_2O, N_2,$ or NH_3	Cause N loss from anaerobic soil
Desulfovibrio	$nH_2 + SO_4^{-2} -- S$ or H_2S	Odor of polluted streams, mud flat
Methane-forming bacteria	$4\ H_2 + CO_2 -- CH_4 + 2\ H_2O$	Sewage disposal plants
Clostridium aceticum	$4\ H_2 + 2\ CO_2 -- CH_3COOH$ $+ 2\ H_2O$	
Photosynthesizers		
Purple sulfur bacteria	$4\ CO_2 + 2\ H_2S + 4\ H_2O \xrightarrow{light}$ $4\ (CH_2O) + 2\ H_2SO_4$	$H_2(A)$ = various electron donors
Nonsulfur purple	$CO_2 + 2\ H_2(A) \xrightarrow{light}$ $(CH_2O) + H_2O + 2\ (A)$	
Algae	$CO_2 + 2\ H_2O \xrightarrow{light}$ $(CH_2O) + 1/2\ O_2$[b]	Plant photosynthesis

a = Anaerobic respiration, with the use of nitrate instead of O_2, is also common for the oxidation of the usual organic substrates by heterotrophs. This metabolism bears no resemblance to autotrophy, as the energy is used for biosynthesis from organic compounds rather than from CO_2.

b = The O_2 is derived directly from H_2O, and not from CO_2.

Table A.4-2.   Relationship Between Representative Microbial Processes and Redox Potential (Berry, Francis, and Bollag, 1987)

Process	Reaction (electron donor + electron acceptor)	Physiological Type	Redox Potential (mV)
Respiration	$OM + O_2 \text{ --- } CO_2$	Aerobes	700 to 500
Denitrification	$OM + NO_3^- \text{ --- } N_2 + CO_2$	Facultative anaerobes	300
Fermentation	OM --- organic acids (mostly acetate, propionate, and butyrate)	Facultative or obligate anaerobes	
Dissimilatory sulfate reduction	$OM \text{ (or } H_2) + SO_4^{-2}$ --- $H_2S + CO_2$	Obligate anaerobes	-200
Proton reduction	$OM (C_4 \text{ to } C_8 \text{ FA}) + H_2 + $ acetate (propionate) + CO_2	Obligate anaerobes	
Methanogenesis	$CO_2 + H_2 \text{ --- } CH_4$ Acetate --- $CO_2 + CH_4$	Obligate anaerobes	<-200 <-200

OM = Organic matter
FA = Fatty acid

Table A.4-3.   Aromatic Hydrocarbons Known to be Oxidized by Microorganisms (Gibson, 1977)

Monocyclic
  Benzene
  Toluene
  Xylenes
  Tri- and Tetramethylbenzenes
  Alkylbenzenes (Linear and Branched)
  Cycloalkylbenzenes

Dicyclic
  Naphthalene
  Methylnaphthalenes (Mono and Di)

Tricyclic
  Phenanthrene
  Anthracene

Polycyclic
  Pyrene
  Benzo(a)pyrene
  Benzo(a)anthracene
  Dibenzo(a)anthracene
  Benzperylene

Table A.4-4. Fuel Components/Hydrocarbons and Microorganisms Capable of
Biodegrading/Biotransforming Them

Fuel Component/ Hydrocarbon	Microorganisms
Acrylonitrile	Mixed culture of yeast, mold, protozoa, bacteria; activated sludge[e]
Alkanes	Pseudomonas[p],Arthrobacter, Acinetobacter, yeasts, Penicillium sp., Cunninghamella blakesleeana, Absidiaglauca, Mucor sp.[f]
n-Alkanes (C_1 to C_4) gaseous	Mycobacterium ketoglutamicum[f]
n-Alkanes (C_3 to C_{16}	Mycobacterium rhodochrous[g]
n-Alkanes (C_8 to C_{16})	Mycobacterium fortuitum, M. smegmatis[g]
n-Alkanes (C_{12} to C_{16})	Mycobacterium marinum, M. tuberculosis[g] Corynebacterium[f]
n-Alkanes (C_5 to C_{16})	(Arthrobacter, Acinetobacter, Pseudomonas putida, yeasts)[f]
n-Alkanes (C_{10} to C_{14})	Corynebacterium[g]
n-Alkanes (C_8 to C_{20})	Acinetobacter[q]
n-Alkanes (C_{11} to C_{19})	Prototheca zopfii [l], Pseudomonas spp[k,m]
Alkanes (straight chain)	Pseudomonas putida[f]
Alkenes (C_6 to C_{12})	Pseudomonas oleovorans[ab]
Anthracene	Stream bacteria[e], (Flavobacterium, Beijerinckiasp., Cunninghamella elegans)[h] (Pseudomonas/Alcaligenes sp, Acinetobacter sp., Arthrobacter sp.)[k]
Aromatics	Pseudomonas sp.[l]
Benzene	Pseudomonasputida[o,h,ae], sewage sludge[e], stabilization pond microbes[e], P. rhodochrous[f],P. aeruginosa[f] methanogens[r,s], anaerobes[t], Acinetobacter sp.[ae], Methylosinus trichosporium OB3b[ag], Nocardia sp.[ah],
Benzo(a)anthracene	Beijerinckia sp.[c,g], Cunninghamella elegans[e,u] Pseudomonas sp.[f]
Benzo(a)pyrene	(Candidalipolytica, C. tropicalis, C. guilliermondii, C. maltosa, Debaryomyces hansenii)[a], Bacillus megaterium[b] Beijerinckia sp.[c,g], Cunninghamella elegans[a,u,ap], Pseudomonas sp.[a], Neurospora crassa[ai], Saccharomyces cerevisiae[aj]

Table A.4-4.   continued

Fuel Component/ Hydrocarbon	Microorganisms
Biphenyl	(*Candida lipolytica, C. tropicalis, C. Guilliermondii, C. maltosa, Debaryomyces hansenii*)[a], (*Beijerinckia* B8/36, *Oscillatoria* sp., *Pseudomonas putida*[e,ae], *Cunninghamella elegans*[h], (*Moraxella* sp., *Pseudomonas* sp., *Flavobacterium* sp.)[ad], *Beijerinckia* sp.[ae], *Oscillatoria* sp.[ap]
n-Butane	*Mycobacterium smegmatis, Pseudobacterium subluteum, Pseudomonas fluorescens, Actinomyces candidus*[g], (*Arthrobacter, Brevibacterium*[f]
Chlorobenzene	*Pseudomonas putida*[ae]
Cresols	*Methylosinum trichosporium* OB3b[ag]
p-Cresol	*Pseudomonas* sp.[al]
Cyclohexane	*Xanthobacter* sp., *Nocardia* sp.[h]
Cyclohexanol	*Xanthobacter autotrophicus*[ak], (*Acinetobacter, Nocardia globerula*)[h]
Cyclohexanone	*Xanthobacter autotrophicus*[ak]
Decane	*Corynebacterium*[f]
Dibenzanthracene	Activated sludge[a]
Dodecane	(*Arthrobacter, Acinetobacter, Pseudomonas putida*, yeasts[f]
Ethane	*Methylosinus trichosporium*[f], *Pseudomonas methanica*[g], *P. putida*[h]
Ethylbenzene	*Pseudomonas putida*[ae,o]
Fluoranthene	Sewage sludge[e] *Pseudomonas* spp[m]
n-Heptane	*Pseudomonas aeruginosa*[g], (*Arthrobacter, Acinetobacter, Pseudomonas putida*, yeasts)[f]
n-Hexane	*Mycobacterium smegmatis*[g]

Table A.4-4.   continued

Fuel Component/ Hydrocarbon	Microorganisms
Hexadecane	*Acinetobacter*sp.[f], *(Candida petrophilum, Pseudomonas aeruginosa, Arthrobacter* sp.)[f], *Micrococcus cerificans (Candida parapsilosis, C. tropicalis, C. guilliermondii, C. lipolytica, Trichosporon* sp., *Rhodosporidium toruloides)*[l],*Prototheca zopfii (alga)*[l], *(Pseudomonas putida,* yeasts)[f], *Nocardia* sp.[ac], *(Pichia, Debaryomyces, Torulopsis, Candida*[an]
Jet fuels	*Cladosporium, Hormodendrum*
Kerosene	*Torulopsis, Candida tropicalis, Corynebacterium hydrocarboclastus (Candida parapsilosis, C. guilliermondii, C. lipolytica, Trichosporon* sp., *Rhohosporidium toruloides)*[l], *Cladosporium resinae*[ao]
Kerosene, Jet fuel, Paraffin wax	*Aspergillus, Botrytis, Candida, Cladosporium, Debaromyces, Endomyces, Fusarium, Hansenula, Monilia, Penicillium, Actinomyces, Micromonospora, Nocardia, Proactinomyces, Streptomyces,*
Methane	*Pseudomonas methanica*[g]
2-Methylhexane	*Pseudomonas aeruginosa*[g]
Octadecane	*Micrococcus cerificans*[g]
Naphthalene	*Pseudomonas* sp.[p], *(Candida lipolytica, C. tropicalis, C. Guilliermondii, C. maltosa, Debaryomyces hansenii)*[a], *Cunninghamella bainieri*[c,h], *Cunninghamella elegans*[c,e,h] *(Agnenellum, Oscillatoria, Anabaena, Cunninghamella elegans, Microcoleus*sp., *Nostoc* sp., *Coccochloris* sp., *Aphanocapsa* sp., *Chlorella* sp., *Dunaliella* sp., *Chlamydamonas* sp., *Cylindriotheca* sp., *Amphora* sp.,*
Flavobacterium,	*Alcaligenes, Corynebacterium, Nocardia, Aeromonas,* stream bacteria)[e] *Pseudomonas rathonis*[g], *(Bacillus naphthalinicum nonliquifaciciens, Pseudomonas desmolyticum, P. fluorescens,*

Table A.4-4.   continued

Fuel Component/ Hydrocarbon	Microorganisms
	P. putida biotype B)[h], *Pseudomonas oleovorans*[g], *P. putida*[f,v], (Mucorales: *Cunninghamella elegans, C. echinulata, C. japonica, Syncephalastrum* sp., *S. racemosum, Mucor* sp., *M. hiemalis, Neurospora crassa, Claviceps paspali, Psilocybe strictipes, P. subaeruginascens, P. cubensis, P. stuntzii)*[aa] *(Pseudomonas* NCIB 9816, *P.* sp. 53/1 and 53/2, *P. desmolyticum, Nocardia* strain R, *Nocardia* sp. NRRL 3385)[ae] Cyanobacteria[ap]
Octane	*Pseudomonas putida*[f,v], *Corynebacterium* sp. 7EIC[f], *Pseudomonas*[g]
Paraffins	*Trichosporon pullulans Nocardia* sp.[l]
n-Pentane	*Mycobacterium smegmatis*[g]
Phenanthrene	*Beijerinckia*[e], *(Pseudomonas putida, Cunninghamella elegans)*[h], *Pseudomonas* spp.[m], *Flavobacterium*[h,w]
Phenol	*(Pseudomonas, Vibrio, Spirillum, Bacillus, Flavobacterium, Chromobacter, Nocardia, Chlamydamonas ulvaensis, Phoridium fuveolarum, Scenedesmus basiliensis, Euglena gracilus, Corynebacterium* sp.)[e] *(Pseudomonas putida*, yeasts)[m], *(Azotobacter* sp., *Pseudomonas putida* CB-173 (ATCC 31800)[ab] *Acinetobacter calcoaceticus*[af]
Pristane	*(Corynebacterium* sp, *Brevibacterium erythrogenes)*[f]
n-Propane	*Mycobacterium smegmatis, M. rubrum, M. rubrum* var. *propanicum, M. carotenum, Pseudomonas puntotropha, (Pseudobacterium subluteum, Pseudomonas methanica)*[g], *(Cunninghamella elegans, Penicillium onatum)*[f]
1-Propanol > 2-propanol	*(Nocardia paraffinica, Brevibacterium* sp)[f]

Table A.4-4.   continued

Fuel Component/ Hydrocarbon	Microorganisms
Pyrene	Stabilization pond organisms[e] *(Pseudomonas/Alcaligenes* sp., *Acinetobacter* sp., *Arthrobacter* sp)[k]
Tetradecane	*Micrococcus cerificans*[g] *(Arthrobacter, Acinetobacter, Pseudomonas putida*, yeasts)[f]
Toluene	*Bacillus* sp.[e], *Pseudomonas putida*[e,f,o,m,ae], *Cunninghamella elegans*[h], *(P. aeruginosa, P. mildenberger)*[f] methanogens[r,s], anaerobes[s,v,t], *Methylosinus trichosporium* OB3b[ag], *(Pseudomonas* sp., *Achromobacter* sp.)[ah], *Pseudomonas aeruginosa*[am]
n-Undecane	*Mycobacterium* sp.[g]
p- and *m*-Xylene	*Pseudomonas putida*[f,ae], methanogens[n], anaerobes[v,t]

References:

a = (Cerniglia and Crow, 1981)
b = (Poglazova, Fedoseeva, Khesina, Meissel, and Shabad, 1967)
c = (Gibson, Mahadevan, Jerina, Yagi, and Yeh, 1975)
d = (Magor, Warburton, Trower, and Griffin, 1986)
e = (Kobayashi and Rittmann, 1982)
f = (Hou, 1982)
g = (Zajic, 1964)
h = (Cerniglia and Gibson, 1977)
i = (Ahearn, Meyers, and Standard, 1971)
j = (Jamison, Raymond, and Hudson, 1975)
k = (Stetzenbach and Sinclair, 1986)
l = (Boehn and Pore, 1984)
m = (Ghisalba, 1983)
n = (Reinhard, Goodman, and Barker, 1984)
o = (Gibson, Koch, and Kallio, 1968)
p = (Solanas, Pares, Bayona, and Albaiges, 1984)
q = (Garvey, Stewart, and Yall, 1985)
r = (Grbic-Galic and Vogel, 1986)
s = (Grbic-Galic and Vogel, 1987)
t = (Battermann and Werner, 1984)
u = (Dodge and Gibson, 1980)
v = (Jain and Sayler, 1987)
w = (Foght and Westlake, 1985)
x = (Rees, Wilson, and Wilson, 1985)

Table A.4-4.   references continued

---

y = (Zeyer, Kuhn, and Schwarzenback, 1986)
aa = (Cerniglia, Hebert, Szaniszlo, and Gibson, 1978)
ab = (Roberts, Koff, and Karr, 1988)
ac = (Mulkins-Phillips and Stewart, 1974b)
ad = (Stucki and Alexander, 1987)
ae = (Knox, Canter, Kincannon, Stover, and Ward, 1968)
af = (Fewson, 1981)
ag = (Higgins, Best, and Hammond, 1980)
ah = (Claus and Walker, 1964)
ai = (Lin and Kapoor, 1979)
aj = (Wiseman, Lim, and Woods, 1978)
ak = (Magor, Warburton, Trower, and Griffin, 1986)
al = (Dagley and Patel, 1957)
am = (Kitagawa, 1956)
an = (Scheda and Bos, 1966)
ao = (Atlas, 1977)
ap = (Atlas, 1981)

Table A.4-5.   Growth of Microorganisms on Components of Gasoline (Jamison, Raymond, and Hudson, 1976)

Compound	Nocardia	Pseudo-monas	Acineto-bacter	Micro-coccus	Flavo-bacterium	Unclas-sified
General Compound Classes						
n-Alkanes	+	-	-	+	-	-
Cyclic alkanes	-	-	-	-	-	-
Alkyl-substituted cyclic alkane	-	-	-	-	-	-
Monomethylalkanes	+	-	-	+	+	-
Dimethylalkanes	-	-	-	-	+	-
Trimethylalkanes	+	+	+	-	-	+
Aromatics	-	+	+	+	-	+
Specific Compounds						
n-Butane	-	-	-	-	-	-
n-Pentane	* + -	- + *	-	- +	-	- +
n-Hexane	* +	- + *	- +	- +	-	-
n-Heptane	* + -	- + *	-	-	-	- +
n-Octane	* + -	- + *	-	-	-	- + *
n-cis-Butene-2	-	-	-	-	-	-
n-Pentane-2	-	-	-	-	-	-
2,3-Dimethyl-butene-1	-	-	-	-	-	-
Cyclopentane	-	-	-	-	-	-
Cyclohexane	-	-	-	-	-	-
Methylcyclo-pentane	-	-	-	-	-	-
2-Methylbutane	- + *	- + *	-	-	-	-
2-Methylpentane	- + *	-	- +	+	-	- +

Table A.4-5.   continued

Compound	Nocardia	Pseudo-monas	Acineto-bacter	Micro-coccus	Flavo-bacterium	Unclas-sified
3-Methylpentane	-	-	-	-	-	-
3-Methylhexane	-	-	-	-	-	-
2-Methylheptane	-	-	-	-	-	-
3-Methylheptane	* + -	- +	-	-	-	-
2,2-Dimethyl-butane	-	-	-	-	-	-
2,3-Dimethyl-butane	-	-	-	-	-	-
2,2-Dimethyl-pentane	-	-	-	-	-	-
2,4-Dimethyl-pentane	-	-	-	-	-	-
2,3-Dimethyl-pentane	-	-	-	-	-	-
2,3-Dimethyl-hexane	-	-	- +	- +	+	- +
1,2-Dimethyl-cyclohexane	-	-	-	-	-	-
2,2,4-Trimethylpentane (isooctane)	* + -	- +	* + -	-	-	* + -
2,3,4-Trimethylpentane (isooctane)	-	-	-	-	-	-
2,3,3-Trimethylpentane (isooctane)	-	-	-	-	-	-
Ethylcyclohexane	-	-	-	-	-	-
Benzene	- + *	-	+ *	-	-	- + *
Ethylbenzene	+ - *	-	+ *	- +	-	- +
Toluene	- +	+ *	+	+ -	-	- + *
o-Xylene	-	-	-	-	-	-
m-Xylene	-	+ -	-	-	-	-
p-Xylene	-	+ -	-	-	-	-
Gasoline	+ *	+ *	+ *	+	+ -	+ *

Growth on the Specific Compounds:

+ - most isolates were +

- + most isolates were -

* some isolates exhibited moderate to heavy growth

Table A.4-6.   Biodegradation of Gasoline Components by Mixed Normal Microflora (Jamison, Raymond, and Hudson, 1976)

Component	Percent Biodegraded above Control	Component	Percent Biodegraded above Control
n-Propane	0	2,3-Dimethylpentane	0
n-Butane	0	2,5-Dimethylhexane	20
n-Pentane	70	2,4-Dimethylhexane	0
n-Hexane	46	2,3-Dimethylhexane	19
n-Heptane	49	3,4-Dimethylhexane	84
n-Octane	54	2,2-Dimethylhexane	75
Olefins--C_4	0	2,2-Dimethylheptane	62

Table A.4-6.   continued

Component	Percent Biodegraded above Control	Component	Percent Biodegraded above Control
Olefins--C_5	16	1,1-Dimethylcyclopentane	25
Olefins--C_6	18	1,2- and 1,3-Dimethylcyclopentane	78
Isobutane	0	1,3- and 1,4-Dimethylcyclohexane	0
Cyclopentane	0	1,2-Dimethylcyclohexane	26
Cyclohexane	45	2,2,3-Trimethylbutane	62
Methylcyclopentane	10	2,2,4-Trimethylpentane	13
Methylcyclohexane	75	2,2,3-Trimethylpentane	54
2-Methylbutane	0	2,3,4-Trimethylpentane	13
2-Methylpentane	6	2,3,3-Trimethylpentane	16
3-Methylpentane	7	2,2,5-Trimethylpentane	23
2-Methylhexane	23	1,2,4-Trimethylcyclopentane	0
3-Methylhexane	0	Ethylpentane	0
2-Methylheptane	38	Ethylcyclopentane	31
3-Methylheptane	45	Ethylcyclohexane	95
4-Methylheptane	48	Benzene	100
2,2-Dimethylbutane	25	Ethylbenzene	100
2,3-Dimethylbutane	0	Toluene	100
2,2-Dimethylpentane	9	o-Xylene	100
2,4-Dimethylpentane	11	m-Xylene	100
3,3-Dimethylpentane	45	p-Xylene	100
		Heavy ends	87

4.1.1 Aerobic Degradation

Microorganisms have evolved catabolic enzyme systems for metabolism of naturally occurring aromatic compounds (Gibson, 1978). In the oxidation of aromatic hydrocarbons, oxygen is the key to the hydroxylation and fission of the aromatic ring. Two hydroxyl groups must be present on the aromatic nucleus for enzymatic fission of the ring to occur, and these may be ortho (adjacent) or para (opposite on the ring) to each other. Subsequent metabolic sequences will then vary, depending upon the organism and the site of ring cleavage. Man-made molecules are degraded by these enzymes, if they are structurally similar to the naturally occurring compounds.

Bacteria incorporate two atoms of oxygen into the hydrocarbons to form dihydrodiol intermediates (Gibson, 1978). The hydroxyl groups are cis-dihydrodiols; i.e., they have a cis-stereochemistry. Oxidation of the dihydrodiols leads to the formation of catechols, which are substrates for enzymatic cleavage of the aromatic ring. In contrast, certain strains of fungi and higher organisms (Eukaryotes) incorporate one atom of molecular oxygen into aromatic hydrocarbons to form arene oxides, which can undergo the enzymatic addition of water to yield trans-dihydrodiols. These differences are illustrated in Figure A.4-1.

For most compounds, the most rapid and complete degradation occurs aerobically (Environmental Protection Agency, 1985b). It can be generalized that for the degradation of petroleum hydrocarbons, aromatics, halogenated aromatics, polyaromatic hydrocarbons, phenols, halophenols, biphenyls, organophosphates, and most pesticides and herbicides, aerobic bioreclamation techniques are most suitable. Aerobic degradation with methane gas as the primary substrate appears promising for some low molecular-weight halogenated hydrocarbons.

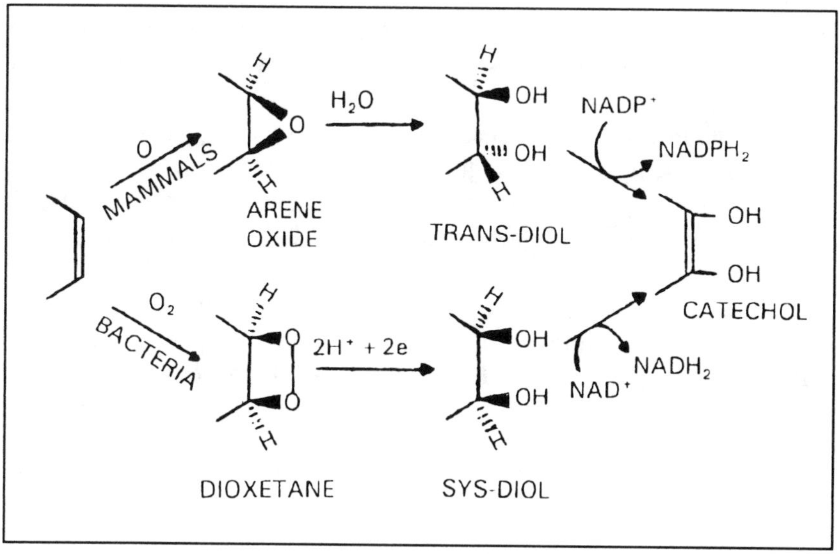

Figure A.4-1. Differences Between the Reactions Used by Eukaryotic and Procaryotic Organisms to Initiate the Oxidation of Aromatic Hydrocarbons (Gibson, 1977)

The sequence of hydrocarbon degradation in an oil spill is also likely to be determined by the ecological succession of the degrading microorganisms (Bartha and Atlas, 1977). n-Alkane degraders with rapid growth rates would out-compete the slow-growing decomposers of the more recalcitrant hydrocarbons for the nutritional resources until the n-alkanes are depleted. These organisms would then be replaced by microbes with slower growth rates but greater metabolic flexibility to degrade the more recalcitrant hydrocarbons (Fredericks, 1966).

The diversity of catabolic pathways for the degradation of hydrocarbons among different species and even strains of microorganisms makes it difficult to summarize all of the varied mechanisms and reactions that can occur (Hornick, Fisher, and Paolini, 1983). However, some of the major degradative pathways can be discussed.

4.1.1.1 Degradation of Straight Chain Alkanes

The n-Alkanes are the most widely and readily utilized hydrocarbons, with those between C_{10} and C_{25} being most suitable as substrates for microorganisms (Bartha and Atlas, 1977). The process is similar to the degradation of fatty acids.

The straight chain alkanes are oxidized by Pseudomonas spp. after 20 days at 20°C, being converted into the isoprenoids, pristane and phytane, the polycyclic sterane, and 17 alpha (H), 21 beta (H)-hopane series, all of which are resistant to biodegradation (Gibson, Koch, and Kallio, 1968). The n-alkanes with shorter chains (from C_5 to C_9) are more easily used as a source of carbon and energy by microorganisms than those with longer chain lengths (from C_{10} to C_{14}) (Williams, Cumins, Gardener, Palmier, and Rubidge, 1981). Biodegradation of n-alkanes with molecular weights up to $n-C_{44}$ has been demonstrated (Haines and Alexander, 1974).

Oxidation of C_{10} to C_{14} alkanes is obtained with Corynebacterium (Zajic, 1964). Alkanes are metabolized by terminal oxidation, alpha-oxidation, and diterminal oxidation. Fatty acids formed as by-products may be metabolized further by beta-oxidation. C_{12}-C_{16} are oxidized by Corynebacterium and further converted to the corresponding ketones or to esters of lower aliphatic acids (Hou, 1982z). Alkanes (C_5 to C_{16}) have been utilized by selected strains of Arthrobacter, Acinetobacter, Pseudomonas putida, and yeasts. Heptane, dodecane, tridecane, tetradecane, and hexadecane have been reported as growth substrates.

n-Alkane biodegradation is a predominantly bacterial activity (Song and Bartha, 1986). See Section 4.1.3 for the products formed from the microbial oxidation of alkanes.

1. Methane

The ability to oxidize small amounts of methane is widely spread among soil microbes. It is oxidized by Pseudomonas methanica (Zajic, 1964). Methane is also oxidized by Mycobacterium fortuitum and M. smegmatis (Lukins, 1962).

2. Ethane, Propane, Butane

These can be cooxidized by Pseudomonas methanica, with oxygen as a limiting factor (Zajic, 1964). Butane is utilized for growth by Pseudobacterium subluteum, Pseudomonas fluorescens, and Actinomyces candidus (Telegina, 1963).

Oxidation of alkanes or aliphatic hydrocarbons is classified as being terminal or diterminal (Zajic, 1964). These terms indicate that the initial breach occurs at one of the terminal carbon atoms:

A. Terminal Oxidation of Alkanes:

$$RCH_2CH_3 \longrightarrow [RCH_2 \cdot CH_2]$$
$$\downarrow O_2$$
$$RCH_2CH_2OH$$
$$\downarrow$$
$$RCH_2COOH$$

B. Alpha Oxidation (variation of monoterminal oxidation):

$$RCH_2CH_3 \longrightarrow [RCH_2 \cdot CH_3] \longleftrightarrow [R\text{-}CHCH_3]$$
$$\downarrow O_2$$
$$RCHOHCH_3$$
$$\underset{\overset{\|}{O}}{RCCH_3}$$
methyl ketones

C. Diterminal Oxidation

$$RCH_2CH_3 \xrightarrow{O_2} RCH_2CH_2OH$$
$$\downarrow O_2 \quad \text{Main path}$$
$$RCH_2COOH \longrightarrow \text{beta oxidation}$$
$$\downarrow O_2 \quad \text{Minor path}$$
$$RCH_2C(CH_3)_x COOH \xrightarrow{O_2} HOOC(CH_2)_x COOH$$
ω-Hydroxymonoic acid   Dioic acid

Monoterminal oxidation proceeds by the formation of a free radical and then the corresponding alcohol, which is readily oxidized to its respective aldehyde or aliphatic acid (Zajic, 1964). The terminal methyl group is enzymatically oxidized by incorporation of molecular oxygen by a monooxygenase, producing a primary alcohol, with further oxidation to a monocarboxylic acid (Atlas, 1981; Hornick, Fisher, and Paolini, 1983). Beta-oxidation of the carboxylic acid results in formation of fatty acids and acetyl coenzyme A, with eventual liberation of carbon dioxide. Some fatty acids are toxic and have been found to accumulate during hydrocarbon biodegradation. Fatty acids produced from certain chain length alkanes may be directly incorporated into membrane lipids, instead of going through the beta-oxidation pathway (Dunlap and Perry, 1967).

Alpha-oxidation of alkanes to methyl ketones may proceed with the same initial breech, but at the alpha-carbon (Zajic, 1964). Alpha-alcohols are formed, which are futher oxidized to methyl ketones. Alpha-oxidation has been observed with propane, butane, pentane, and hexane.

Diterminal (omega) oxidation (Kester and Foster, 1963) of C_{10} to C_{14} alkanes has been obtained with a culture of Corynebacterium. A breach occurs first at one terminal carbon atom and is followed by a second breach at the omega-carbon atom, synthesizing first a primary alcohol and then an aliphatic acid. The second breach at the omega carbon gives an omega-hydroxymonoic acid, which is oxidized to the corresponding dioic acid. Certain species of Pseudomonas oxidize the fatty acids formed from heptane and hexane by classical beta-oxidation (Heringa, Huybregtse, and van der Linden, 1961). Decarboxylation of fatty acids does not occur in these particular systems.

Subterminal oxidation occurs with C_3 to C_6, and longer, alkanes with formation of a secondary alcohol and subsequent ketone; these may or may not be metabolized. However, this is probably not the primary metabolic pathway for most n-alkane-utilizing microorganisms.

At chain lengths greater than C_6, the degradability generally increases until about C_{11} to C_{12} (Hornick, Fisher, and Paolini, 1983). As the alkane chain length increases, the molecule becomes less soluble in water. However, at chain lengths of C_{11} to C_{12} and above, the liquid n-alkanes are "accommodated" in water at a higher concentration than would be extrapolated from the solubility of a series of lower alkanes. This may be due to a change in the structure of the water molecules surrounding the alkanes. During the microbial utilization of long chain liquid alkanes, microbial attachment to droplets of alkane is seen with "transport" of these long chain alkanes through the cell membrane. This is also thought to occur with solid long chain alkanes, although, in both cases, the actual mechanism is unknown. Entirely separate transport mechanisms may exist for gaseous alkanes, short chain liquid alkanes, "accommodated" intermediate chain liquid alkanes, and long chain solid alkanes (Perry, 1968).

4.1.1.2 Degradation of Branched and Cyclic Alkanes

Only a few microorganisms utilize these hydrocarbons, possibly because the oxidation enzymes are not capable of handling branched-chain substrates (Hou, 1982). The fatty acids resulting from oxidation of single-branched alkanes are incorporated into the cell lipid. Subsequent oxidation is usally then by the beta-oxidation pathway.

Multiple-branched alkanes, such as pristane, can be converted to succinyl-CoA by Corynebacterium sp. and Brevibacterium erythrogenes.

Isoalkanes (branched alkanes) are degradable by a large number of microorganisms, although they are generally inferior to n-alkanes as growth substrates, especially if the branching is extensive or creates quaternary carbon atoms (Hornick, Fisher, and Paolini, 1983). A small number of methyl or ethyl side chains does not drastically decrease the degradability, but complex branched chains, and especially, terminal branching, are harder to degrade. The position of the side chain also has an effect on the degradability. 1-Phenylalkanes are more degradable than interior substituted ones, and the further the side group is into the molecule, the slower the degradation.

Cycloalkanes are more resistant to microbial attack than straight chain alkanes and are highly toxic (Atlas, 1981). This is probably due to the absence of an exposed terminal methyl group for the initial oxidation (Hornick, Fisher, and Paolini, 1983). Complex alicyclic compounds, such as hopanes, are among the most persistent components of petroleum spillages in the environment. Up to six-membered condensed ring structures have been reported to be subject to microbial degradation. Bacterial degradation of aromatic hydrocarbons normally involves the formation of a diol followed by cleavage and formation of a cis diacid, while that of fungi forms a trans-diol. Light aromatic hydrocarbons are subject to microbial degradation in a dissolved state. Condensed ring aromatic structures are subject to microbial degradation by a similar metabolic pathway as monocyclic structures; however, these hydrocarbons are relatively resistant to enzymic attack. Structures with four or more condensed rings have been shown to be attacked by cooxidation or as a result of commensalism.

Cooxidation by mixed microbial communities appears to be very important in the degradation of cycloalkanes, since only a few species of bacteria have been shown to use cyclohexane as a sole carbon source (Hornick, Fisher, and Paolini, 1983). Cooxidation changes the cycloalkane to a cycloalkanone, using a variety of oxygenases, such as peroxidase or polyphenoloxidase (Beam and Perry, 1974). A commensalistic symbiosis is then postulated to occur, with the cycloalkanones produced by cooxidation being used as a sole carbon source by a wide range of microorganisms (Donoghue, Griffin, Norris, and Trudgill, 1976). Alkyl side chains on cycloalkanes are degraded by the normal alkane oxidation mechanism before degradation of the cycloalkane itself, depending upon the size of the side chain.

4.1.1.3 Degradation of Alkenes

Unsaturated 1-alkenes are generally oxidized at the saturated end of the molecule by the same mechanism as used for alkanes (Hornick, Fisher, and Paolini, 1983). Some microorganisms, such as the yeast Candida lipolytica, attack at the double bond and convert the alkene into an alkane-1,2-diol. Other minor pathways have been shown to proceed via an epoxide, which eventually is converted into a fatty acid. For similar amounts of degradation, the chain length of 1-alkenes must be longer than the corresponding alkane. Many microorganisms will not grow on 1-alkenes less than C_{12}. Since 2-alkenes are more readily attacked than 1-alkenes, the presence of a terminal methyl group at each end of the molecule appears to make the molecule degradable for more organisms.

4.1.1.4 Degradation of Aromatic Compounds

Five phases can be distinguished for aerobic and anaerobic metabolism of aromatics (Evans, 1977).

1. Entry into the cell--this can be by free diffusion or with specific transport mechanisms
2. Manipulations of the side chains and formation of substrates for ring cleavage
3. Ring cleavage
4. Conversion of the products of ring cleavage into amphiboli intermediates
5. Utilization of the amphibolic intermediates

The side groups of the ring are first modified by hydroxylation, demethylation, or decarboxylation (which are generally enzymatic reactions) to produce one or two basic molecules, which are then cleaved by the second group of enzymes and further degraded to molecules utilizable by the cell (Hornick, Fisher, and Paolini, 1983). The most common ring cleavage mechanism is the "ortho" pathway. This is followed by a series of enzymatic reactions, with the final products being low molecular weight organic acids and aldehydes that are readily incorporated into the tricarboxylic acid cycle. Polycyclic aromatic hydrocarbons, such as anthracene and phenanthrene, are also degraded by the "ortho" cleavage pathway. The other major pathway is the "meta" cleavage mechanism, where the aromatic ring is cleaved by a dioxygenase to form a keto acid or an aldehydro acid.

When an alkyl chain is present on a PAH, it is removed by beta-oxidation, if it is larger than an ethyl group (Hornick, Fisher, and Paolini, 1983). Ring cleavage usually can occur when methyl side chains are present; however, when certain locations on the ring are substituted, the resulting compound is very resistant to degradation (McKenna and Heath, 1976; Gibson, 1976).

In one study, aromatic compounds were found to be susceptible to aerobic, but not anaerobic, biodegradation (Rittmann, Bouwer, Schreiner, and McCarty, 1980). However, halogenated aliphatic compounds evaluated were degradable only under anaerobic conditions and not aerobic conditions (Roberts, McCarty, Reinhard, and Schreiner, 1980). Degradation of chlorinated aromatics only under aerobic conditions suggests the need for mixed-function oxidase systems to bring about dehalogenation and ring cleavage of these compounds (Bitton and Gerba, 1985).

4.1.1.4.1 Degradation of Mononuclear Aromatic Hydrocarbons. Groundwater spiked with benzene, toluene, and xylenes (BTX) was introduced below a shallow water table, and the migration of contaminants was monitored (Barker and Patrick, 1985). The BTX migrated slightly slower than the groundwater due to sorptive retardation. Essentially, all the injected mass of BTX was lost within 434 days due to biodegradation. Rates of mass loss were highest for m- and p-xylene, lower for o-xylene and toluene, and lowest for benzene, which was the only component to persist beyond 270 days. Laboratory biodegradation experiments produced similar rates. However, phenolic and acidic breakdown products observed in laboratory experiments were not found in the field plume. BTX is rapidly degraded under aerobic condition, but persists in conditions of low dissolved oxygen. Washed cell suspensions of P. putida, grown with toluene as a sole source of carbon, were able to oxidize benzene, toluene, and ethylbenzene at equal rates (Gibson, Koch, and Kallio, 1968).

1. Phenol

Phenol is rapidly degraded in aerobically incubated soil (Baker and Mayfield, 1980). Substituted phenols incubated under aerobic conditions are sorbed irreversibly by clays and soils and are transformed into polymerized species (Sawhney and Kozloski, 1984). Phenol can be utilized as the sole carbon source via catechol by Pseudomonas putida, yeasts, and other organisms. (Ghisalba, 1983). Acinetobacter calcoaceticus can degrade phenol and many aromatic compounds (Fewson, 1967).

2. Benzene

The availability of dissolved oxygen is a dominant factor in the biodegradation of benzene (Barker and Patrick, 1985). Pseudomonas rhodochrous and Pseudomonas aeruginosa metabolize benzene through catechol and cis, cis-muconic acid (Hou, 1982).

3. Toluene

A major factor in the biodegradation of toluene is the availability of dissolved oxygen (Barker and Patrick, 1985). Turnover time, the amount of time required to remove the concentration of substrate present, for toluene is greater than 10,000 hr; therefore, biodegradation of this compound in aquifers would be expected to be a very slow process (Swindoll, Aelion, Dobbins, Jiang, Long, and Pfaender, 1988).

Toluene can undergo two types of attack: 1) immediate hydroxylation of the benzene nucleus, followed by ring cleavage or 2) oxidation of the methyl group, followed by hydroxylation and cleavage of the ring (Fewson, 1981). Toluene can be converted to 3-methylcatechol by a Pseudomonas sp. and an Achromobacter sp. (Claus and Walker, 1964).

P. aeruginosa can oxidize toluene (Hou, 1982). Extracts of this organism grown on xylene can also oxidize toluene. The compound is also oxidized by other species of Pseudomonas (e.g., P. mildenberger) and Achromobacter.

Some soil bacteria (P. putida, etc.) can utilize toluene as the sole carbon source (Ghisalba, 1983). More than 90 percent of toluene added to core samples from depths of 1.2, 3.0, and 5.0 m was degraded in one week (McNabb, Smith, and Wilson, 1981). There was no degradation of toluene in material that was autoclaved prior to addition of the compound.

4. Xylene (o,m,p)

A dominant factor in the biodegradation of xylenes is the availability of dissolved oxygen (Barker and Patrick, 1985).

P. putida can directly oxidize the aromatic ring of p- and m-xylenes (Hou, 1982). These authors were not able to isolate microorganisms that grow on o-xylenes. The oxidation of o-xylene to o-toluene was demonstrated in a Nocardia sp. only with the cooxidation technique.

5. Alkylbenzenes

Long-chain alkylbenzenes are oxidized at the terminal methyl group (Hou, 1982). A Nocardia species grows on n-decylbenzene, n-dodecylbenzene, n-octadecylbenzene, n-nonylbenzene, and n-dodecylbenzene. As the alkyl chain length grows and the substituent becomes the major part of the molecule, these compounds are more realistically regarded as substituted alkanes, rather than substituted benzenes. n-Alkylbenzenes are also oxidized by yeasts.

4.1.1.4.2 Degradation of Polynuclear Aromatic Hydrocarbons (PNAs) or Polycyclic Aromatic Hydrocarbons (PAHs). Polycyclic aromatic hydrocarbons are made up of two or more fused benzene rings in linear, angular, or cluster arrangements and contain only carbon and hydrogen (Edwards, 1983). The quantities of PAHs that are formed in nature are very small in comparison with those from anthropogenic sources. Most PAHs are practically insoluble in water. Molecular weights range from 178 to 300. Some are strongly, and some weakly, carcinogenic or mutagenic. The well-known PAHs, such as benzo(a)pyrene (BaP), are innocuous by themselves, but can be biologically activated by enzymes to form epoxides that are carcinogenic and mutagenic (Levin, Wood, Wislocki, Chang, Kapitulnik, Mah, Yagi, Jerina, and Conney, 1978). Conversely, there are some green plants that contain a substance (ellagic acid) that can destroy the diol epoxide form of BaP, inactivating its carcinogenic and mutagenic potential (Sayer,

Yagi, Wood, Conney, and Jerina, 1982). Some of the PAHs of concern are anthracene, benzo(a)pyrene (BaP), benzo(a)anthracene, fluoranthene, phenanthrene, perylene, pyrene, and fluorene. Biphenyl and naphthalene, diaromatics, will be considered as PAHs in this report.

The toxicity of PAHs to microorganisms is related to their water solubility (Sims and Overcash, 1983). Degradation of some of the less water soluble, such as benzo(a)anthracene and BaP, occurs only when the PAHs are mixed with soil, water, and a substance to stimulate growth of oxygenase-active organisms (Groenewegen and Stolp, 1981). The biotransformation process for PAHs with more than three rings appears to be cometabolism (Sims and Overcash, 1983). Benzo(a)anthracene and pyrene are probably slow to be degraded (Ghisalba, 1983). Generally, PAHs with four or more rings can be degraded by microbes by metabolism via hydroxylation and ring fission. When the soils become acclimated to PAHs, their ability to degrade these compounds is enhanced.

PAH mineralization in sediments is related to the length of incubation time, temperature, molecular size of the substrate and previous exposure to PAH or related contaminants (Sherrill and Sayler, 1982). The capacity of bacteria to degrade BaP increases with BaP content in the soil, and microflora of soil contaminated with BaP are more active in metabolizing BaP than those in "clean" soil (Shabad, Cohan, Ilnitsky, Khesina, Shcherbak, and Smirnov, 1971). On the other hand, a PAH-degrading bacterial population added to various sediment systems did not significantly enhance PAH (naphthalene, phenanthrene, benzo(a)pyrene) mineralization rates (Sherrill and Sayler, 1982).

There is a correlation between PAH contamination and the rate of mineralization of naphthalene (Kerr and Capone, 1986). Rates of PAH mineralization in all environments appear to be primarily controlled by the amount of pollutant present. There is an inverse relationship between ambient concentrations and mineralization/transformation rates (Shiaris and Jambard-Sweet, 1984). The rates are generally not affected by ambient salinity regimes (Kerr and Capone, 1986). However, rates have been shown to be substantially modified by changes in salinity outside ranges normally experienced in the coastal environments tested. The rates are also affected by seasonal fluctuations.

Anthracene, fluoranthene, and phenanthrene are utilized by Pseudomonas spp. as a sole carbon source (Ghisalba, 1983). Napthalene, biphenyl, and BaP can be metabolized by a number of different species of yeast, some of which are reported in high numbers in oil-polluted soils (Cerniglia and Crow, 1981). A filamentous fungus has been noted to degrade BaP and benz(a)anthracene (Dodge and Gibson, 1980; Cerniglia and Gibson, 1979).

Di- and tricyclic aromatic hydrocarbons are extremely insoluble, but it appears that degradation of compounds, such as phenanthrene, may be more rapid in natural environments than was once imagined (Fewson, 1981). The rate is related to previous pollution at the site, the length of time allowed for biodegradation, the temperature, and the molecular weight of the hydrocarbon (Sherill and Sayler, 1980). It has been suggested that naphthalene and phenanthrene are utilized only in the soluble form, since the growth rates of bacteria are related to the solubilities of the hydrocarbons on which they are growing (Wodzinski and Coyle, 1974). Microorganisms, however, could adhere to particles of insoluble substrates (Cox and Williams, 1980). Addition of chemical surfactants could increase the rate of utilization of naphthalene.

1. Alkylnaphthalenes

Degradation of alkylnaphthalenes, by Pseudomonas spp. depends upon the position, number, and type of the substituents on the molecule (Solanas and Pares, 1984). Pure cultures of Pseudomonas spp. preferentially degraded the less substituted aromatics in a light crude oil residue, providing indirect evidence that the organisms are attacking the ring system rather than the alkyl chains. Certain strains of bacteria, notably Pseudomonas spp., metabolize aromatic hydrocarbons with fused benzene rings through cis-dihydrodiol intermediates at different sites of the molecule (Jerina, Selander, Yagi, Wells, Davey, Mahadevan, and Gibson, 1976). It has been, therefore, assumed that alkyl substituents may hinder the initial oxygenative attack of the molecule by the corresponding enzymatic system. However, in some cases, the oxidation rates seem to be enhanced, such as with meta-substituted naphthalenes (Solanas and Pares, 1984). Degradation of naphthalene by Pseudomonas involves the formation of 1,2-dihydroxynaphthalene as the first stable metabolite (Davies and Evans, 1976). Oxidation becomes difficult when a substituent is present in these ortho (1,2) positions, unless it is a methyl group adjacent to the sites of oxidative attack.

2. Naphthalene

A Nocardia sp. can biotransform naphthalene and substituted naphthalenes to their corresponding diols (Hou, 1982). Naphthalene metabolism can be found in many genera from the major fungal taxa.

Candida lipolytica, C. guilliermondii, C. tropicalis, C. maltosa, and Debaryomyces hansenii are able to metabolize this compound (Cerniglia and Crow, 1981). C. lipolytica oxidizes naphthalene to 1-naphthol, 2-naphthol, 4-hydroxy-1-tetralone, and trans-1,2-dihydroxy-1, 2-dihydronaphthalene. The primary metabolite is 1-naphthol.

Cunninghamella bainieri oxidizes naphthalene through trans-1,2-dihydroxy- 1,2-dihydronaphthalene (Gibson and Mahadevan, 1975). Cunninghamella elegans oxidizes naphthalene by a sequence of reactions resulting in six metabolites: 1-naphthol, 4-hydroxy-1-tetralone, 1,4-naphthoquinone, 1,2-naphthoquinone, 2-naphthol, and trans-1,2-dihydroxy-1,2-dihydronaphthalene.

Naphthalene is oxidized by species of Cunninghamella, Syncephalastrum, and Mucor to 2-naphthol, 4-hydroxy-1-tetralone, trans-naphthalene dihydrodiol, 1,2-naphthoquinone, 1,4-naphthoquinone, and predominantly 1-naphthol (Cerniglia, Hebert, Szaniszlo, and Gibson, 1978). Neurospora crassa, Claviceps paspali, and Psilocybe strains also showed a similar degradative capacity.

Metabolism of naphthalene by Pseudomonas rathonis and Pseudomonas oleovorans with optimal synthesis of salicylic acid was accomplished by addition of 0.4 percent $Al(OH)_3$ (Zajic, 1964). Inorganic boron compounds also increase the yields of salicylic acid.

Algae also oxidize naphthalene (Naval Civil Engineering Laboratory, 1986). Methyl-, dimethyl-, and trimethyl-naphthalenes were more toxic than naphthalene to the freshwater alga, Selenastrum capricornutum, while dibenzofuran, fluorene, phenanthrene, and dibenzothiophene were the most toxic to this organism (Hsieh, Tomson, and Ward, 1980). In general, compounds with higher boiling points were more toxic.

Turnover time for naphthalene was greater than 10,000 hr; therefore, biodegradation of this compound in aquifers would be expected to be a very slow process (Swindoll, Aelion, Dobbins, Jiang, Long, and Pfaender, 1988). However, naphthalene has been reported to be degraded rapidly in aerobic groundwaters contaminated by polynuclear aromatic hydrocarbons or in groundwater near oil and gas beds (Slavnia, 1965). Naphthalene can be slightly sorbed onto sediments (Erlich, Goerlitz, Godsy, and Hult, 1982). It was biodegraded in an aquifer recharged with reclaimed water from wastewater treatment after an initial lag (Roberts, McCarty, Reinhard, and Schreiner, 1980z). These results demonstrate the importance of acclimation of the organisms to the contaminant.

3. Biphenyl

There is evidence that fungi metabolize biphenyl to metabolites similar to those formed by mammalian systems; i.e., the trans-configuration (Hou, 1982). Candida lipolytica, C. guilliermondii, C. tropicalis, C. maltosa, and Debaryomyces hansenii are able to metabolize this compound (Cerniglia and Crow, 1981). C. lipolytica oxidizes biphenyl to 2-, 3-, and 4-hydroxybiphenyl, 4,4'-dihydroxybiphenyl, and 3-methoxy-4-hydroxybiphenyl, with 4-hydroxybiphenyl as the main metabolite.

Strains of Moraxella sp., Pseudomonas sp., and Flavobacterium sp. able to grow on biphenyl were isolated from sewage (Stucki and Alexander, 1987).

4. Benzo(a)pyrene (BaP or BP)

After growth with succinate plus biphenyl, a mutant strain of Beijerinckia contains an enzyme system that can oxidize benzo(a)pyrene to a mixture of vicinal dihydrodiols, mainly cis-9, 10-dihydroxy-9, 10-dihydrobenzo(a)pyrene (Gibson and Mahadevan, 1975).

Fungi oxidize this compound by a mechanism similar to that observed in mammalian systems (Hou, 1982); e.g., it is oxidized by Cunninghanella elegans to polar products with trans-configuration. Candida lipolytica, C. guilliermondii, C. tropicalis, C. maltosa, and Debaryomyces hansenii are able to metabolize this compound (Cerniglia and Crow, 1981). C. lipolytica oxidizes benzo(a)pyrene to 3-hydroxybenzo(a)pyrene and 9-hydroxybenzo(a)pyrene. Neurospora crassa and Saccharomyces cerevisiae have inducible hydroxylases that can attack this compound (Knackmuss, 1981). Cunninghamella elegans also degrades this compound, with formation of a carcinogen.

Bacillus megaterium strains accumulate BP, which is stored in the cytoplasm and the lipid inclusions of the cells but not in the form of BP-protein complexes (Poglazova, Fedoseeva, Khesina, Meissel, and Shabad, 1967). The chemical remains unaltered in some of the strains but gradually disappears or is reduced in others. When the organisms are grown on a medium that does not contain aromatic hydrocarbons, they eventually lose their ability to destroy BP, but this ability can be restored and further enhanced by adding the compound to the medium.

5. Benzo(a)anthracene (BaA)

A mutant strain of Beijerinckia is able to oxidize benzo(a)anthracene after growing on succinate in the presence of biphenyl (Gibson and Mahadevan, 1975). This compound is metabolized to four dihydrodiols, primarily cis-1,2-dihydroxy-1,2-dihydrobenzo(a)anthracene. Fungi, e.g., Cunninghamella elegans, oxidize this compound to polar products with trans-configuration (Hou, 1982).

6. Anthracene, Phenanthrene, and Pyrene

A Pseudomonas/Alcaligenes sp., an Acinetobacter sp., and an Arthrobacter sp. grew on anthracene or pyrene as the sole carbon source (Stetzenbach and Sinclair, 1986). Pyrene degradation was correlated with the presence of oxygen with no decrease in concentration observed under anaerobic or microaerophilic conditions.

The metabolic pathways for phenanthrene and anthracene are similar to the sequence for naphthalene (Hou, 1982). A Beijerinckia sp. oxidizes these two compounds. An Aeromonas sp. uses an alternative pathway for phenanthrene metabolism.

4.1.1.5 Degradation of Carboxylic Acids Including Fatty Acids

P. putida grown on aviation turbine fuel oxidized straight chain fatty acids from acetate (C_2) to palmitic acid (C_{16}) (Williams, Cumins, Gardener, Palmier, and Rubidge, 1981). No odd-numbered fatty acids were detected, indicating that beta-oxidation is the major pathway for fatty acid dissimilation.

Certain species of Pseudomonas oxidize the fatty acids formed from heptane and hexane by classical beta-oxidation (Zajic, 1964).

4.1.1.6 Degradation of Alcohols

Methanol and tertiary butyl alcohol (TBA) may each be present at 5 percent in some gasolines (Novak, Goldsmith, Benoit, and O'Brien, 1985). Methanol is readily biodegradable in both aerobic and anaerobic environments (Novak, Goldsmith, Benoit, and O'Brien, 1985; Lettinga, DeZeeuw, and Ouborg, 1981). It is degraded in soil samples to 31 m, especially in the saturated region (Novak, Goldsmith, Benoit, and O'Brien, 1985). Many bacteria and yeasts that can catabolize methanol under aerobic or anaerobic conditions and at low and high concentrations have been isolated from deep soil (to 100 ft) (Benoit, Novak, Goldsmith, and Chadduck, 1985).

TBA is more refractory. It is slowly degraded in anaerobic aquifers, with the degradation rate increasing with increasing concentration. At 1 mg/l TBA, it appears the microbial population receives insufficient energy to cause a population increase and utilization rates remain slow. Rates are faster at the highest concentration. Several species that can degrade tertiary butyl alcohol under aerobic conditions have been isolated (Benoit, Novak, Goldsmith, and Chadduck, 1985).

4.1.1.7 Degradation of Alicyclic Hydrocarbons

Pure strains do not grow on these compounds (Hou, 1982). Instead, these materials have been found to be cooxidized by microorganisms growing on other substrates (see Section 5.1.1.2).

n-Alkyl-substituted alicyclic hydrocarbons are more susceptible to microbial attack than the unsubstituted parent compounds; e.g., methylcyclohexane and methylcyclopentane (Hou, 1982).

Cyclohexane and Oxygenates. Complete biodegradation of cyclohexane can be achieved with commensalism by using two organisms--an n-alkane oxidizer that converts cyclohexane into either cyclohexanol or cyclohexanone during growth on the n-alkane and another that can grow on cyclohexanol or cyclohexanone (Beam and Perry, 1974; de Klerk and van der Linden, 1974). Acinetobacter and Nocardia globerula grow rapidly with cyclohexanol as the sole source of carbon. Microbial attack on cyclohexane could be initiated by hydroxylation.

In one experiment, a Pseudomonas strain furnished biotin to allow a Nocardia sp. to grow on cyclohexane (Hou, 1982).

Xanthobacter autotrophicus cannot utilize cyclohexane but can grow with a limited range of substituted cycloakanes, including cyclohexanol and cyclohexanone (Magor, Warburton, Trower, and Griffin, 1986). Another species of Xanthobacter can grow on cyclohexane. Both pathways produce adipic acid.

Cyclohexane carboxylic acid and cyclohexane propionic acid are more suitable to microbial growth than cyclohexane acetic acid and cyclohexane butyric acid (Hou, 1982). Effective cleavage of an alicyclic ring occurs only when the side chain contains an odd number of carbon atoms.

4.1.2 Anaerobic Degradation

It is now believed that three major groups of microorganisms are essential for complete mineralization of organic carbon to carbon dioxide and methane in anoxic sites that are without light and are low in electron acceptors other than carbon dioxide (Berry, Francis, and Bollag, 1987). These three groups are the fermenters, the proton reducers, and the methanogens (Boone and Bryant, 1980; McInerney and Bryant, 1981; McInerney, Bryant, Hespell, and Costerton, 1981).

There are some compounds, most notably the lower molecular weight halogenated hydrocarbons, that will degrade only anaerobically (Environmental Protection Agency, 1985b). Aromatic compounds consisting of either a homocyclic [e.g., benzoate, Figure A.4-2 (a)] or a heterocyclic [e.g., nicotinate, Figure A.4-2 (b)] aromatic nucleus can be metabolized by microorganisms under anaerobic conditions (Berry, Francis, and Bollag, 1987).

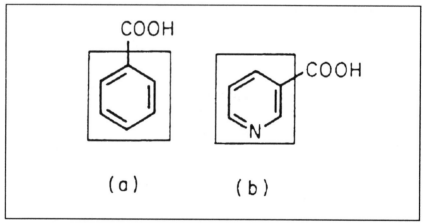

Figure A.4-2. Homocyclic Aromatic "Benzenoid" Nucleus (Enclosed) of Benzoate (a) and Heterocyclic Aromatic "Pyridine" Nucleus (Enclosed) of Nicotinate (b) (Berry, Francis, and Bollag, 1987)

The chemistry of the anaerobic breakdown of aromatics involves an initial ring hydrogenation step (ring reduction followed by a ring hydration ring cleavage reaction sequence) (Berry, Francis, and Bollag, 1987). This pathway is believed to be common to all microorganisms involved in benzenoid metabolism, including the denitrifiers, the sulfate reducers, and the fermenters.

4.1.2.1 Anaerobic Respiration

Anaerobic respiration in soil involves biological oxidation-reduction reactions in which inorganic compounds (nitrate, sulfate, and carbonate ions; manganic and ferric ions) serve as the ultimate electron acceptor, instead of molecular oxygen. Nitrate may be used by nitrate reducers and sulfate by sulfate reducers as a terminal electron acceptor (Environmental Protection Agency, 1985b). The organic waste serves as the electron donor or energy source. If oxygen is depleted during decomposition of a particular waste, the system will be dominated by facultative anaerobic bacteria, which are able to adapt from the aerobic to the anaerobic conditions, as necessary. Obligate anaerobes, on the other hand, cannot tolerate oxygen and are inhibited or killed by exposure to it.

Without oxygen, oxygenases are inactive, and only those aromatic compounds with oxygen-containing functional groups (phenols and benzoates) are mineralized (Zeyer, Kuhn, and Schwarzenback, 1986). Under anaerobic conditions, the aromatic ring may be first reduced to a substituted cyclohexane before hydrolytic ring cleavage. In some cases, removal or modification of a substituent must occur before reduction of the ring. The mechanisms used with aromatic hydrocarbons that have no activating groups to facilitate hydration of the ring are still unknown. However, it does appear that there is an oxidation of the ring in these instances. Apparently, attachment of an oxygen atom to the ring structure facilitates ring catabolism under anoxic conditions (Berry, Francis, and Bollag, 1987).

It appears that many aromatic compounds can be cleaved under strict anaerobic conditions, producing carbon dioxide and methane (Healy and Young, 1979). This decomposition of organic matter to carbon dioxide and methane involves interactions of heterotrophic bacteria and several fastidiously anaerobic bacteria: hydrolytic bacteria that catabolize major components of biomass, such as saccharides, proteins, and lipids; hydrogen-producing, acetogenic bacteria that catabolize products from the activity of the first group, such as fatty acids and neutral end products; homoacetogenic bacteria that catabolize multicarbon compounds to acetic acid; and methanogenic bacteria (Kobayashi and Rittmann, 1982). Both fastidious and facultative anaerobes are important.

If conditions do not favor methanogens, certain classes of organic compounds, such as phenols, cresols, and xylenes, can be degraded by bacteria that respire nitrate or sulfate (Wilson, Leach, Henson, and Jones, 1986). Anaerobic degradation of aromatic hydrocarbons might also be facilitated in nature by the presence of other substrates and oxidants, including oxidized organics, metal organic compounds, and perhaps even water (Kochi, Tang, and Bernath, 1972).

Petroleum can be microbially degraded anaerobically by the reduction of sulfates and nitrates (Shelton and Hunter, 1975). An alkane dehydrogenase is proposed to be the initial enzyme involved with production of an alkene as the first intermediate compound. The fatty acids produced from the alkanes can be fermented under anaerobic conditions (Rosenfeld, 1947). Alkanes shorter than C_9 can be degraded anaerobically, while some of the longer chain alkanes may be transformed into napthalenes and other polycyclic aromatic hydrocarbons. Anaerobic degradation of aromatic compounds can occur with both the aromatic ring and side chains of substituted aromatic compounds as carbon sources, for many anaerobic microorganisms (Balba and Evans, 1977).

4.1.2.1.1 Denitrification.

Microorganisms that carry out nitrate respiratory metabolism (e.g., the denitrifiers) are facultative and appear to prefer oxygen as their electron acceptor (Gottschalk, 1979). Under aerobic conditions, this group of microorganisms uses a wide range of organic compounds as carbon and energy sources. In many instances the same range of organic carbon is used under denitrifying conditions. Active nitrate-respiring microorganisms are found in a variety of anoxic environments, including soils, lakes, rivers, and oceans (Berry, Francis, and Bollag, 1987).

Catabolism of aromatic compounds can occur under anoxic conditions and in the presence of nitrate (Braun and Gibson, 1984). Pseudomonas sp. strain PN-1 can use p-hydroxybenzoate, benzoate, and m-hydroxybenzoate, but not phenol, to grow under both aerobic and nitrate-reducing conditions. This organism appears to have an oxygenase enzyme system that does not require oxygen as an inducer. Some facultative microorganisms retain low levels of oxygenase activity when grown in the presence of aromatic compounds, even under anaerobic conditions, while others do not (Taylor, Campbell, and Chinoy, 1970). A nitrate-respiring P. stutzeri is capable of using phenol as a substrate (Ehrlich, Godsy, Goerlitz, and Hult, 1983).

Microorganisms can degrade o-, m-, and p-xylenes and toluene under denitrifying conditions, such as may occur in lake sediments, sludge digestors, and groundwater infiltration zones from landfills and polluted rivers (Zeyer, Kuhn, and Schwarzenback, 1986; Kuhn, Colberg, Schnoor, Wanner, Zehnder, and Schwarzenbach, 1985).

4.1.2.1.2 Sulfate Reduction. Those microorganisms that carry out dissimilatory sulfate reduction to obtain energy for growth are strict anaerobes (Berry, Francis, and Bollag, 1987). For these bacteria, organic carbon serves as a source of both carbon and energy. Reducible sulfur compounds (e.g., sulfate, thiosulfate) serve as terminal electron acceptors. Dissimilatory sulfate-reducing bacteria (sulfidogens) are most commonly associated with aquatic environments (i.e., marine and freshwater sediments) (Laanbroek and Pfennig, 1981; Lovley and Klug, 1983), although sulfate reducers can also be found in soil.

Organisms, such as Desulfovibrio, utilize sulfate as an electron acceptor, reducing it to sulfide, and by using organic acids as electron donors (Environmental Protection Agency, 1985b). A newly recognized sulfate-reducing organism, Desulfonema magnum, was isolated from marine sediment and is capable of mineralization of various fatty acids and benzoate (but not ethanol, cyclohexane carboxylate, or glucose) to carbon dioxide in the presence of reducible sulfur compounds (Tiedgi, Sexstone, Parkin, Revsbech, and Shelton, 1984).

An interesting syntrophic association between Pseudomonas aeruginosa and Desulfovibrio vulgaris links sulfate respiration to the utilization of benzoate (Balba and Evans, 1980). D. vulgaris seems to produce organic acids, which are used by P.aeruginosa as electron acceptors while it metabolizes the benzoate.

4.1.2.1.3 Methanogenesis. During biodegradation, certain anaerobic bacteria commonly produce short-chain organic acids that can be further broken down to methane, carbon dioxide, and inorganic substances by other bacterial forms (Freeze and Cherry, 1979). Methane bacteria are an example of obligate anaerobes that ferment organic acids to methane. This suggests that such chemicals that enter an anaerobic environment may not be refractory and can possibly be mineralized to carbon dioxide and methane (Healy and Young, 1979). However, organic compounds occurring in a highly reduced state, such as hydrocarbons, are degraded only slowly in an anaerobic environment (Pettyjohn and Hounslow, 1983). In addition, degradation of petroleum hydrocarbons, straight-chain and branched alkanes, and alkenes, is not possible under methanogenic conditions (Environmental Protection Agency, 1985b).

Bacterial methanogenesis is a process common to many anaerobic environments (Berry, Francis, and Bollag, 1987). This strictly anaerobic process is frequently associated with the decomposition of organic matter in ecosystems, such as anoxic muds and sediments, the rumen and intestinal tract of animals, and anaerobic sewage sludge digesters

(Stanier, Doudoroff, and Adelberg, 1970; Lynch and Poole, 1979). Methane bacteria are able to use only a few simple compounds to support growth (Berry, Francis, and Bollag, 1987):

$$CO_2 + 4H_2 \dashrightarrow CH_4 + 2H_2O$$

$$4HCOOH \dashrightarrow CH_4 + 3CO_2 + 2H_2O$$

$$4CH_3OH \dashrightarrow 3CH_4 + CO_2 + 2H_2O$$

$$CH_3COOH \dashrightarrow CH_4 + CO_2$$

The importance of fastidious anaerobic consortia of organisms is illustrated by the types of detoxification reactions known to occur in the animal rumen, the best known of all anaerobic systems (Prins, 1978). These reactions include reductive dechlorination (or dehalogenation), possibly a limiting factor in degradation of certain compounds; nitrosamine degradation, a removal mechanism for a suspected carcinogen; reduction of epoxide groups in various compounds to olefins; reduction of nitro groups, as found in nitrophenol; and breakdown of aromatic structures (Kobayashi and Rittmann, 1982).

Methanogens are an essential component of anaerobic consortia degrading aromatics to methane (Ferry and Wolfe, 1976), but they appear to serve as electron sinks for other organisms rather than themselves attacking the primary substrates (Zeikus, 1977). Acetate and carbon dioxide plus hydrogen are probably the most important substrates for methane bacteria in natural ecosystems (Gottschalk, 1979). Since these organisms can use only simple compounds to support growth, they must rely on syntrophic associations with fermenters, which degrade complex organic compounds (i.e., aromatic compounds) into usable substrates. These associations are generally obligatory and may be similar from one methanogenic habitat to another (Suflita and Miller, 1985). It has been observed that the presence of --Cl or --NO_2 groups on phenol can inhibit methane production (Boyd, Shelton, Berry, and Tiedje, 1983).

Many phenolic compounds in a creosote waste were degraded to carbon dioxide and methane by anaerobic bacteria in an aquifer (Ehrlich, Goerlitz, Godsy, and Hult, 1982). However, polynuclear aromatic hydrocarbons, such as naphthalene, were not degraded under the same conditions.

Acetate and formate are important intermediate products in methanogenic fermentation and can indicate the presence of these organisms (McInerney and Bryant, 1981; McInerney, Bryant, Hespell, and Costerton, 1981).

4.1.2.2 Fermentation

During fermentation, obligate and facultative anaerobic bacteria use organic compounds as both electron donors and electron acceptors (Thibault and Elliott, 1979). This produces incompletely oxidized organic compounds, such as organic acids and alcohols.

Mixed cultures are often important in the degradative process (Berry, Francis, and Bollag, 1987). Interrelationships exist among various members of the anaerobic community. Neither Pelobacter acidigallici nov. sp. nor Acetobacterium woodii, a demethylating microorganism, was able to degrade the aromatic ring when cultured separately (Bache and Pfennig, 1981; Schink and Pfennig, 1982). However, a coculture completely metabolized syringic acid to acetate and carbon dioxide. In addition, if the isolates were cocultured with Methanosarcina barkeri, any metabolizable aromatic substrate could be completely mineralized to carbon dioxide and methane.

4.1.2.3 Specific Compounds

4.1.2.3.1 Mononuclear Aromatic Hydrocarbons.

Laboratory studies confirmed field tests that alkylbenzenes in ground water could be transformed both aerobically and anaerobically (Wilson, Bledsoe, Armstrong, and Sammons, 1986). Anaerobic degradation may be a useful adjunct to the aerobic degradation in heavily contaminated areas of groundwater and soil where the oxygen supply has been depleted.

1. Phenol

Phenol, hydroquinone, and p-cresol are converted to CO_2 and CH_4 under different reducing conditions, and the rate improves with acclimation (Young and Bossert, 1984). Phenolic compounds have been anaerobically biodegraded in near-surface groundwater (Ehrlich, Goerlitz, Godsy, and Hult, 1982). Methane and methane-producing bacteria were found only in water from the contaminated zone. However, it has been reported that under anaerobic conditions, transformation of phenols into polymerized species may be inhibited, and they may possibly leach through the soil more readily (Sawhney and Kozloski, 1984).

2. Benzene

Benzene can be biodegraded without the presence of molecular oxygen (Grbic-Galic and Young, 1985; Reinhard, Goodman, and Barker, 1984; Kuhn, Colberg, Schnoor, Wanner, Zehnder, and Schwarzenback, 1985). The anaerobic transformation of benzene might be a fermentation in which the substrates are partially reduced but also partially oxidized (Grbic-Galic and Vogel, 1987). Reduction results in the production of saturated alicyclic rings. Benzene degradation by mixed methanogenic cultures may lead to carbon dioxide and methane with intermediates of phenol, cyclohexane, and propanoic acid. The oxygen for ring hydroxylation might be derived from water. No anaerobic transformation of the compound was observed under methanogenic conditions by enrichment cultures from anaerobic sewage sludge, freshwater sediments, or marine sediments (Schink, 1985). However, after nitrate was injected into a contaminated aquifer, the pollutant was degraded anaerobically (Battermann and Werner, 1984).

Benzene is transformed by methanogenic cultures acclimated to lignin-derived aromatic acids under strictly anaerobic conditions with several intermediates, including demethylation products, aromatic alcohols, aldehydes and acids, cresols, phenol, alicyclic rings, and aliphatic acids (Grbic-Galic and Vogel, 1987). It is ultimately converted to carbon dioxide and methane.

3. Toluene

Toluene can also be biodegraded without the presence of molecular oxygen (Grbic-Galic and Young, 1985; Reinhard, Goodman, and Barker, 1984; Kuhn, Colberg, Schnoor, Wanner, Zehnder, and Schwarzenback, 1985). The anaerobic transformation of toluene may be a fermentation in which the substrates are partly oxidized and partly reduced (Grbic-Galic and Vogel, 1987). Reduction produces saturated alicyclic rings, while oxidation of the methyl group may give a primary product of benzyl alcohol, which, in turn, may be converted to benzaldehyde and benzoic acid.

Microorganisms adapted to growth on m-xylene in the absence of molecular oxygen with nitrate as an electron acceptor are also able to degrade toluene under denitrifying conditions (Zeyer, Kuhn, and Schwarzenback, 1986).

No anaerobic transformation of the compound was observed under methanogenic conditions by enrichment cultures from anaerobic sewage sludge, freshwater sediments, or marine sediments (Schink, 1985). However, after nitrate was injected into a contaminated aquifer, the pollutant was degraded anaerobically (Battermann and Werner, 1984). Methanogenic alluvium from the floodplain of the South Canadian River, which receives leachate from a landfill, showed toluene degradation by an order of magnitude after 11 months (Rees, Wilson, and Wilson, 1985).

Toluene is transformed by methanogenic cultures acclimated to lignin-derived aromatic acids under strictly anaerobic conditions, with intermediates, including demethylation products, aromatic alcohols, aldehydes and acids, cresols, phenol, alicyclic rings, and aliphatic acids (Grbic-Galic and Vogel, 1987). It is ultimately converted to carbon dioxide and methane.

4. Xylene

Xylenes can be biodegraded without the presence of oxygen (Grbic-Galic and Young, 1985; Reinhard, Goodman, and Barker, 1984; Kuhn, Colberg, Schnoor, Wanner, Zehnder, and Schwarzenback, 1985). Up to 0.4 mM 1,3-dimethylbenzene (m-xylene) was rapidly mineralized in a laboratory aquifer column operated in the absence of molecular oxygen with nitrate as an electron acceptor (Zeyer, Kuhn, and Schwarzenback, 1986).

No anaerobic transformation of the compound was observed under methanogenic conditions by enrichment cultures from anaerobic sewage sludge, freshwater sediments, or marine sediments (Schink, 1985). However, after nitrate was injected into a contaminated aquifer, the pollutant was degraded anaerobically (Battermann and Werner, 1984). Other workers also reported that all three xylene isomers could be degraded under anoxic denitrifying conditions in a laboratory column system simulating saturated flow conditions typical for a river water/groundwater infiltration system (Wilson and Rees, 1985). The three isomers were also preferentially removed over other petroleum products from a methanogenic landfill leachate (Reinhard, Goodman, and Barker, 1984).

m-Xylene can be mineralized in the absence of oxygen by reducing the redox potential, $E^{'}$, of the inflowing medium with sulfide to -0.11 V (Zeyer, Kuhn, and Schwarzenback, 1986).

5. Benzoate

Rhodopseudomonas palustris photometabolizes benzoate anaerobically (Atlas and Schofield, 1975). Pseudomonas PN-1 and P. stutzeri grow anaerobically on benzoate-nitrate-mineral salts medium.

4.1.2.3.2 Polycyclic Aromatic Hydrocarbons (PAHs)

1. Naphthalene

No anaerobic transformation of the compound was observed under methanogenic conditions by enrichment cultures from anaerobic sewage sludge, freshwater sediments, or marine sediments (Schink, 1985). Polynuclear aromatic hydrocarbons, such as naphthalene, were not degraded to carbon dioxide and methane by anaerobic bacteria in an aquifer (Ehrlich, Goerlitz, Godsy, and Hult, 1982).

There was no evidence of anaerobic degradation of naphthalene in groundwater samples contaminated by wood creosoting products, although it disappeared at a faster rate in the aquifer than if only dilution were occurring (Ehrlich, Goerlitz, Godsy, and Hult, 1982).

2. Pyrene

Pyrene degradation did not occur under anaerobic or microaerophilic conditions.

4.1.2.3.3 Branched-Chain Alkanes and Alkenes

4.1.2.3.4 Straight-Chain Aliphatics

4.1.2.3.5 Alcohols.
Many bacteria and yeasts that can catabolize methanol under anaerobic conditions and at low and high concentrations have been isolated from soil as deep as 100 ft (Benoit, Novak, Goldsmith, and Chadduck, 1985).

4.1.2.3.6 Alicyclic Hydrocarbons

1. Cyclohexane

A strain of Acinetobacter anitratum utilizes cyclohexane carboxylic acid through 2-oxocyclohexane carboxylic acid (Hou, 1982). Some anaerobic photosynthetic strains, as well as aerobic nonphotosynthetic strains, also metabolize cyclohexane carboxylic acid by this pathway.

4.1.3 End Products

Aerobic breakdown of organic molecules may cause accumulation of organic acid intermediates that reduce soil pH and inhibit biological activity. These effects can be handled with regular reinoculation and use of chemical pH control agents, such as lime. For efficient biodegradation, it is important to have a provision for the removal of toxic wastes and by-products (Texas Research Institute, Inc., 1982). This can be accomplished in two ways: 1) having a diverse microbial population so the by-products are consumed and 2) creating a flow through the system to remove toxins.

Table A.4-7 shows some of the end products formed from the microbial oxidation of hydrocarbons by specific microorganisms (Zajic, 1964).

1-Naphthol is a major product of cyanobacterial metabolism of naphthalene (Cerniglia, van Baalen, and Gibson, 1980a). It is also produced by the filamentous fungus, Cunninghamella elegans, from oxidation of naphthalene (Dagley, 1981). The fate of 1-naphthol in natural environments is of particular interest because it appears to be very toxic (Fewson, 1981). This compound can be totally degraded by some microorganisms (Bollag, Czaplicki, and Minard, 1975), and a soil pseudomonad has been isolated, which can grow on it as a sole source of carbon and energy (Walker, Janes, Spokes, and van Berkum, 1975). 1-Naphthol can also undergo other transformations, such as polymerization by the extracellular enzyme laccase (Fewson, 1981).

The role of phenol oxidases in determining the fate of xenobiotics may have been underestimated (Bollag, Sjoblad, and Minard, 1977). Polymerization reactions and enzymic oxidative coupling are probably important in the covalent bonding of phenolic compounds to soil organic polymers. The degree to which phenols become bound to soil humic molecules as a result of enzymatically mediated oxidative coupling reactions may be affected by substituent groups on the aromatic ring (Berry and Boyd, 1984). For instance, electron donating groups, such as methoxy $(-OCH^3)$ (common substituents on lignin-derived polyphenols), would facilitate this reactivity.

It is not completely known at this time what the effect of microbial degradation of complex hydrocarbons may be on the environment (Texas Research Institute, Inc. 1982). It has been established that complete oxidation of many carcinogenic hydrocarbons is not required to render them noncarcinogenic. However, an EPA study reported that, with moderate temperatures and the presence of inorganic nitrogen and phosphorus, naturally occurring freshwater microorganisms are able to form mutagenic biodegradation products from crude oil that are bactericidal to Escherichia coli K-12 (Morrison and Cummings, 1982). The public health significance of this finding is undetermined.

4.2 HEAVY METALS

In the last 200 years, microorganisms have been adapting to the changes in the distribution of elements at the surface of the earth, as a result of industrialization (Wood and Wang, 1983). Several strategies for resistance to metal ion toxicity have been identified in these organisms:

1. The development of energy-driven efflux pumps that keep toxic element levels low in the interior of the cell (e.g., for Cd(II) and As(V))
2. Oxidation (e.g., AsO_3^{-2} to AsO_4^{-3}) or reduction (e.g., Hg^{+2} to Hg^0), which can enzymatically and intracellularly convert a more toxic form of an element to a less toxic form
3. Biosynthesis of intracellular polymers that serve as traps for the removal of metal ions from solution (e.g., for Cd, Ca, Ni, and Cu)
4. The binding of metal ions to cell surfaces
5. The precipitation of insoluble metal complexes (e.g., metal sulfides and metal oxides) at cell surfaces
6. Biomethylation and transport through cell membranes by diffusion-controlled processes.

Microorganisms that have short generation times and, consequently, increased evolution rates, have adapted themselves to deal with high concentrations of metal ions. Microorganisms are evolving strategies to maintain low intracellular concentrations of toxic pollutants. Some resist high concentrations through their evolution under extreme environmental conditions (Brock, 1978). Others have achieved resistance to the recently polluted environment through acquisition of extrachromosomal molecules (plasmids) (Silver, 1983).

Table A.4-7.   Products Formed from the Oxidation of Hydrocarbons by Certain Microorganisms (Zajic, 1964)

Substrate	Microbe	Product
Alkanes	*Pseudomonas* spp.[h]	pristane, phytane, sterane, 17 alpha (H), 21 beta (H)-hopane series
	Pseudomonas oleovorans[k]	carboxylic acids
(C_{12} to C_{16})	*Corynebacterium*[a]	ketones, esters of aliphatic acids
(multiple-branched; e.g., pristane)	(*Brevibacterium erythrogenes Corynebacterium* sp.)[a]	succinyl-CoA
(C_{10} to C_{14})	*Corynebacterium*[l]	fatty acids (metabolyzable by beta-oxidation)
Alkenes (C_6 to C_{12})	*Pseudomonas oleovorans*[k]	1,2-epoxides
Benzo(a)anthracene (BaA)	*Cunninghamella elegans*[c]	trans-BA 3,4-, 8,9-, and 10,11-dihydrodiols
	Beijerinck (mutant, acclimated)[g]	dihydrols, primarily c-1,1-dihydroxy-1,2-dihydrobenzo(a)anthracene
Benzo(a)pyrene (BaP, BP)	*Cunninghamella elegans*[b]	trans-7,8-dihydroxy-7,8-dihydro-BP; trans-9,10-dihydroxy-9,10-dihydro-BP; BP-3- and 9-phenols; and BP-1,6- and 3,6-quinones. BP-7,8- and BP-9, 10-diols are further oxidized to metabolites known to be carcinogenic, tumorigenic, and mutagenic to experimental animals
	Candida lipolytica[f]	3-hydroxybenzo(a)pyrene 9-hydroxybenzo(a)pyrene
	Beijerinckia (mutant, acclimated)[g]	vicinal dihydrodiols, mainly cis-9,10-dihydroxy-9,10-dihydro-benzo(a)pyrene
	Neurospora crassa[o]	mainly 3-hydroxybenzo-(a)pyrene

Table A.4-7.   continued

Substrate	Microbe	Product
	Saccharomyces cerevisiae[p]	7,8-dihydroxy-7,8-dihydrobenzo(a)pyrene, 9-hydroxybenzo(a)-pyrene, 3-hydroxybenzo-(a)pyrene
Biphenyl	Candida lipolytica	2-,3-,and 4-hydroxybi-phenyl, 4,4'-dihydroxy-biphenyl, 3-methoxy-4-hydroxybiphenyl
	Cunninghamella elegans[m]	2-,3-,4-hydroxybiphenyl and 4,4'-dihydroxy-biphenyl, glucuronides, sulphates
n-Butane	M. smegmatis	2-butanone
Decane	Corynebacterium[a]	1-decanol, 1,10-decanediol
Ethane	Methylosinus trichosporium	acetone
n-Heptane	Pseudomonas aeruginosa	n-hexanoic acid,
Hexadecane	M. cerificans Arthrobacter sp.[a]	cetyl palmitate ketones
n-Hexane	M. smegmatis	2-hexanol, 2-hexanone
Kerosene, heavy naphthas, aromatic naphthas, petroleum pitches, tars, and asphalts	Fusarium moniliforme[l]	giberellin
Kerosene	55 strains[l]	amino acids
2-Methylhexane	P. aeruginosa	2-methylhexanoic acid, 5-methylhexanoic acid
Naphthalene	Nocardia sp.	diols of substituted naphthalene
	Cunninghamella elegans	1-naphthol, 2-napththol trans-1,2-dihydroxy-1, 2-dihydronaphthalenes

Table A.4-7.   continued

Substrate	Microbe	Product
	Oscillatoria sp. strain JCM[h]	1-naphthol, cis-1,2-dihydroxy-1,2-dihydro-naphthalene, and 4-hydroxy-1-tetralone
	Candida lipolytica[b]	1-naphthol, 2-naphthol, 4-hydroxy-1-tetralone, trans-1,2-dihydroxy-1,2-dihydronaphthalene
	Candida elegans[c]	1-naphthol, 4-hydroxy-1-tetralone, 1,4-naphthoquinone, 1,2-naphthoquinone, 2-naphthol, trans-1,2-dihydroxy-1,2-dihydro-naphthalene
	Cunninghamella elegans, C. echinulata, C. japonica, Syncephalastrum sp., S. racemosum, Mucor sp., M. hiemalis[j]	mainly 1-naphthol; also 2-naphthol, 4-hydroxy-1-tetralone, trans-na-phthalene dihydrodiol, 1,2-naphthoquinone, 1,4-naphthoquinone
	Lake sediments[l]	naph-cis-1,2-dihydroxy-1,2-dihydronaphthalene, 1-naphthol, salicylic acid, catechol
Octadecane	M. cerificans	octadecyl stearate
n-Octane	Pseudomonas	n-octanol
n-Pentane	M. smegmatis	2-pentanone
n-Propane	Mycobacterium smegmatis	acetone
Phenanthrene		1-hydroxy-2-naphthoic acid[a]
Tetradecane	Micrococcus cerificans	myristyl palmitate
Toluene	Pseudomonas sp., Achromobacter sp.[n]	3-methylcatechol

Table A.4-7.   continued

Substrate	Microbe	Product
Tridecane	*Pseudomonas aeruginosa*[a]	tridecane-1-ol, undecan-1-ol
n-Undecane	*Mycobacterium* sp.	undecanoic acid, 1,11 undecandioic acid

Additional references:

a = (Hou, 1982)
b = (Cerniglia and Gibson, 1980)
c = (Dodge and Gibson, 1980)
d = (Cerniglia, Wyss, and Van Baalen, 1980)
e = (Lawlor, Shiaris, and Jambard-Sweet, 1986)
f = (Cerniglia and Crow 1981)
g = (Gibson and Mahadevan, 1975)
h = (Gibson, Koch, and Kallio, 1968)
i = (Zajic, 1964)
j = (Cerniglia, Hebert, Szaniszlo, and Gibson, 1978)
k = (Roberts, Koff, and Karr, 1988)
l = (Heitkamp, Freeman, and Cerniglia, 1987)
m = (Dodge, et al., 1979)
n = (Claus and Walker, 1964)
o = (Lin and Kapoor, 1979)
p = (Wiseman, Lim, and Woods, 1978)

Metal uptake by a microorganism is metabolism dependent (Hornick, Fisher, and Paolini, 1983). What the organism does not require can be precipitated intracellularly and stored. Microbes are also capable of producing organic compounds, such as citric acid and oxalic acid, which act as binding or chelating agents. Production of hydrogen sulfide by microbes is of great importance in that heavy metals form insoluble sulfides. Both bacteria and yeast have exhibited hydrogen sulfide production and have created a more tolerant environment for more sensitive organisms.

The availability of metal ions for transport into cells is restricted by their abundance and solubility in water (Wood and Wang, 1983). Solubility is greatly influenced by pH, temperature, standard reduction potential (E^0), the presence of competing anions and cations, and the presence of surface-active substances, such as particulates and macromolecules, including proteins, humic acids, and clays.

Both the pH and the E^0 can vary widely from the outside of the living cell to the inside of that cell (Wood and Wang, 1983). Most metal ions function as Lewis acids (electron acceptors), but depending upon pH, oxidation state, and complexation, metal complexes also can function as bases. Living cells are not at equilibrium with the external environment, and, therefore, a kinetic approach to metal ion transport, binding, toxicity, and resistance to toxicity is much more meaningful than a thermodynamic approach.

Metal interactions in biology can be divided into three classes: ions in fast exchange with biological ligands (e.g., Na^+, K^+, Ca^{+2}, Mg^{+2}, and H^+), ions in intermediary exchange with biological ligands (e.g., Fe^{+2} and Mn^{+2}), and ions in slow exchange with biological ligands (e.g., Fe^{+3}, Zn^{+2}, Ni^{+2}, and Cu^{+2}) (Williams, 1983). Living cells have membranes that act as initial barriers to metal ion uptake (Wood and Wang, 1983). Prokaryotes select those ions in fast exchange. Eukaryotes use spatial partitioning of metals. Once the cell buffering capacity for essential metal ions is exceeded, toxicity becomes evident. Toxicity occurs at much lower concentrations for nonessential metals than for essential.

Insoluble metal complexes can be precipitated at the cell surface through the activities of membrane-associated sulfate reductases (Galun, Keller, Malki, Feldstein, Gallun, Siegal, and Siegal, 1983) or through the biosynthesis of oxidizing agents, such as oxygen or hydrogen peroxide (Wood, 1983). The reduction of sulfate to sulfide and the diffusion of oxygen and hydrogen peroxide through the cell membrane provide highly reactive means by which metals can be complexed and precipitated. A green alga, Cyanidium caldarium, can grow in acidic conditions and at high temperatures (Lovelock, 1979). It can remove 68 percent of the iron, 50 percent of the copper, 41 percent of the nickel, 53 percent of the aluminum, and 76 percent of the chromium from solution. Anaerobic cultures growing in the dark produce hydrogen sulfide gas for sulfide precipitation of metals.

Some microorganisms can use biomethylation to eliminate heavy metals, such as mercury and tin, and metalloids, such as arsenic and selenium (Wood and Wang, 1983). The synthesis of less polar organometallic compounds from polar inorganic ions has certain advantages for cellular elimination by diffusion-controlled processes (Wood, Cheh, Dizikes, Ridley, Rackow, ad Lakowicz, 1978). Mechanisms for B_{12}-dependent synthesis of metal alkyls (requiring the presence of vitamin B_{12}) have been discovered for the metals Hg, Pb, Tl,

Pd, Pt, Au, Sn, and Cr and for the metalloids As and Se (Craig and Wood, 1981).

Heavy metal contaminants can be altered by transformation to volatile forms (Arthur D. Little, Inc., 1976). Soil reactions can transform them to less toxic forms or make them unavailable to plants. Modifications in soil conditions can promote transformation of mercury and arsenic (e.g., increasing organic material, moisture, temperature, and pH of the soil for mercury and decreasing the oxygen level for arsenic).

There is a possible sequence of activating measures to increase transformation of these metals (Arthur D. Little, Inc., 1976). The first steps would be to add organic material, decrease oxygen, increase temperature, and decrease pH. Logistically, lowering the pH should come first, since ferrous sulfate should be added before irrigation begins. The organic matter should be incorporated into the soil next, immediately preceding irrigation; this will help produce anaerobic conditions. A PVC sheeting can then be applied. These measures should be kept in operation for about 2 to 3 years. Monitoring will indicate the actual time required. After the level of one toxicant is acceptable, conditions can be changed to transform and remove another. Finally, efforts must be made to decrease the availability of heavy metals and other undegradable toxicants left in in the soil, e.g., increasing the pH of the soil with an application of limestone. Leaching of metals must be prevented throughout the treatment process.

Figure A.4-3 presents a possible sequence for soil manipulation.

Bacterial resistance to heavy metals in the environment can result in bioaccumulation, biotransformation, changes in ecological diversity, and coselection of resistance factors for antibiotics (Sterritt and Lester, 1980). Microorganisms can be used for the removal of heavy metals from industrial effluents and as indicator organisms in bioassays (Anderson and Abdelghani, 1980). A simple, rapid method for the determination of bacterial resistance to a wide range of metals has been developed through modification of the antibiotic susceptibility test (Thompson and Watling, 1983).

Table A.4-8 lists bacteria, fungi, and algae; the metals they can remove; and the methods they use for removal (Monroe, 1985).

4.2.1 Specific Elements

The following are the elements present in the fuels addressed in this review and their potential for transformation by soil microbes (JRB and Associates, Inc., 1984b):

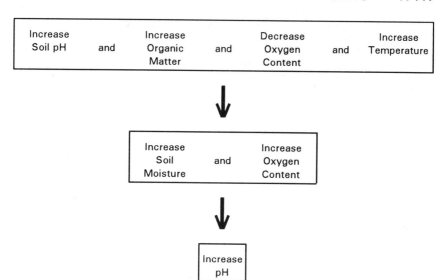

Figure A.4-3. Possible Sequencing of Soil Manipulation (Arthur D. Little, Inc., 1976).

4.2.1.1 Arsenic (As)

Arsenic is apparently oxidized by aerobic heterotrophic organisms, and many heterotrophs can also reduce arsenate (Alexander, 1977). Methylation of arsenicals is an important process in soils, and trimethylarsine is an important gaseous product (Woolson, 1977). Soil contaminated with arsenic must be managed to minimize volatilization through microbial reduction. Addition of organic material and maintenance of aerobic conditions can help stimulate the oxidation of arsenite to arsenate (Sims and Bass, 1984). Further treatment with ferrous sulfate will form highly insoluble $FeAsSO_4$.

Resistance of microbes to arsenic occurs through the evolution of cellular exclusion mechanisms (Wood and Wang, 1983). Resistance in Staphylococcus aureus and Escherichia coli to arsenate and arsenite is induced by an operon-like system (Novick, Murphy, Gryczan, Baron, and Edellman, 1979). An operon is a DNA region that codes for several enzymes in a reaction pathway.

Table A.4-8.   Microbial Mechanisms for Metal Extracting/Concentrating/Recovery (Monroe, 1985)

Microorganism	Metal Removed	Method
Thiobacillus, Sulfolobus	Iron, sulfur	Oxidation
Sphaerotilus, Leptothrix, Hyphomicrobrium, Gallionella	Iron, manganese	Oxidation
Spirogyra, Oscillatoria, Rhizoclonium, Chara	Molybdenum, selenium, uranium, radium	Oxidation
Desulfovibrio spp.	Mercury, lead	Reduction
Scenedesmus, Synechococcus, Oscillatoria, Chlamydomonas, Euglena	Nickel	Surface ion-exchange
Saccharomyces cerevisiae, Rhizopus arrhizus	Uranium, cesium, radium	Surface ion-exchange
Penicillium digitatum	Uranium	Surface ion-exchange
Ustilago sphaerogena	Iron	Surface chelation
Aspergillus niger	Aluminum	Surface chelation
Cynanidium caldarium	Iron, copper, nickel, aluminum, chromium	Surface precipitation
Staphylococcus aureus, Escherichia coli	Cadmium, zinc, arsenate, arsenite, antimony	Chemosmotic efflux
Pseudomonas aeruginosa	Uranium, cesium, radium	Intracellular trap
Synechococcus	Nickel, copper, cadmium	Intracellular trap
Clostridium cochlearium	Mercury	Biomethylation
Pseudomonas spp.	Tin	Biomethylation

Arsenic behaves much like phosphorus in soils in that its adsorption increases as iron oxide content increases, and that iron and aluminum hydrous oxides specifically adsorb the metal (Hornick, 1983). Arsenate, however, can be reduced to arsenite, but arsenate is the most common form in soils. Generally, both forms are strongly retained in soils.

4.2.1.2 Cadmium (Cd)

Growth of soil bacteria is retarded by cadmium, and the soil microbial community structure is affected (JRB and Associates, Inc., 1984b). However, bacterial populations in contaminated sites have been found to be able to adapt to the heavy metal contamination (Tripp, Barkay, and Olson, 1983). Soil microbial biomass may contribute to the soil cadmium-binding capacity and affect cadmium availability, with dead cells sorbing more cadmium than live cells (Kurek, Czaban, and Bollag, 1982). Since cadmium exists in nature only in the valence state of +2, microbial oxidation or reduction of this element is unlikely.

Intracellular traps can be biosynthesized as a temporary measure for organisms to remove metal ions (e.g., synthesis of metallothionen and removal of cadmium by this sulfhydryl-containing protein) (Williams, 1981b), and to prevent metals from reaching toxic levels (Wood and Wang, 1983). Mutants with intracellular trapping mechanisms tend to bioconcentrate the toxic metal intracellularly to about 200 times over the external concentration. This strategy works well for some organisms but is not as effective as the extracellular binding or precipitation of metals.

Cellular exclusion mechanisms are also responsible for resistance of microorganisms to cadmium (Wood and Wang, 1983). This resistance is mediated by a plasmid. Resistant cells of <u>Staphylococcus aureus</u> have a very efficient chemosmotic efflux system specific for Cd^{+2} ions, as a result of two separate plasmid genes.

Cadmium is complexed by organic matter, oxides of iron and manganese, and chlorides (Chaney and Hornick, 1978). In alkaline, low organic matter, sandy soils, precipitation of cadmium compounds occurs. The soil pH is the most important factor governing cadmium solubility and resultant availability, with the solubility increasing as pH decreases.

4.2.1.3 Chromium (Cr)

Chromium should be amenable to oxidation or reduction by microbes, since it commonly exists in two oxidation states, Cr III and Cr VI (Zajic, 1969). Chromium VI is more toxic and mutagenic than chromium III

(Ross, Sjogren, and Bartlett, 1981). Gram negative bacteria are also more sensitive to chromium VI than gram positive organisms. It should not be assumed that chromium III is harmless to the soil microflora at high levels.

The most soluble, mobile, and toxic form of chromium in soils occurs in the hexavalent state as chromate or dichromate (Hornick, 1983). When aerobic conditions exist, the hexavalent form is rapidly reduced to trivalent chromium, which forms insoluble hydroxides and oxides and is unable to leach. Liming soil to pH 6.5 with the presence of an alkaline oil waste in the soil will maintain the soil pH near neutral (Hornick, Fisher, and Paolini, 1983). A near neutral soil pH will also prohibit the formation of dichromates (Hornick, 1983).

Solid phase chromium should not be readily mobilized during the retention of dredge slurries in disposal areas, possibly because of the slow oxidation of reduced chromium hydroxide (Lu and Chen, 1977). Little chromium is released from dispersed sediments under oxidizing conditions. Reduced chromium (Cr III) is generally highly insoluble at pH values above 5.5, unless complexed with soluble organic compounds. Chromium has also not been noted to oxidize to more soluble chromium (VI) forms under short-term oxidizing conditions.

The green alga, Cyanidium caldarium may be effective for selective removal of chromium from polluted wastewaters (Wood and Wang, 1983).

4.2.1.4 Iron (Fe)

The concentration of free metal ions can be controlled by the biosynthesis of ligands in the form of small molecules with high stability constants, such as with the removal of iron (Wood and Wang, 1983). The cell may expend energy to pump the metal ion out of the cell, it may synthesize ligands that bind metals strongly at the cell surface, or it may use the activities of surface-bound enzymes to precipitate metals extracellularly.

4.2.1.5 Lead (Pb)

Inorganic lead is toxic to a broad range of microorganisms, including cyanobacteria, marine algae, fungi, and protozoa (JRB and Associates, Inc., 1984b). Lead and its compounds also affect microbial activity in soil, including inhibition of nitrogen mineralization, stimulation of nitrification, and the synthesis of soil enzymes. Species diversity is lower in lead-contaminated soils. A low pH of 5 or 6 increases the toxicity of lead, while higher pH and other abiotic factors (phosphate and carbonate ions, clay minerals, particulate humic acid, and soluble organics) reduce toxicity. Lead can be methylated by some microbes, thereby increasing its volatility and potential loss from the soil.

Lead solubility is determined by the amount of sulfate, phosphate, hydroxides, carbonates, and organic matter present in a soil system (Hornick, 1983). The formation of lead sulfate at very low pH, lead phosphate and hydroxides at intermediate pH, and lead carbonates at a calcareous pH limits the mobility of lead in soils. As the soil pH and available phosphorus in a soil decrease, soluble lead increases.

4.2.1.6 Mercury (Hg)

Microbes oxidize, reduce, methylate, and demethylate mercury (JRB and Associates, Inc., 1984b). Serratia marcescens had the highest rate of mercury transformation at pH 8 over the range tested (Mason, Anderson, and Shariat, 1979). Both aerobic and anaerobic heterotrophic bacteria were resistant to 14 ppm Hg^{+2} (Callister and Winfrey, 1983). Mercury-resistant populations of oil-degrading bacteria have been isolated (Walker and Colwell, 1974a). Petroleum biodegradation proceeded when mercury in the soil was present in the low ppm concentration range, but was absent at 85 ppm (Walker and Colwell, 1976a). Mercury methylation was highest in anaerobically incubated surface sediments (Callister and Winfrey, 1983). Sediment-bound mercury remained available for methylation over 7 days.

Mercury methylation and demethylation are usually ascribed to different bacteria; however, the anaerobic Clostridium cochlearium T-2 was found to acquire demethylating capabilities, in addition to its methylating (Pan-Hou, Hosono, and Imura, 1980). This trait is probably on a plasmid. Mercury resistance is an inducible trait, and the genetic material coding for it is transposable (Silver and Kinscherf, 1982). Biomethylation of Hg(II) salts to CH_3Hg^+ by a B_{12}-dependent strain of C. cochlearium allows detoxification and gives the organism an advantage in mercury-contaminated systems (Wood and Wang, 1983). Mercuric and organomercurial strains of bacteria have been isolated from a variety of ecosystems, such as soil, water, and marine sediments (Friello and Chakrabarty, 1980; Vonk and Sjipesteijn, 1973). These organisms catalyze both the forward and reverse conversion of CH_3Hg^+ to Hg^{+2} and then to Hg^0 (Wood, 1983).

Mercury reacts in soils with chlorides and sulfur to form insoluble HgS, $HgCl_3$, and $HgCl_4^{-2}$ (Hornick, 1983). Mercury can be chelated by organic matter as $HgCl_4^{-2}$ or can be absorbed by sesquioxide surfaces (CAST, 1976). Through chemical and microbial degradation, mercury can be volatilized or associated with clay particles and organic matter. Mercury is not very mobile in the soil profile due to its strong sorption reactions with soil constituents.

An enzyme capable of metabolizing phenylmercuric acetate has been isolated from soil microorganisms (Tonomura and Kanzaki, 1969). The enzyme(s) is capable of cleaving the carbon-mercury bond and requires both a sulfhydryl compound and NADH to carry out the reaction.

4.2.1.7 Nickel (Ni)

Nickel inhibits soil nitrification, carbon mineralization, and the activities of acid and alkaline phosphatase and arylsulfatase (JRB and Associates, Inc., 1984b). Toxicity to fungi is reduced by the presence of clay or an increase in pH to 7.0 (Babich and Stotzky, 1983a; 1983b). Survival of certain bacteria and yeasts was improved by raising the pH from 4.9 to 7.7. The presence of ions such as Mg^{+2}, Zn^{+2}, S^{-2}, PO^{-3}, and PO_4; alkaline pH; and type and amount of clay minerals greatly reduce the microbial toxicity of nickel. However, other ions (e.g., potassium, sodium, calcium, and iron), amino acids, tryptone, casamino acids, yeast extract, and chelating agents (citrate, EDTA, DPA, NTA) do not reduce Ni toxicity.

Both nickel binding and nickel toxicity are very pH dependent (Wood and Wang, 1983). The optimum pH for binding is between 8 and 8.5. Orientation of ligands at the cell surface must be important, since only surface-active substances, such as humic acids, could compete effectively for nickel binding (Galun, Keller, Malki, Feldstein, Galun, Siegal, and Siegal, 1983).

Nickel availability in soils is governed by iron and manganese hydrous oxides and by organic chelates that complex nickel less strongly than copper (CAST, 1976). Nickel differs from copper and zinc in that it is more available from organic sources than inorganic sources. Acid soils increase the solubility of nickel in the soil solution.

Green algae are much more resistant to high concentrations of Ni^{+2} than are the blue-green algae (Wood and Wang, 1983). Cyanobacteria and brown and green algae all bioconcentrate nickel (Fuge, 1973; Karata, Yoichi, and Fumio, 1980; Ballester and Castelvi, 1980; Hirschberg, Skane, and Thorsby, 1977). The former are more sensitive to nickel toxicity than the green algae, indicating different transport mechanisms for prokaryotes and eukaryotes (Galun, Keller, Malki, Feldstein, Galun, Siegal, and Siegal, 1983).   Nickel-tolerant mutants of the cyanobacterium, Synechoccus, can tolerate up to 20 x 10^{-5} M nickel sulfate by synthesizing an intracellular polymer that removes nickel from solution (Simon and Weathers 1976).   This intracellular trap prevents nickel toxicity.   The green alga, C. caldarium, may be effective in treating waters polluted with nickel, so effluents can meet federal standards (Wood and Wang, 1983).

4.2.1.8 Selenium (Se)

Normal soils contain between 0.1 and 2.0 ug/g selenium (Girling, 1984). Selenium behaves similarly to sulfur in the soil solution, existing as selenates and selenite (Hornick, 1983).   Selenate, the predominant form in alkaline soils, is quite soluble as $CaSeO_4$ and can move readily in these soils. Addition of lime will increase the availability of selenium (Gissel-Nielsen, 1971).

Most of the selenium in the soil is biologically unavailable (Girling, 1984). Some bacteria and fungi reduce biologically available selenium to elemental insoluble forms, while others produce volatile organic forms of selenium that are lost to the atmosphere. Some bacteria are able to oxidize colloidal selenium to selenate or selenite so it becomes biologically available.   Soil microflora are capable of several transformations of selenium, such as oxidation, reduction, and methylation (JRB and Associates, Inc., 1984b). The oxidized form, selenate, is very toxic (Alexander, 1977).   Methylation is greatly accelerated when a readily available carbon source, such as glucose, is added to the soil. Selenium salts can be biologically methylated to volatile organic products (Challenger and North, 1933). The fungus, Scopulariopsis brevicaulis, can produce dimethylselenide from inorganic selenite or selenate. Eleven strains of fungi have been found capable of producing dimethylselenide, including strains of Penicillium, Fusarium, Cephalosporium, and Scopulariopsis (Barkes and Fleming, 1974). Strains of Bacillus spp. are able to oxidize around 1.5 percent of the total selenium added to selenite and trace amounts of selenate (Shrift, 1973). Inorganic selenium can be reduced to elemental selenium. Isolates of bacteria, fungi, and actinomycetes from soils containing high selenium contain as much as 0.18 percent (dry wt.) selenium inside their cells (Koval'skii, 1968). Their resistance to the high selenium levels depends upon their ability to reduce the soil selenium to the biologically inert elemental form.

Waste material containing significant amounts of selenium may present environmental problems if the selenium can be transformed into a form that is biologically available (Girling, 1984).

4.2.1.9 Silver (Ag)

Silver is known to be bacteriocidal (Woodward, 1963); however, soil treated with 100 ppm Ag had about twice as many Ag^+-reducing bacteria than untreated soil (Klein and Molise, 1975; Sokol and Klein, 1975).

Several bacteria have been found that precipitate silver as Ag_2S at the cell surface (Wood and Wang, 1983).

APPENDIX SECTION 5

Enhancement of Biodegradation

5.1 OPTIMIZATION OF SOIL BIODEGRADATION

Some of the information in this section may duplicate material covered in Section 5.2, Optimization of Groundwater Biodegradation, and Section 5.3, Optimization of Freshwater, Estaurine, and Marine Biodegradation; however, it is presented here under a separate heading, with other related information, to accommodate those readers who may specifically wish to address treatment of soil contamination only.

5.1.1 Biological Enhancement

5.1.1.1 Seeding of Microorganisms

The seeding of microorganisms has been used in a number of different environments to degrade organics (Environmental Protection Agency, 1985b). Microbial seeding does not appear to be very effective in reducing oil contamination in the sea (Atlas and Bartha, 1973e; Gutnick and Rosenberg, 1979; McDowell, Bourgeois, and Zitrides, 1980), but it has been argued that seeding and nutrient supplementation would work in contained environments like oil tankers (Gutnick and Rosenberg, 1979).

Seeding with hydrocarbonoclastic microorganisms appears to be a special application to severely environmentally distressed soil (Cook and Westlake, 1974), such as in Arctic or near-Arctic climates, where the activity and growth rate of indigenous organisms may be limited (Cook and Westlake, 1974; Hunt, et al., 1973). When an addition of 10^4 microorganisms/g dry soil was tested in such an environment (with addition of N and P and adjustment of the pH to 7), microbial activity over the controls was increased by at least a factor of four in 40 days.

However, there have been varying degrees of success with seeding microorganisms onto petroleum-contaminated soils (Knox, Canter, Kincannon, Stover, and Ward, 1968). Other examples have not been as successful. Addition of 10^8 cells/g soil of two hydrocarbon-degrading isolates did not significantly influence oil concentrations (Lehtomakei and Niemela, 1975), and application of 10^6 oil-utilizing bacteria/cm^2 resulted in only a slight additional degradation of the C_{20} to C_{25} group of n-saturated compounds (Jobson, Cook, and Westlake, 1972). The seed organisms in the latter case were species of Flavobacterium and Cytophaga (41 percent), Pseudomonas (34 percent), Xanthomonas (10 percent), Alcaligenes (9 percent), and Arthrobacter (5 percent). Even when oil-degrading bacteria were added to soil of the boreal region of the arctic, there was no increase in the changes of the recovered oil; however, this may have been due to insufficient application (Westlake, Jobson, and Cook, 1978).

Microorganisms able to degrade organic pollutants in culture sometimes may fail to function when inoculated into natural environments because the concentration of the hydrocarbon in nature may be too low to support growth or because the organisms may be susceptible to toxins or predators in the environments (Alexander, 1986). The organisms may use other organic compounds in preference to the pollutant, or they may be unable to move through the soil to sites containing the chemical.

Use of microorganisms from sewage effluent is a low-cost, fairly effective method for the removal of water-soluble biodegradable organics (Wentsel, et al., 1981). Formaldehyde was successfully removed at levels less than 2000 ppm with such a population mixed with a Pseudomonas sp. Attempts to acclimate an activated sludge culture to groundwater contaminated by several priority pollutants and at least 70 other organics were only minimally successful (Shuckrow and Pajak, 1981). However, coupling an activated sludge process with treatment by granular activated carbon proved beneficial. Microbial growth on the activated carbon may also play a role in the removal of contaminants (Werner, 1982).

Generally, the organisms may be applied in liquid suspension or with a solid carrier (Sims and Bass, 1984). Run-on and run-off controls may be necessary. The potential achievable level of treatment is high.

Nevertheless, relatively long periods of time may be required to complete treatment, and excessive precipitation may wash out the inoculum, necessitating retreatment.

Seed microorganisms would have to be freeze dried or frozen to maintain viability (Atlas, 1977). Growing cultures are metabolically active for immediate biodegradation, but would require large volumes and would be difficult to transport to a location. Frequent transferring also raises the danger of contamination or mutation. Mixing of organisms should not be done until just prior to application to prevent competition within the mixture. Freeze-dried or frozen cultures could be mixed well in advance without this problem. Freeze-dried cultures would occupy small volumes for easy transport. However, freeze-dried cultures are quite expensive to prepare and take some time to become metabolically active again, especially in the suboptimal environmental condition of the environments in which they would be used. An alternative method would be to store the culture in a freeze-dried state and then initiate growth in a laboratory just before actually seeding. Then the organisms can be added in the appropriate concentrations to the environment.

The size of the inoculum required depends largely upon the size of the spill, how it is dispersed, and on the growth rate of the seed microorganisms. Some workers report that the size of the inoculum is of little importance, as long as a minimal inoculation of the oil occurs (Miget, 1973). Of greater importance is whether the oil will support the growth of the seed organism in the given environment. However, other researchers concluded that too low a level of application (10^6 bacterial cells/cm^2) was responsible for the only slight increase in the rate of oil utilization (C_{20} to C_{25}) obtained after seeding, rather than the inability of the added bacteria to survive under the natural field conditions (Jobson, McLaughlin, Cook, and Westlake, 1974).

Migration of microorganisms through soil can occur via two processes: convective transport and molecular "diffusion" (Ahlert and Kosson, 1983). Convective transport involves addition of significant quantities of an aqueous nutrient feed solution that causes movement of organisms with the feed and distribution throughout the soil column. Microbiological "diffusion" is analogous to molecular or surface diffusion. It occurs as a result of the life/death cycle and the natural movements of microorganisms. If the selective organisms added to in situ systems have to depend only on growth to diffuse through the subsurface, it will take too long for bioaugmentation to be cost effective (Kobayashi and Rittmann, 1982). However, the organisms can be spread through the soil more rapidly by convective transport when the soil is flooded with nutrient solution and the water is continuously pumped from the aquifer, establishing a closed path that pulls the liquid and organisms through the soil.

5.1.1.1.1 Commensals. In a process that uses both cooxidation and commensalism, complete biodegradation of cyclohexane can be achieved in a two-step process using two organisms--an n-alkane oxidizer, such as a pseudomonad, that first converts cyclohexane into either cyclohexanol or cyclohexanone during growth on the n-alkane (e.g., n-heptane) and another strain (e.g., a pseudomonad) that can grow on cyclohexanol or cyclohexanone (de Klerk and van der Linden, 1974; Donoghue, Griffin, Norris, and Trudgill, 1976). The same general results have been reported for the oxidation of other unsubstituted cycloparaffinic hydrocarbons (Beam and Perry, 1974).

5.1.1.1.2 Acclimated Microorganisms. Acclimated microbes and nutrients can be employed to stimulate degradation (Quince and Gardner). However, before a cleanup system employing acclimated bacteria can be implemented, a laboratory investigation of the kinetics of biodegradation for the acclimated bacteria, the potential for inhibition under various conditions, the oxygen and nutrient requirements, and the effects of temperature should be evaluated (Sommers, Gilmore, Wildung, and Beck, 1981). Emulsifiers may be necessary to increase the solubility of the contaminant. The hydrogeologic data that are needed are formation porosity, hydraulic gradient, depth to water, permeability, groundwater velocity and direction, and recharge/discharge information.

Many systems using acclimated bacteria recharge the effluent from biological treatment to the aquifer to create a closed loop of recovery, treatment, and recharge (Quince and Gardner, 1982). This flushes contaminants out of the soil rapidly and establishes hydrodynamic control separating the contaminated zone from the rest of the aquifer.

The systems that have used acclimated bacteria to restore contaminated aquifers typically have relied on biological wastewater treatment techniques, such as activated sludge, aerated lagoons, trickling filters, aerobic digestion, composting, and waste stabilization (Lee and Ward, 1985). These bacteria can be added to the aquifer and can act in situ to degrade the contaminant. The recharge water can be adjusted to provide optimal conditions for the growth of the acclimated bacteria and of the indigenous populations, before returning it to the soil. The bacteria, air, and nutrients can be injected into the subsurface after levels of organics have been reduced by treatment in clarifiers and through air stripping (Quince and Gardner, 1982) or by treatment with clarification, adsorption onto granular activated carbon, air stripping, and then recharge (Ohneck and Gardner, 1982). However, an experiment showed that indigenous bacteria were able to metabolize contaminants at the same or greater rates than a hydrocarbon-degrading bacteria inoculum.

Acclimated organisms have been used effectively in remedial actions for cleaning up contaminated aquifers (Lee and Ward, 1984; Quince and Gardner, 1982). An acrylonitrile spill was handled by removing the groundwater, treating it with mutant bacteria in a small reactor, and recharging it (Polybac Corporation, 1983). Acclimated organisms were used to clean up contaminated groundwater at a site where ethylene glycol and propyl acetate had been spilled (Lee and Ward, 1985). Initial treatment by air stripping and clarification was followed by addition of acclimated bacteria, nutrients, and air to the groundwater. The total organic carbon in the water was reduced from 40,000 to 1 ppm. In another instance, groundwater from a highly polluted aquifer near a hazardous waste dump site was treated in a batch study with organisms from several wastewater treatment systems receiving the organics found in the groundwater (phenols, creosols, dichlorobenzene, and others). After 3 weeks of acclimation and stabilization, these organisms were able to reduce the organic levels by more than 80 percent within 24 hr.

5.1.1.1.3 Mutant Microorganisms. Genetically engineered microorganisms have been used in cases of environmental contamination. In two instances of mutant bacteria being added to groundwater contaminated by acrylonitrile, levels of the contaminant fell from 1000 ppm to the limits of detection in 1 month in one case (Walton and Dobbs, 1980) and from 100 ppm to 1 ppm in 3 months in the other (Polybac Corporation, 1983). However, in both cases, there was no conclusive evidence that the added organisms contributed more to the degradation in the field than the normal flora of the site.

Genetic Engineering

For those compounds that are recalcitrant to attack by natural microbes, there is the theoretical possibility of constructing strains that can degrade them by the introduction into one bacterium of a number of enzyme activities from different bacteria, which under natural circumstances might have little or no chance to exchange genetic information (Williams, 1981a).

Microorganisms that are potentially useful for treating organic wastes must first be isolated or enriched from natural microbial communities in the environment (Pierce, 1982a). Genetic manipulation of these strains requires extensive knowledge of the biochemistry of the microbial pathway under investigation (Leisinger, 1983). Information of the rate-limiting steps in the degradative pathways, on the substrate specificities of the relevant enzymes, and on the types of regulatory mechanisms involved in gene expression are prerequisites for a rational approach in strain construction. The diversity of substrate utilization of the organism must be determined; growth may be possible on more organic compounds than were used for initial isolation (Pierce, 1982a). This allows the construction of a profile of catabolic activity, which facilitates the physiological, enzymatic, and genetic characterization needed for rational improvement of the strain. Those with high substrate diversity have complex regulatory mechanisms with considerable interaction between plasmid and chromosomal genes. Appropriate hosts and cloning vectors must be carefully selected. The natural origin, as opposed to laboratory or clinical strains, of the isolated organisms may be of value when the modified strains are then used in waste treatment.

Catabolic plasmids in microorganisms occur naturally and some of the common and well-studied catabolic plasmids are listed in Table A.5-1 (Jain and Sayler, 1987).

The degradation potential of microbes can be altered by genetic manipulation (Zitrides, 1978; McDowell, Bourgeois, and Zitrides, 1980). Radiation can be used to increase the genetic variability of an adapted microbial population. Selected strains of bacteria are chosen for their known ability to degrade similar compounds and exposed to successively increasing concentrations of substrate. The fastest growing strains are irradiated. The genetic alterations should increase the growth rate and fix the desired biochemical capability. Adaptation of the strains at high concentrations of a preferred substrate can induce increased production of the specific enzymes required to degrade that substrate (Thibault and Elliott, 1979). However, this is probably achieved at the expense of the ability to produce other enzymes. Therefore, an adapted mutant put back into the environment will only be able to compete against indigenous organisms when the preferred substrate is available.

Table A.5-1.   Some Catabolic Pathways Encoded by Naturally Occurring Plasmids in Microorganisms (Jain and Sayler, 1987)

Growth Substrate	Plasmid	Microorganism
Toluene, m-Xylene, p-Xylene	TOL (pWWO); pDK1	*Pseudomonas putida*
Naphthalene	NAH7	*P. putida*
Salicylate	SAL	*P. putida*
Camphor	CAM	*P. putida*
n-Octane	OCT	*P. putida*
4-Chlorobiphenyl	pKF1; pSS50	*Acinetobacter* sp. P6; *Alcaligenes* sp. A5
3-Chlorobenzoate	pAC25	*Pseudomonas* sp. B13
2,4-Dichlorophen-oxyacetate	pJP1	*A. paradoxus*
Styrene	—	*P. fluorescens*

Interstrain and interspecies genetic engineering promises to develop organisms with extraordinary abilities to degrade xenobiotic compounds (Chakrabarty, 1982).   However, more information is needed on the ability of genetically engineered organisms to survive, grow, and function in the soil environment (Liang, Sinclair, Mallory, and Alexander, 1982).   Other practitioners are skeptical because of the importance of the soil environment in determining the microbial activity and hence the success of applying exogenous organisms (Sims and Bass, 1984), especially since many compounds that can be degraded under laboratory conditions continue to persist in the environment (Pierce, 1982d).

An example of potential genetic engineering technology occurs with Pseudomonas putida (Hou, 1982; Jain and Sayler, 1987).   Naturally occurring strains of the organism have the ability to degrade octanes (OCT plasmid), camphors (CAM plasmid), naphthalenes (NAH plasmid), toluene, and m- and p-xylenes, but no single strain is able to degrade all these pollutants simultaneously.   By genetic and molecular techniques, it has been possible to transfer the plasmids NAH, TOL, and CAM-OCT to produce a multiplasmid Pseudomonas putida (MPP), thereby, constructing an organism that can transform, simultaneously, some linear alkanes and aromatic and polyaromatic hydrocarbons (Thibault and Elliott, 1979).   This organism might be useful in the cleanup of oil spills or of wastes from industrial pollution.   The creation of improved strains in this manner would accomplish the in situ treatment and removal of several environmental pollutants at the same time.

A number of genetically engineered strains have been developed. One successful application of the technique has been the development of a bacterial strain of Pseudomonas cepacia, capable of degrading 2,4,5-trichlorophenoxyacetic acid (2,4,5-T) (Kellogg, Chatterjee, and Chakrabarty, 1981). After the compound was degraded in soil, the numbers dropped until more of the compound was added. The population increased until the material was degraded. No appreciable effects on the number or types of indigenous bacteria were noted. This organism still needs to be field tested.

A Flavobacterium sp. marine isolate degrades a variety of 2- and 3-ring PAHs (Foght and Westlake, 1985). It can use naphthalene or phenanthrene as sole carbon sources. Partial oxidation of anthracene and incomplete metabolism of dibenzothiophene occurs. This organism contains three large plasmids, which were initially refractory to isolation and demonstrated inconsistent stability in vitro. Complete curing was not achieved, but phenotypic variants, which showed plasmid involvement in PAH degradation, were generated.

Acinetobacter (P-7) can utilize a variety of growth substrates (Garvey, Stewart, and Yall, 1985). Growth was observed with n-alkanes (C_8 to C_{20} carbon chain length), camphor, phenethyl alcohol, and crude oil. This organism harbors a plasmid, pYGI. When cured of the plasmid, the ability to grow on decane, nonane, or octane was lost. Growth on larger carbon chain-length n-alkanes (C_{11} to C_{20}) was normal. Chromosomally mediated n-alkane utilization can be divided into three classes: C_{11} to C_{12}, C_{13} to C_{14}, and C_{15} to C_{20}.

Plasmids, such as TOL and NAH, mediate bacterial degradation of small aromatic compounds (Foght and Westlake, 1983). However, there is no evidence that plasmids play a role in degradation of larger, polycyclic aromatic hydrocarbons, such as phenanthrene, benzo(a)pyrene, or dibenzothiophene. Almost all aromatic degraders have at least one plasmid, which can be divided into two groups: those with limited ability to degrade aromatics (e.g., only biphenyls) have small plasmids; those with extensive capability to degrade PAHs have larger plasmids. Aromatic-degrading strains identified include members of the Flavobacterium, Acinetobacter, Pseudomonas, and Aeromonas genera. There is no evidence for any PAH-degrading isolates being capable of degrading saturates (e.g., n-alkanes) or vice versa.

Mutant bacterial formulations have been developed by Polybac Corp. to degrade complex organic materials, including industrial surfactants, crude and refined petroleum products, pesticides and herbicides, and solvents. Polybac maintains an up-to-date library of information on the relative biodegradability of a wide range of organic chemicals by various mutant strains. This allows selection of the proper formulation of strains for a specific waste mixture to be treated.

Cytochrome P450 systems catalyze the monooxygenation of a broad range of xenobiotic compounds (Chen, Dey, Kalb, Sanglard, Sutter, Turi, and Loper, 1987). The individual P450 enzymes provide the specificity for a broad range of oxidative reactions. P450 systems are being engineered in yeasts for the oxidative detoxication and biodegradation of environmentally stable organic pollutants. Because of the highly substituted structure of the persistent polychlorinated aromatics, the P450 monooxygenases are likely candidates as catalysts for their specific oxidation or reductive dechlorination.

The hydroxylation of n-octane by Corynebacterium sp. 7EIC involves cytochrome P450 (Hou, 1982). Only n-alkanes with more than 14 carbon atoms are inducers of the synthesis of cytochrome P-450, the hydroxylase system, the NADPH-cytochrome c reductase, and cytochrome b.

Two organisms, Saccharomyces cerevisiae (baker's yeast) and Candida tropicalis ATCC750, a yeast capable of n-alkane assimilation, have been widely used in genetic studies (Chen, Dey, Kalb, Sanglard, Sutter, Turi, and Loper, 1987). S. cerevisiae is highly characterized genetically and has become the eukaryotic microorganism of choice for the application of molecular genetic techniques. C. tropicalis ATCC750 is representative of a group of yeasts capable of growth on n-alkanes in petroleum as a carbon and energy source. Such yeasts already express lipophilic properties that may be useful in genetically engineering cells for the uptake and metabolism of hydrophobic hazardous compounds. Access to the DNA sequence of the C.tropicalis gene for P450alk will facilitate testing of its expression as a foreign gene in S. cerevisiae. Experiments using these methods form important steps toward genetically engineering other specific P450s in yeasts for the degradation of hazardous wastes.

There are four major areas that may benefit from genetic engineering: stabilization; enhanced activity; multiple degradative activities; and health, safety, and environmental concerns (Pierce, 1982d).

1. Stabilization

The extrachromosomal-DNA (plasmids) responsible for degrading some hydrocarbons can be lost, if the organisms are not grown in the presence of these compounds (Pierce, Facklam, and Rice, 1980). As a result, the microbes lose the ability to degrade the material (Fisher, Appleton, and Pemberton, 1978). Selective pressure must be present to maintain these plasmid-encoded traits. Genetic engineering can stabilize these traits by constructing hybrid plasmids (Bassford, et al., 1980). This can involve insertion of desired genetic elements into a region where replication is under control of a conserved element.

2. Enhanced Activity

By taking a plasmid containing a degradative trait, which has a low number of copies of itself in a cell, and hybridizing it with another plasmid that has a large number of itself present, more degrading enzyme may be produced (Pierce, 1982d). Optimization of the location of a promoter of protein systhesis will also increase the amount of protein produced (Shine and Dalgarno, 1975).

3. Multiple Degradative Activities

In nature, strains of Pseudomonas will contain only one degradative plasmid, since they belong to the same incompatibility group (Korfhagen, Sutton, and Jacoby, 1978). Incompatible plasmids are mutually exclusive, and an organism will contain only one from a given group. Fusion of the CAM, TOL, and OCT plasmids led to the construction of novel microorganisms with multiple degradative traits. Recombinant-DNA techniques could also be used to insert several elements into a single plasmid. Combinations can result in an organism with degradative traits different from either parent (Vandenbergh, Olsen, and Coloruotolo, 1981). Plasmid-assisted molecular breeding is also being used to produce novel degradative traits.

4. Health, Safety, and Environmental Concerns

The potential for adverse effects should be considered for wild-type and engineered strains (Pierce, 1982d). It is unlikely that recombinant-DNA techniques, as applied to microbial degradation, will result in additional biohazards. Strains may be constructed with a minimized potential for adverse effects, such as use of nontransmissible plasmids, reduction of the plasmid to the smallest size possible (removal of superfluous genetic information and cryptic genes), development of a host microorganism that will survive only under defined conditions, and construction of containment systems to limit the potential for escape.

Limiting Factors of Genetic Engineering

1. Compatibility and Coordination

Plasmid incompatibility must be overcome (Pierce, 1982d). The host must have the ability to synthesize the protein for the new degradative trait. A promoter must also be present to allow adequate genetic control of protein synthesis. The pathway or reactions encoded by the inserted plasmid must be compatible with the biochemical pathways encoded by the host chromosome. There is evidence that degradative plasmids are responsible for only the first several biochemical reactions of the degradative pathway, while the remainder is encoded by chromosomal genes (Fisher, Appleton, and Pemberton, 1978). There are also three pathways in pseudomonads for the degradation of aromatic compounds (ortho and meta cleavage and gentisate pathways), which further complicates the coordination of biochemical pathways and their regulation mechanisms (Doelle, 1975).

2. Environmental Conditions

Genetically engineered microorganisms may not be successful in the field (Pierce, 1982d). These microbes would have to compete with resident microflora. If there are already 10^8 to 10^9 bacteria/g soil, a very large number of the new organisms would have to be added for them to become established in the community. If environmental parameters, such as temperature, pH, substrate concentration, oxygen tension, are not optimal for the added organisms, there will be reduced activity (e.g., a bacterium that degrades n-alkanes optimally at 30°C in an aqueous medium would probably perform poorly in an oil spill in the North Sea).

Microorganisms that are tolerant of multiple kinds of stresses (e.g., abiotic stress, starvation, and biological antagonism) have a higher potential for survival in the soil after genetic manipulation in the laboratory than organisms with less tolerance and versatility (Liang, Sinclair, Mallory, and Alexander, 1982). Genetic engineering is trying to produce bacteria capable of complete degradation of xenobiotic compounds by using zymogenous organisms, such as Pseudomonas (Chakrabarty, 1982). However, strains of Arthrobacter, more slowly growing autochthonous organisms, have also demonstrated xenobiotic degrading capabilities, and these have also shown potential for treating hazardous waste-contaminated soils (Stanlake and Finn, 1982; Edgehill and Finn, 1983).

There have been concerns that genetically engineered organisms may continue to persist in the environment (Stotsky and Krasovsky, 1981). It is hard to predict the fate of these organisms in such complex ecosystems as soils and aquatic environments. Unless the new genetic material confers a wider substrate range, or increases the organism's ability to detoxify its environment, it would seem energetically disadvantageous for the organism to maintain the genes. Eventually, the organism or its new genes would be selected against. Such an organism may decrease to unmeasureable numbers, if its target substrate is depleted (Chakrabarty, 1982). However, others argue that more stress-tolerant organisms could persist for extended periods (Stanlake and Finn, 1982; Edgehill and Finn, 1983).

Most of the bacteria typically used in microbial genetic work are eutrophs of the families Enterobacteriaceae and Pseudomonadaceae, which may not be able to attack substrates in the parts-per-billion range that are often found in environmental samples (Johnston and Robinson, 1982). Environmental stress, such as unsuitable water availability or the presence of toxicants may also affect an introduced microorganism. However, genetic engineering holds great promise, especially in treatment facilities where conditions can be controlled.

Since microbial growth doubles for about every 10°C increase in temperature, biodecontaminations would be slowed considerably in cool weather (Thibault and Elliott, 1979). Mutant organisms could be developed to provide the optimal degradation at any given temperature. Section 5.1.2.2 discusses the temperature dependency of the biodegradation of crude oil by a commercially available mutant bacterial formulation (PETROBAC[R] Mutant Bacterial Hydrocarbon Degrader). Biodegradation does occur at a temperature as low as 5°C.

It may not be feasible to engineer strains that are resistant to heavy metals, since it is more than likely that the toxic conditions found in industrial wastewaters will cause plasmid losses from organisms, even though they function well in the laboratory (Wood and Wang, 1983). It looks more promising to employ organisms, such as the green alga, Cyanidium caldarium, that naturally tolerate extreme conditions of pH and temperature.

Because of the large amount of information necessary before rational and reproducible experiments in strain construction can be performed, this approach has not yet been applied on a wide scale in biodegradation research (Leisinger, 1983). Assembling degradative pathways from different bacterial strains in one cell by genetic manipulation is a laborious technique by which to obtain organisms with novel degradative capacities. This method will find its application in cases where a precise strategy can be formulated and where obstacles to degradation, as, for example, the intracellular accumulation of dead-end metabolites from problem compounds, cannot be overcome by cocultivation of strains with different degradative pathways.

The problem with the use of genetically selected microorganisms to degrade contaminants is that no conclusive evidence has been found that commercially available organisms are effective in establishing themselves or significantly enhancing biodegradation of pollutants in aeration basins or natural environments having an active native microbial population (Johnston and Robinson, 1982). In general, acclimated or genetically engineered organisms will not survive or offer significant advantage in treatment of hazardous wastes unless environmental parameters can be controlled to promote survival of the added organisms (Lee, Wilson, and Ward, 1987).

The addition of adapted mutant microbes has not been completely successful but has great potential (Knox, Canter, Kincannon, Stover, and Ward, 1968). A number of companies are involved in the production of microbial strains to be used to treat abandoned hazardous waste sites and chemical spills.

5.1.1.2 Use of Analog Enrichment for Cometabolism

Not all of the organic compounds at a contaminated site may be readily degraded (Atlas, 1981). It may be possible to induce oxidation of compounds that might otherwise not be degraded by addition of a second substrate that will promote cooxidation. Uncharacterized dissolved organic carbon may also play an important role in controlling the rate and extent of biodegradation of organic compounds present at low concentrations (Alexander, 1986).

Various organisms can be used in combination with chemical analogues of specific organic compounds to promote cooxidation of the latter. Some of these combinations, and the resultant metabolic products, are described below and summarized in Table A.5-2.

Sucrose has been used as a cometabolite in the degradation of 4,6-dinitro-o-cresol (DNOC), a phenolic priority pollutant, in wastewater, by use of an anaerobic recycle fluidized-bed reactor as a pretreatment stage, followed by an activated-sludge reactor as the aerobic treatment stage (Slonim, Lien, Eckenfelder, and Roth, 1985). There appeared to be a relationship between the sucrose concentration and

Table A.5-2.   Chemical Analogues/Growth Substrates and Microorganisms for
Cooxidation

Organic Compound	Analogue/ Growth Substrate	Organism	Products
Ethane, propane, butane[a]	Methane	*Pseudomonas methanica*	
2-Methylheptane[b]	Hexane	*Pseudomonas* sp.	Aldehydes, ketones, etc.
o-Xylene[b]	Hexane	*Pseudomonas* sp.	o-Toluic acid
Ethylcyclohexane[b]	Hexane	*Pseudomonas* sp.	Cyclohexane acid
o-Xylene[b]	Hexadecane	*Nocardia* sp.	o-Toluic acid
p-Xylene[b]	Hexadecane	*Nocardia* sp.	p-Toluic acid
Ethylcyclohexane[b]	Hexadecane	*Nocardia* sp.	Cyclohexane acid
Benzo(a)pyrene[c]	Phenanthrene		
Pyrene, 3,4-benzpyrene, 1,2-benzanthracene, 1,3,5,6-dibenzanthracene[d]	Phenanthrene, naphthalene		
4,6-Dinitro-o-cresol[e]	Sucrose		
Cycloparaffins[f]	Propane	*Mycobacterium vaccae*	Cycloalkanones
Cycloalkanes[f]	n-Heptane	*Pseudomonas aeruginosa*	Alcohols
n-Butylcyclohexane	n-Octadecane-grown	*Nocardia* sp.	Cyclohexane acetic acid

a = (Foster, 1962)
b = (Jamison, Raymond, and Hudson, 1976)
c = (Sims and Overcash, 1981)
d = (McKenna, 1977)
e = (Slonim, Lien, Eckenfelder, and Roth, 1985)
f = (Hou, 1982)

degradation of the compound. A ratio of sucrose to DNOC of 2:1 or higher resulted in a 95 to 100 percent conversion of DNOC, while a lower ratio did not permit cometabolism or degradation of DNOC.

The rate of cometabolism of benzo(a)pyrene was significantly increased with enrichment of the soil with phenanthrene as an analog (Sims and Overcash, 1981). Similar results were obtained by using naphthalene, and especially phenanthrene, as growth substrates for the nongrowth substrates of pyrene, 3,4-benzpyrene, 1,2-benzanthracene, 1,3,5,6-dibenzanthracene, with approximately 35 percent of the nongrowth substrate remaining after 4 weeks (McKenna, 1977).

It was discovered that Pseudomonas methanica cometabolically oxidized ethane, propane, and butane while the organism metabolized its growth substrate, methane (Foster, 1962). Soil aerobic organisms that can grow on aliphatic hydrocarbons, such as natural gas or propane, could possibly be added to the soil while methane or propane is pumped into the ground for degradation of a variety of chlorinated solvents, including TCE (Rich, Bluestone, and Cannon, 1986).

When two components of gasoline were combined, one for growth and one as a cooxidizable substrate, a Pseudomonas sp. was able to cooxidize 2-methylheptane, o-xylene, and ethylcyclohexane with hexane as the growth substrate (Jamison, Raymond, and Hudson, 1976). Nocardia sp. cooxidized primarily o-xylene and p-xylene. o-Xylene was oxidized to o-toluic acid, ethylcyclohexane to cyclohexane acid, p-xylene to p-toluic acid, and 2-methylheptane to a mixture of products, including ketones and aldehydes.

Bacteria capable of utilizing or cooxidizing phenanthrene were enumerated by spreading water and sediment samples on nutrient agar plates containing PHE as sole carbon source and plates containing PHE and one alternative carbon source (Shiaris and Cooney, 1981). Colonies producing on the first plates were classified as utilizers, and those producing clear zones only on the second were considered cooxidizers. The alternative substrates were effective in this order:

yeast extract/peptone > glucose > benzoic acid > oil/kerosene

Compounds with two or more chlorine atoms can be transformed anaerobically in the laboratory through reductive halogenation by addition of a primary substrate--acetate, methanol, or isopropanol--to increase the concentration and activity of methane-forming bacteria (Rich, Bluestone, and Cannon, 1986).

An alkane-utilizing strain of Mycobacterium vaccae (JOB5) cooxidizes a variety of alicyclic hydrocarbons to the corresponding ketones (Hou, 1982). A strain of Pseudomonas aeruginosa grown on n-heptane cooxidizes cycloalkanes to their corresponding alcohols. Cycloalkanes can be metabolized readily by mixed cultures of n-alkane utilizers, which cometabolize cycloalkanes, and other microorgnisms, which are able to utilize the alicyclic oxidation products. A Nocardia sp. produces biotin, which is utilized by a Pseudomonas strain to grow on cyclohexane.

Methane-grown microorganisms will stationary-oxidize alicyclic hydrocarbons. Stationary transformation is a term used to describe metabolism of nongrowth substrates by microorganisms in the absence of growth substrates. This is to be distinguished from cooxidation, in which the microorganisms require a growth substrate in order to metabolize the nongrowth compound.

The ability to use aromatic compounds can be an induced phenomenon in bacteria (Claus and Walker, 1964). Toluene-grown strains of Pseudomonas and Achromobacter oxidized, without lag, benzene, catechol, 3-methyl-catechol, benzyl alcohol, and, more slowly, o- and m-cresol, but not benzaldehyde or benzoic acid. The mutual adaptations to use benzene and toluene suggest that enzymes with similar activities may be involved in the metabolism of the two compounds. n-Octadecane grown Nocardia sp. was found to cooxidize n-butylcyclohexane to cyclohexane acetic acid (Hou, 1982).

Diauxie Effect

The opposite of cometabolism is a sparing, or diauxic, phenomenon, which occurs when a compound cannot be degraded in the presence of another compound (Atlas, 1981). The metabolic pathways of degradation are not altered, but the enzymes necessary for metabolic attack of a particular hydrocarbon may not be produced. This can lead to persistence of particular hydrocarbons in a petroleum mixture. Both processes of cooxidation and sparing can occur within the context of a petroleum spillage.

When a microorganism with a broad substrate range is offered more than one type of organic substrate, it will not attack the substrates simultaneously, but rather in a definite sequence (Bartha and Atlas, 1977). A diauxie effect, where the presence of one compound will inhibit the degradation of another, may determine whether or not the hydrocarbon components of an oil spill are degraded, and if so, in what order. A Brevibacterium erythrogenes strain was capable of utilizing pristane and other branched alkanes only in the absence of n-alkanes (Pirnik, Atlas, and Bartha, 1974). B. erythrogenes utilizes n- alkanes by a monoterminal beta-oxidation sequence, but degrades isoalkanes by diterminal oxidation. The common phenomenon that the n-alkane components of an oil spill disappear before the isoalkanes and other hydrocarbon classes show substantial biodegradative change strongly suggests that such diauxic regulatory mechanisms apply not only to some pure cultures, but most likely also to the mixed microbial community of the environment (Bartha and Atlas, 1977).

The presence of nonhydrocarbon substrates may repress the inductive synthesis of enzymes required for hydrocarbon oxidation (van Eyk and Bartels, 1968). Addition of glucose to lake water repressed hexadecane utilization by its microbial community in a diauxic manner (Bartha and Atlas, 1977). In another study, the rate of mineralization of organic compounds in trace concentrations was found to be enhanced by the addition or inorganic nutrients, arginine, or yeast extract, but reduced by addition of glucose (Rubin and Alexander, 1983). This effect should be considered when selecting a substrate as a cometabolite.

5.1.1.3 Application of Cell-Free Enzymes

Enzymatic methods show promise for removing aromatic compounds from high strength industrial wastewater (Maloney, Manem, Mallevialle, and Fiessinger, 1985). Enzymatic oxidative coupling may be useful in eliminating aromatics that are not well removed in biological or physical water treatment. Wastewaters containing aromatic compounds are treated with horseradish peroxidase and hydrogen peroxide (Alberti and Klibanov, 1981). The resulting high molecular weight compounds are less soluble in water and can be removed by sedimentation or filtration.

One concern in the use of this technique for drinking water treatment is the nature of the products of the oxidative coupling (Maloney, Manem, Mallevialle, and Fiessinger, 1985). Biphenyls accounted for 3 percent of the initial carbon concentration (Schwartz and Hutchinson, 1981). They would be incomplete polymerization products and may be the predominant ones. They may not be removed by sedimentation or filtration and may pass through the treatment process in their altered (polymerized) form.

Another problem in drinking water treatment is the presence of competition or interfering compounds (Maloney, Manem, Mallevialle, and Fiessinger, 1985). Raw water supplies usually have background organic carbon composed mainly of humic acids (McCarty, 1980). It has been suggested that humic acids may deactivate peroxidase (Pflug, 1980).

The peroxidase-peroxide system is effective in eliminating chlorinated phenols from drinking water supplies, but does not remove their breakdown products from the water (Maloney, Manem, Mallevialle, and Fiessinger, 1985). Further work is necessary to determine if these by-products present a potential risk for human health and if they are removed in other unit processes. It is of interest that extracellular peroxidases have been found in soil (Kaufman, 1983). These could presumably be involved in a vast array of soil metabolic reactions affecting xenobiotic residues.

The Cetus Company developed a novel multienzyme process for the oxidation of propylene (Hou, 1982). The first enzyme reaction converts olefin to halohydrin in the presence of halide, hydrogen peroxide, and haloperoxidase. The latter can be obtained from horseradish, seaweed, or Caldariomyces. In the second reaction, propylene halohydrin is transformed to propylene oxide by halohydrin epoxidase or by whole cells of Flavobacterium sp.

Cell-free enzymes for treating hazardous waste constituents are not currently in bulk production (Munnecke, Johnson, Talbot, and Barik, 1982). Only eight companies accounted for 90 percent of worldwide production of industrial enzymes in 1981; five of the firms, in Western Europe. Only 16 enzymes (primarily amylases, proteases, oxidases, and isomerases) accounted for 99 percent of the 1981 market. This suggests that specialized enzyme production, even on a large scale, may be quite expensive. Current prices for bulk enzyme materials range in price from $1.45 to $164/lb. If the enzyme can be produced through chemical synthesis, it will be much less expensive than if it is produced by microorganisms in fermenters.

5.1.1.4 Addition of Antibiotics

5.1.1.5 Effect of Biostimulation on Counts

The following accounts indicate what favorable influence biostimulation can have on the total counts or hydrocarbon-degrading organisms at actual field locations.

- After biostimulation at a site contaminated with gasoline, bacterial levels increased up to 6 million times the initial levels (Minugh, Patry, Keech, and Leek, 1983).

- From 10^2 to 10^5 gasoline-utilizing organisms/ml were found to be present in contaminated groundwater when preliminary tests were made before a bioreclamation operation began. The microbial population responded to the addition of nutrients and oxygen with a 10- to 1000-fold increase in the numbers of gasoline-utilizing and total bacteria in the vicinity of the spill, with levels of hydrocarbon utilizers in excess of 10^6/ml in several wells. The microbial response was an order of magnitude greater in the sand than the groundwater.

- Aeration of the groundwater contaminated with methylene chloride, n-butyl alcohol, dimethyl aniline, and acetone (temperature 12 to 14°C) in a monitoring well with a small sparger and the subsequent addition of nutrients resulted in an increase of bacteria from 1.8×10^3/ml to 1.6×10^6/ml in a 7-day period (Jhaveri and Mazzacca, 1985).

- A natural flora of gasoline-utilizing organisms were present at levels of 10^3/ml (Jamison, Raymond, and Hudson, 1975) in an area contaminated with over 3000 barrels of high-octane gasoline. This population could be increased a 1000-fold by supplementing the groundwater with air, inorganic nitrogen, and phosphate salts.

- At a contaminated site in Millville, NJ, a population of 10^2 to 10^5 gasoline-utilizing organisms/ml groundwater responded to the addition of nutrients and oxygen with a 10- to 1,000-fold increase in their numbers (Raymond, 1978). In several wells, levels of hydrocarbon-utilizers exceeded 10^6/ml.

- After biostimulation at a LaGrange, OR, site, bacterial levels increased up to 6 million times the initial levels (Minugh, Patry, Keech, and Leek, 1983).

- After the biostimulation program ended at Ambler, PA, the numbers of gasoline-utilizing bacteria declined, suggesting a depletion of nutrients and gasoline (Raymond, Jamison, and Hudson, 1976).

- In the solvent contamination at the Biocraft Laboratories, Waldwick, NJ, the wells had populations of 10^3 to 10^4 colonies/ml prior to biostimulation; addition of nitrogen and phosphorus increased the numbers of resident organisms as high as 4 times that of the control level (Lee and Ward, 1985).

- At a site contaminated with over 3000 barrels of high-octane gasoline, the natural flora of gasoline-utilizing organisms were present at levels of 10^3/ml (Jamison, Raymond, and Hudson, 1975). This population could be increased a 1000-fold by supplementing the groundwater with air, inorganic nitrogen, and phosphate salts.

5.1.2 Optimization of Soil Factors

Biodegradation of contaminants in the soil can be enhanced by making environmental factors optimum for the required reactions. Table A.5-3 lists the site and soil properties that should be identified to predict potential migration of a contaminant and to indicate the factors that will have to be adjusted to achieve optimum biodegradation (JRB and Associates, Inc., 1984b).

The inherent capacity of soil to degrade toxicants by chemical and biological mechanisms can be maximized by identification of the soil conditions that promote the degradation of each toxicant and manipulation of the soil environment to bring about these conditions (Arthur D. Little, Inc., 1976). Although each toxicant, in general, has a unique set of ideal soil conditions for degradation, for some compounds these ideal conditions overlap, and more than one toxic substance can be the focus of soil manipulation at one time. For other compounds, the ideal conditions do not overlap and are sometimes even contradictory; these materials must be treated in series.

Table A.5-4 lists the soil factors that may have to be modified during the use of various treatment technologies (Sims and Bass, 1984). The most important soil factors that affect biodegradation are water content, temperature, soil pH, oxygen supply, available nutrients, oxidation-reduction potential, and soil texture and structure (Hornick, 1983).

5.1.2.1 Soil Moisture

Bacterial activity is highest in the presence of moisture (JRB and Associates, Inc., 1984b). The greatest diversity and activity of microorganisms and the highest population densities are consistently observed in the sandy water-bearing strata, whereas, the dense, dry-clay layer zones have the least microbiological activity (Fredrickson and Hicks, 1987). The aerobic biodegradation of simple or complex organic material in soil is commonly greatest at 50 to 70 percent of the soil water-holding (field) capacity (Pramer and Bartha, 1972; Zitrides, 1983). Inhibition at lower values is due to inadequate water activity, and higher values interfere with soil aeration. Excess moisture, extremely dry conditions, pooling, or flooding should be avoided (Zitrides, 1983). Biodecontamination programs should not be conducted during heavy rains or drought. However, an observed lack of inhibition at 30 percent of the

Table A.5-3.   Important Site and Soil Characteristics for <u>In Situ</u> Treatment (JRB and Associates, Inc., 1984b)

Site location/topography and slope

Soil type, and extent

Soil profile properties
    Boundary characteristics
    Depth
**   Texture
    Amount and type of coarse fragments
**   Structure
    Color
    Degree of mottling
**   Bulk density
    Clay content
    Type of clay
**   Cation exchange capacity
 *   Organic matter content
 *   pH
 *   Eh
 *   Aeration status

Hydraulic properties and conditions
    Soil water characteristic curve
    Field capacity/permanent wilting point
**   Water holding capacity
**   Permeability (under saturated and a range of unsaturated conditions)
 *   Infiltration rates
    Depth to impermeable layer or bedrock
 *   Depth to groundwater, including seasonal variations
    Flooding frequency
 *   Run-off potential

Geological and hydrogeological factors
    Subsurface geological features
 *   Groundwater flow patterns and characteristics

Meteorological and climatological data
    Wind velocity and direction
    Temperature
    Precipitation
    Water budget

---

 *   Factors that may be managed to enhance soil treatment
**   Factors that may be managed to enhance soil treatment with shallow depth contamination.

Table A.5-4.   Soil Modification Requirements for Treatment Technologies (Sims and Bass, 1984)

Technology	Oxygen Content	Moisture Content	Nutrient Content	pH	Temperature
EXTRACTION	-	-	-	x	x
IMMOBILIZATION					
Sorption (heavy metals)					
Agricultural products	-	-	-	x	-
Activated carbon	-	-	-	x	-
Tetren	-	-	-	x	-
Sorption (organics)					
Soil moisture	-	x	-	-	-
Agricultural Products	-	-	-	-	-
Activated carbon	-	-	-	-	-
Ion exchange					
Clay	-	-	-	x	-
Synthetic resins	-	-	-	x	-
Zeolites	-	-	-	x	-
Precipitation					
Sulfides	x	x	-	x	-
Carbonates, phosphates, and hydroxides	x	x	-	x	-
DEGRADATION					
Oxidation					
Soil-catalyzed reactions	x	-	-	x	-
Oxidizing agents	x	-	-	x	-
Reduction					
Reducing agents	x	x	-	x	-
Chromium	x	-	-	x	-
Selenium	x	-	-	x	-
PCBs and Dioxins	-	x	-	-	x
Polymerization	-	-	-	-	-
Modification of soil properties (for biodegradation)					
Soil moisture	-	x	-	-	-
Soil oxygen--aerobic	x	-	-	-	-
Soil oxygen--anaerobic	x	x	-	-	-
Soil pH	-	-	-	x	-
Nutrients	-	-	x	-	-

Table A.5-4.   continued

Technology	Oxygen Content	Moisture Content	Nutrient Content	pH	Temperature
Nonspecific org. amendments	-	-	-	-	x
Analog enrichment for cometabolism	-	-	-	-	x
Exogenous acclimated or mutant microorganisms	-	-	x	-	x
Cell-free enzymes	-	-	-	-	x
Photolysis					
Proton donors	-	-	-	-	-
Enhance volatilization	-	x	-	-	-
ATTENUATION					
Metals	-	-	-	-	-
Organics	-	-	-	-	-
REDUCTION OF VOLATILES					
Soil vapor volume	-	x	-	-	-
Soil cooling	-	-	-	-	x

x = required
- = not required

field capacity suggests that the moisture requirement for maximum activity on hydrophobic petroleum may be different than the optimal moisture levels for the biodegradation of hydrophilic substrates (Dibble and Bartha, 1979a).

Rainfall is useful to biodegradation. It dissolves contaminants and acts as a carrier as it percolates through the soil on its way to the groundwater (Dietz, 1980). Rainwater also keeps the contaminated soil moist, and microorganisms will utilize the oxygen dissolved in interstitial water droplets (Thibault and Elliot, 1980).

Many organisms are capable of metabolic activity at water potentials lower than -15 bar (Soil Science Society of America, 1981). The lower limit for all bacterial activity is probably about -80 bar, but some organisms cease activities at -5 bar. Although many microbial functions continue in soils at -15 bars or drier, optimum biochemical activity is usually observed at soil water potentials of -0.1 to 1.0 bar (Sommers,

Gilmore, Wildung, and Beck, 1981). The kinds of microorganisms that are metabolically active in the soil will be affected. Degradation rates are highest at soil water potential between 0 and -1 bar. When natural precipitation cannot maintain near optimal soil moisture for microbial activity, irrigation may be necessary (Sims and Bass, 1984). Although degradation of hazardous organic compounds may be accelerated by soil moisture optimization, more rapid treatment of the contaminated soil may be achieved when moisture augmentation is used in combination with other techniques.

Moisture control is widely practiced in agriculture; however, there is little information available on its use to stimulate biological degradation of hazardous materials in soil (Sims and Bass, 1984). Most laboratory studies have been conducted at or near optimal soil moisture. The success of this technology depends upon the biodegradability of the waste constituents and the suitability of the site and soil for moisture control. Its effectiveness may be enhanced by combination with other treatment techniques to increase biological activity. The technology is reliable in that it has been used in agriculture, but retreatment is necessary. Leaching of soluble hazardous compounds may occur, and erosion may also be a problem.

Control of soil moisture content can be practiced to optimize degradative and sorptive processes and may be achieved by several means (Sims and Bass, 1984). Supplemental water may be added to the site (irrigation), excess water may be removed (drainage, well points), or these methods can be combined with other techniques, such as using soil additives, for greater moisture control.

5.1.2.1.1 Irrigation. Soil may be irrigated by subirrigation, surface irrigation, or overhead (sprinkler) irrigation (Fry and Grey, 1971). With subirrigation, water is applied below the ground surface and moves upward by capillary action. If the water has high salinity, salts may accumulate in the surface soil, with an adverse effect on soil microbiological activity. The site must be nearly level and smooth, with either a natural or perched water table, which can be maintained at a desired elevation. The groundwater is regulated by check dams and gates in open ditches, or jointed perforated pipe to maintain the water level in the soil. Use of such systems is limited by the restrictive site criteria. There may be situations in which a subirrigation system may be combined with a drainage system to optimize soil moisture content. However, at a hazardous waste site, raising the water table might result in undesirable groundwater contamination.

With trickle irrigation, filtered water is supplied directly on or below the soil surface through an extensive pipe network with low flow-rate outlets only to areas that require irrigation. It does not give uniform coverage to an area, but with proper management, does reduce percolation and evaporation losses. For most in-place treatment sites, this method would probably not be appropriate, but it may be applicable in an area where only "hot spots" of wastes are being treated.

Surface irrigation includes flood, furrow, or corrugation irrigation. Since the prevention of off-site migration of hazardous constituents to ground- or surface waters is a primary restraint on a treatment technology, surface irrigation should be considered with caution. Contaminated water may also present a hazard to on-site personnel.

In flood irrigation, water covers the surface of a soil in a continuous sheet. Theoretically, water should stay at every point just long enough to apply the desired amount, but this is difficult or impossible to achieve under field conditions.

In furrow irrigation, water is applied in narrow channels or furrows. As the water runs down the furrow, part of it infiltrates the soil. Considerable lateral movement of the water is required to irrigate the soil between furrows. Salts also tend to accumulate in the area between furrows. Furrow irrigation frequently requires extensive land preparation, which usually would not be possible or desirable at a hazardous waste site due to contamination and safety considerations.

In corrugation irrigation, as with furrow irrigation, water is applied in small furrows from a head ditch. However, in this case, the furrows are used only to guide the water, and overflooding of the furrows can occur.

In general, control and uniform application of water is difficult with surface irrigation. Also, soils high in clay content tend to seal when water floods the surface, limiting water infiltration.

The basic sprinkler irrigation system consists of a pump to move water from the source to the site, a pipe or pipes leading from the pump to the sprinkler heads, and the spray nozzles. Sprinkler irrigation has many advantages. Erosion and run-off of irrigation water can be controlled or eliminated, application rates can be adjusted for soils of different textures, even within the same area, and water can be distributed more uniformly. Sprinkler irrigation is also possible on steep, sloping land and irregular terrain. Usually less water is required than with surface flooding methods, and the amount of water applied can be controlled to meet the needs of the in-place treatment technique.

Also, a larger soil surface area is covered, which could facilitate soil washing.

There are several types of sprinkler irrigation systems:

1. Permanent installations with buried main and lateral lines
2. Semipermanent systems with fixed main lines and portable laterals
3. Fully portable systems with portable main lines and laterals,
    as well as a portable pumping plant

The first two types (especially the first) would likely not be appropriate nor cost effective for a hazardous waste site because of the required land disturbance for installation and the limited time period for execution of the treatment.

The fully portable systems may have hand-moved or mechanically moved laterals. Portable systems can be installed in difficult areas, such as forests, in a way that will avoid interfering with trees. Mechanically moved laterals may be side-roll/wheel-move, center-pivot systems, or traveling sprinklers. This equipment is more expensive, but it requires considerably less labor than hand-moved systems. The health and safety of the workers must be considered, as well as the cost in the choice of an appropriate system.

5.1.2.1.2 Drainage. When irrigation is used, controls for erosion and proper drainage due to run-off are necessary (Sims and Bass, 1984). A properly designed drainage system removes excess water or lowers the groundwater level to prevent waterlogging (Fry and Grey, 1971). Open ditches and lateral drains are used for surface drainage, while a system of open ditches and buried tube drains into which water seeps by gravity is used for subsurface drainage. The collected water is conveyed to a suitable disposal point. Subsurface drainage may also be accomplished by pumping from wells to lower the water table. The drainage water to be disposed of off-site must not be contaminated with hazardous substances, and must be collected, stored, treated, or recycled if not acceptable for off-site release.

Surface drains are used where subsurface drainage is impractical (e.g., impermeable soils, excavation difficult) to remove surface water or lower the water table (Donnan and Schwab, 1974). Subsurface drains are used to lower the water table. Construction materials include clay or concrete tile, corrugated metal pipe, and plastic tubing. Selection depends upon strength requirements, chemical compatibility, and cost.

5.1.2.1.3 Well Points. Like subsurface drains, well points can be used to lower the water table in shallow aquifers. They typically consist of a series of riser pipes screened at the bottom and connected to a common header pipe and centrifugal pump. Well point systems are practical up to 10 m (33 ft) and are most effective at 4.5 m (15 ft). However, their effectiveness depends upon site-specific conditions, such as the horizontal and vertical hydraulic conductivity of the aquifer (Ehrenfeld and Bass, 1983).

5.1.2.1.4 Additives. Various additives are available to enhance moisture control; e.g., the water-retaining capacity of the soil can be enhanced by adding water-storing substances (Nimah, Ryan, and Chaudhry, 1983). Water-repelling agents are available to diminish water absorption by soils. Water-repelling soils can be treated with surface-active wetting agents to improve water infiltration and percolation. Surface-active agents also accelerate soil drainage, modify soil structure, disperse clays, and make soil more compactable. There are also evaporation retardants to retain moisture in a soil.

5.1.2.2 Soil Temperature

A temperature gradient exists in the soil (Ahlert and Kosson, 1983). As a result of heat transfer phenomena, temperature responds less to daily weather fluctuations at increased depths. Microorganisms near the surface of the soil column must adapt more readily to temperature fluctuations than those at greater depth.

Soil temperature can be modified by soil moisture control and by the use of mulches of natural or artificial materials (JRB and Associates, Inc., 1984b). Mulches can affect soil temperature in several ways. In general, they reduce diurnal and seasonal fluctuations in soil temperature (Sims and Bass, 1984). In the middle of summer, there is little overall temperature difference between mulched and bare plots, but mulched soil is warmer in spring, winter, and fall, and warms up more slowly in the spring.

Mulches with low thermal conductivities decrease heat flow both into and out of the soil; thus, soil will be cooler during the day and warmer during the night. White paper, plastic, or other types of white mulch increases the reflection of incoming radiation; thereby reducing excessive heating during the day. A transparent plastic mulch transmits solar energy to the soil and produces a greenhouse effect. A black paper or plastic mulch absorbs radiant energy during the day and reduces heat loss at night.   Placing a black covering over the soil to increase the soil temperature during the winter has been suggested as a means of overcoming the problem of slower biodegradation at the lower winter temperatures (Guidin and Syratt, 1975). Humic substances also increase soil temperature by their dark color, which increases the surface soil's heat absorption (Sims and Bass, 1984). Use of film mulch as a means of stimulating waste oil biodegradation by increasing soil temperatures during the winter, however, would preclude tilling of the soil and, thus, decrease its aeration (Dibble and Bartha, 1979a). Some workers believe this would not have an overall beneficial effect and may, in fact, be unnecessary, since it has been reported that the albedo decrease due to oil contamination raised the temperature in the upper 10 to 20 cm of tundra soils by as much as 5°C (Freedman and Hutchinson, 1976).

The type of mulch required determines the application method (Sims and Bass, 1984). Mulches are also used to protect soil surfaces from erosion, reduce water and sediment run-off, prevent surface compaction or crusting, conserve moisture, and help establish plant cover (Soil Conservation Service, 1979).   Commercial machines for spraying mulches are available. Hydromulching is a process in which seed, fertilizer, and mulch are applied as a slurry. To apply plastic mulches, equipment that is towed behind a tractor mechanically applies plastic strips that are sealed at the edges with soil. For treatment of large areas, special machines that glue polyethylene strips together are available (Mulder, 1979). Table C-9 describes the organic materials available for use as mulch and the situations when each would be most suitable.

Irrigation increases the heat capacity of the soil, raises the humidity of the air, lowers air temperature over the soil, and increases thermal conductivity, resulting in a reduction of daily soil temperature variations (Baver, Gardner, and Gardner, 1972). Sprinkle irrigation, for example, has been used for temperature control, specifically frost protection in winter and cooling in summer and for reduction of soil erosion by wind (Schwab, Frevert, Edminster, and Barhes, 1981). Drainage decreases the heat capacity, thus, raising the soil temperature. Elimination of excess water in spring causes a more rapid temperature increase. The addition of humic substances improves soil structure, thus improving soil drainability, resulting indirectly in increasing the insulative capacity of the soil.

Several physical characteristics of the soil surface can be modified to alter soil temperature (Baver, Gardner, and Gardner, 1972). Compaction of the soil surface increases the density and, thus, the thermal conductivity. Tillage, on the other hand, creates a surface mulch that, when dry, reduces heat flow from the surface to the subsurface. The diurnal temperature variation in a cultivated soil is often much greater than in an untilled soil. A loosened soil has more surface area exposed to the sun but is colder at night and more susceptible to frost.

Raising the temperature of a contaminated zone can also be achieved by pumping in heated water or recirculating groundwater through a surface heating unit (Environmental Protection Agency, 1985b).

Since microbial growth doubles for about every 10°C increase in temperature, biodecontaminations can be slowed considerably in cool weather (Thibault and Elliott, 1979). Mutant organisms are being developed to provide the optimal degradation at any given temperature. Figure A.5-1 shows the temperature dependency of the biodegradation of crude oil by a commercially available mutant bacterial formulation (PETROBAC[R] Mutant Bacterial Hydrocarbon Degrader) (Thibault and Elliott, 1979).

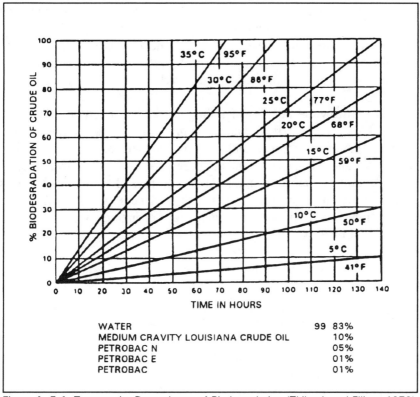

Figure A. 5-1 Temperatire Dependency of Biodegradation (Thibault and Elliott, 1979)

5.1.2.3 Soil pH

5.1.2.3.1 Increasing Soil pH.   Liming is a frequent agricultural practice and is the most common method of controlling pH, while acidification is much less common (Sims and Bass, 1984). Methods have been developed to determine the lime requirement of soils, taking into account the buffering capacity of the soil (McLean, 1982). A lime requirement test may be performed to find the loading rate to use for increasing soil pH. However, there are no readily available guidelines for reducing soil pH, and the acidification requirements for a particular soil have to be determined experimentally in the laboratory, taking into account the buffering capacity of the waste. Thorough mixing is required in the zone of contamination to change the pH. Run-off and minor controls are necessary to control drainages and erosion of the tilled soil. The achievable level of treatment is high, depending upon the wastes, site, and soil. It may be necessary to repeat the process during the treatment.

Liming is the addition to the soil of any calcium or calcium- and magnesium-containing compound capable of reducing acidity (i.e., raising pH) (Sims and Bass, 1984). Lime correctly refers only to calcium oxide, but is commonly used to refer to calcium hydroxide, calcium carbonate, calcium-magnesium carbonate, and calcium silicate slags.

There are several benefits of liming to biological activity (Sims and Bass, 1984). At higher pH values, aluminum and manganese are less soluble; both of these compounds are toxic to most plants. In addition, phosphates and most microelements necessary for plant growth (except molybdenum) are more available at higher pH. Microbial activity is greater at or near neutral pH, which enhances mineralization and degradation processes and nitrogen transformations (e.g., nitrogen fixation and nitrification).

The liming material to use depends upon several factors (Sims and Bass, 1984). Calcitic and dolomitic limestones are the most commonly used materials. However, these must be ground in order to be effective quickly, since the velocity of reaction is dependent upon the surface in contact with the soil. The finer they are ground, the more rapidly they react with the soil. A more finely ground product usually contains a mixture of fine and coarse particles to effect a rapid pH change and still be relatively long-lasting, as well as reasonably priced. Many states require that 75 to 100 percent of the limestone pass an 8- to 10-mesh sieve and that 20 to 80 percent pass anywhere from an 8- to 100-mesh sieve.   Calcium oxide and calcium hydroxide are manufactured as powders and react quickly. Other factors to consider in the selection of a limestone are neutralizing value, magnesium content, and cost per ton applied to the land.

Lime requirement for soil pH adjustment depends on soil factors, such as soil texture, type of clay, organic matter content, and exchangeable aluminum (Follett, Murphy, and Donahue, 1981).   The buffering capacity reflects the soil cation exchange capacity and will directly affect the amount of lime required to adjust soil pH.   The amount of lime required is also a function of the depth of incorporation at the site; i.e., volume of soil to be treated. The amount of lime necessary to effect a pH change in a particular site/soil/waste system is determined by a commercial soil testing laboratory in short-term treatability studies or soil-buffer tests (McLean, 1982).   Lime requirements are also affected by acid-forming fertilizers.

Limestone does not migrate easily in the soil since it is only slightly soluble and must be placed where needed (Sims and Bass, 1984). Plowing or discing surface-applied lime into the soil may, therefore, be required.   The application of fluid lime is becoming more popular, especially when mixed with fluid nitrogen fertilizer. This combination results in fewer passes over the soil, and the lime is available to counteract acidity produced by the nitrogen. Also, limestone has been applied successfully to a pharmaceutical wastewater land treatment facility through a spray irrigation system.

The addition of basic waste to acidic soil increases the pH of the surface layer (4" to 18") but not the subsoil (Brown, 1975).   The reaction neutralizes the buffer capacity of the soil. Basic waste can cause physical damage to the soil system; however, weak organic bases added to the soil may increase the soil buffer capacity and exchange capacity as the bases are degraded.

5.1.2.3.2 Decreasing Soil pH. Ferrous sulfate can be added to the soil to decrease alkalinity (Arthur D. Little, Inc., 1976). Under acidic conditions in soils, solubilities of complexed cations, such as copper (Cu) and zinc (Zn), increase, and iron (Fe), manganese (Mn), and Cu are easily reduced to more soluble forms (JRB and Associates, Inc., 1984b). Acidic wastes may also be used as a treatment process for saline-sodic soils.

5.1.2.4 Oxygen Supply

Biodegradation of most organic contaminants requires approximately two parts of oxygen to completely metabolize one part of organic compound (Wilson, Leach, Henson, and Jones, 1986). The complete oxidation of 1 mg of hydrocarbon to carbon dioxide and water requires 3 to 4 mg of oxygen (Texas Research Institute, Inc., 1982). Less oxygen is needed when microbial biomass is generated or when oxidation is not complete. Typically, about half of the carbon in hydrocarbons is converted to biomass (Green, Lee, and Jones, 1981). The problem is providing the necessary amount to the site where it will be used. Oxygen can be provided to the subsurface through the use of air, pure oxygen, hydrogen peroxide, or ozone (Environmental Protection Agency, 1985b). Fluid and semisolid systems can be aerated by means of pumps, propellers, stirrers, spargers, sprayers, and cascades (Texas Research Institute, Inc., 1982). The advantages and disadvantages of various oxygen supply alternatives are summarized in Table A.5-5 (Environmental Protection Agency, 1985b).

A great deal of water would be needed in fine-textured subsurface materials to supply the necessary oxygen to promote biodegradation (Wilson, Leach, Henson, and Jones, 1986). Oxygen levels can be increased about 5-fold by sparging injection wells with oxygen instead of air. Laboratory studies have shown that hydrocarbon-degrading bacteria can adapt to tolerate hydrogen peroxide equivalent to 200 mg/l oxygen, a 20-fold increase in oxygen over water sparged with air (Lee and Ward, 1987). However, the rate of decomposition of hydrogen peroxide to oxygen must be controlled. Rapid decomposition of only 100 mg/l hydrogen peroxide will exceed the solubility of oxygen in water, resulting in bubble formation, which could lead to gas blockage and loss of permeability.

The flow of oxygen into the system is controlled by oxygen concentration in the carrier and the permeability of the geological material to that carrier (Wilson, Leach, Henson, and Jones, 1986). Air is much less viscous than water (1.8×10^2 and 1.0×10^4 upoise, respectively). Air also has a 20-fold greater oxygen content on a volume basis. If the air- and water-filled porosity are about the same, and the pressure gradients are the same, then air should be about 1000 times more effective than water. Air should be particularly effective for oxygen supply to contaminated regions high in the unsaturated zone.

Table A.5-5.   Oxygen Supply Alternatives (Environmental Protection Agency, 1985b)

Substance	Application Method	Advantages	Disadvantages
Air	In-line	Most economical	Not practical except for trace contamination <10 mg/l COD
	In situ wells	Constant supply of oxygen possible	Wells subject to blow out
Oxygen-enriched air or pure oxygen	In-line	Provides much higher oxygen solubility than air	Not practical except for low levels of contamination <25 mg/l COD
	In situ wells	Constant supply of oxygen possible	Very expensive, wells subject to blow out
Hydrogen peroxide	In-line	Moderate cost, intimate mixing with groundwater, greater oxygen concentrations can be supplied to subsurface (100 mg/l), H_2O_2 provides 50 mg/l oxygen, helps to keep wells free of heavy growth	Chemical decomposes rapidly on contact with soil, and oxygen may bubble out prematurely unless properly stabilized
Ozone	In-line	Chemical oxidation will occur, rendering compounds more biodegradable	Ozone generation is expensive, toxic to microorganisms except at low concentrations, may require additional aeration

Whether the contaminant is above or below the water table, the rate of bioreclamation in hydrocarbon-contaminated zones is effectively controlled by the rate of supply of oxygen (Wilson, Leach, Henson, and Jones, 1986). Table A.5-6 compares the number of times that water in contaminated material below the water table, or air in material above it, must be replaced to totally reclaim subsurface materials of various textures. The calculations assume typical values for the volume occupied by air, water, and hydrocarbons (De Pastrovich, Baradat, Barthal, Chiarelli, and Fussel, 1979). The actual values at a specific site will probably be different. It is also assumed that the oxygen content of the water is 10 mg/l, that of the air 200 mg/l, and that the hydrocarbons are completely metabolized to carbon dioxide.

After the oxygen in the air is consumed during the biological degradation of the contaminant, the remaining air should physically weather (remove volatiles by evaporation) the hydrocarbons (Wilson, Leach, Henson, and Jones, 1986). The extent of weathering depends upon the vapor pressure of the contaminant. Light hydrocarbons, such as gasoline, can be vaporized to a greater extent than they are metabolized with oxygen. The vapor pressure of gasolines varies from 100 to 1000 mm at 100°C. If the vapor pressure is reduced fourfold at typical groundwater temperatures of 10°C, and benzene is typical of the vapors, then the oxygen demand for complete metabolism of the gasoline vapors ranges from twice to 20 times the oxygen content of air. The biological and physical weathering of the hydrocarbon should preferentially remove the more volatile and more water-soluble components, which are the greatest hazards to groundwater quality (De Pastrovich, Baradat, Barthal, Chiarelli, and Fussel, 1979).

There is another treatment approach worth considering. If preliminary remediation has removed any hydrocarbons floating on the water table, and if the geology is favorable, then it might be possible to lower the water table to bring the entire contaminated soil into the unsaturated zone where it is available to permeation by air (Wilson, Leach, Henson, and Jones, 1986).

There are a multitude of chemical, photosynthetic, and electrochemical reactions that produce oxygen, either as a major or minor product (Texas Research Institute, Inc., 1982). The chemical reaction types most often encountered are:

1. *Decomposition of Peroxides, Superoxides (Shanley and Edwards)*

$H_2O_2 \dashrightarrow H_2O + 1/2O_2$ (a good, ecologically sound additive, used
Hydrogen peroxide   extensively in sewage treatment)

$Na_2O_2 + H_2O \dashrightarrow NaOH + H_2O_2$
Sodium peroxide

$NaO_2 + H_2O \dashrightarrow Na_2O_2 + O_2$
Sodium superoxide

Table A.5-6.   Estimated Volumes of Water or Air Required to Renovate Completely Subsurface Material that Originally Contained Hydrocarbons at Residual Saturation (Atlas, 1978c)

Texture	Proportion of the total volume of the subsurface occupied by			Volumes required to meet the oxygen demand of the hydrocarbons	
	Hydrocarbons when drained	Air when drained	Water when flooded	Air	Water
Stone to coarse gravel	0.005	0.4	0.4	250	5000
Gravel to coarse sand	0.008	0.3	0.4	530	8000
Coarse to medium sand	0.015	0.2	0.4	1500	15,000
Medium to fine sand	0.025	0.2	0.4	2500	25,000
Fine sand to silt	0.040	0.2	0.5	4000	32,000

Barium and strontium peroxides are used in the production of oxygenating cakes employed by fishermen for maintaining live bait (Texas Research Institute, Inc., 1982). A typical formulation would contain barium peroxide, manganese dioxide, calcium sulfate, and dental plaster, which releases oxygen slowly when in contact with water. However, use of materials such as these may not be advisable, because of the resulting heavy metal contamination of the water table. Barium peroxide is definitely highly poisonous.

There is also a urea-peroxide addition compound that has been used in conjunction with phosphate solutions to treat plants suffering from oxygen starvation in the root zone (U.S. Patent 3,912,490) (Texas Research Institute, Inc., 1982). The compound is available commercially from Western Europe. It is probably of the inclusion type, one in which H_2O_2 molecules are trapped within channels formed by the crystallization of urea (March, 1968). Since the molecules are held together only by Van der Waal's forces, when dissolved, the solution will behave as a mixture of urea and hydrogen peroxide. By weight, 35 percent of the compound is H_2O_2.

2. Decomposition of Peroxyacids and Salts (Austin American Statesman, 1980)

Peroxy mono- and disulfuric acids, peroxy mono- and diphosphoric acid, and peroxyborates all produce acidic solutions, but the salts may be important for consideration.

The exact mode of degradation of the salt, $KHSO_5$ (potassium monoperoxy sulfate), is uncertain, but it has been used as an aid in the degradation of atrazine (a pesticide)--presumably by virtue of its oxygen-producing ability. Degradation probably results in the formation of H_2O_2 and $KHSO_4$.

Impure salts of peroxy monophosphoric acid (H_3PO_5) might prove useful.

3. Thermal Decomposition of Oxygen-Bearing Salts

$$2NaNO_3 \xrightarrow[\text{heat}]{} 2NaNO_2 + O_2$$

$$2KClO_3 \xrightarrow[\text{heat}]{} 2KCl + 3O_2$$

Generation of oxygen by this method has no particular advantage to treating underground contamination.

Two powerful oxidizing agents that do have potential for in-place treatment are ozone and hydrogen peroxide.

5.1.2.4.1 Ozone

Ozone is an oxygen molecule containing three oxygen atoms (Roberts, Koff, and Karr, 1988). Ozone gas is a very strong oxidizing agent that is very unstable and extremely reactive (Environmental Protection Agency, 1985b). It cannot be shipped or stored; therefore, it must be generated on-site prior to application.

Ozone may be used to degrade recalcitrant compounds directly by creating an oxygenated compound without chemical degradation (Texas Research Institute, Inc., 1982). Ozonation is an oxidation process appropriate for aqueous streams that contain less than 1.0 percent oxidizable compounds (Roberts, Koff, and Karr, 1988). This chemical oxidation can be used on many organic compounds that cannot be easily broken down biologically, including chlorinated hydrocarbons, alcohols, chlorinated aromatics, pesticides, and cyanides (Lee and Ward, 1986).

Ozone can also be used to increase the dissolved oxygen level in the water for enhancing biological activity (Texas Research Institute, Inc., 1982). It can be employed as a pretreatment for wastes to break down refractory organics or to furnish a polishing step after biological or other treatment processes to oxidize untreated organics (Roberts, Koff, and Karr, 1988). The rate of ozone reaction can be controlled by adjusting the pH of the medium (Texas Research Institute, Inc., 1982). At high pH, hydroxyl free-radical reactions dominate over the more rapid direct ozone reactions.

The most effective and cost-effective uses of ozone in soil system decontamination appear to be in the treatment of contaminated water extracted from contaminated soil systems through recovery wells, and in the stimulation of biological activity in saturated soil (Nagel, et al., 1982). Ozone treatment may be very effective for enhancing biological activity, if the organic contaminants are relatively biodegradable. However, if much of the material is relatively biorefractory, the amount of ozone required would greatly increase the cost of the treatment.

Ozone has been used to treat groundwater contaminated with oil products to reduce dissolved organic carbon concentration (Nagel, et al., 1982). Dosages of 1 g ozone/g dissolved organic carbon resulted in residual water ozone concentration of 0.1 to 0.2 ppm. The treated water was then infiltrated into the aquifer through injection wells. There was an increase in dissolved oxygen in the contaminated water. This increased microbial activity in the saturated soil zone, which stimulated microbial degradation of the organic contaminants.

Ozone was used to treat a petroleum contamination in Karlsruhe, Germany, that threatened a drinking water supply (Atlas and Bartha, 1973e). The polluted groundwater was withdrawn, treated with ozone, and infiltrated back into the system via three infiltration wells. About 1 g ozone/g DOC was added to the groundwater, which increased the biodegradability of the petroleum contaminants and added dissolved oxygen. The dissolved oxygen reached equilibrium at about 80 percent of the initial concentration injected. The oxygen consumption peaked at about 40 kg/day during the initial infiltration period. Levels of cyanide, a contaminant identified after the treatment began, also decreased, although biodegradation was not shown to be the cause. Total bacterial counts in the groundwater increased tenfold, but bacteria potentially harmful to man did not increase. The drinking water from this aquifer contained no trace of contaminants after one and a half years of ozone treatment.

Saturated aliphatic compounds that do not contain easily oxidized functional groups are not readily reactive with ozone; for example, saturated aliphatic hydrocarbons, aldehydes, and alcohols (Sims and Bass, 1984). Reactivity of aromatic compounds with ozone is a function of the number and type of substituents. Substituents that withdraw electrons from the ring deactivate the ring toward ozone; for example, halogen, nitro, sulfonic acid, carbonyl, and carboxyl groups. Substituents that release electrons activate the ring toward ozone; for example, alkyl, methoxyl, and hydroxyl.

The following reactivity patterns with ozone are:

1.   phenol, xylene > toluene > benzene
2.   pentachlorophenol < dichloro-, trichloro-, tetrachlorophenol

The relatively rapid decomposition rates of ozone in aqueous systems, especially in the presence of certain chemical contaminants or other agents that catalyze its decomposition to oxygen, preclude its effective application to subsurface waste deposits (Amdurer, Fellman, and Abdelhamid, 1985). The half-life of ozone in groundwater is less than ½ hour (Ellis and Payne, 1984) (about 18 minutes) (Environmental Protection Agency, 1985b). Since the flow rates of water are likely to be in inches/hour or less, it is unlikely that effective doses of ozone could be delivered very far for chemical oxidation. However, it has been used successfully to supply oxygen for microbial biodegradation (Rice, 1984).

5.1.2.4.2 Hydrogen Peroxide.   Hydrogen peroxide is a weaker oxidizing agent than ozone; however, it is considerably more stable in water (Amdurer, Fellman, and Abdelhamid, 1985). It is used to degrade recalcitrant compounds and modify the mobility of some metals (Sims and Bass, 1984). It can be used to increase oxygen levels in the soil. This can increase microbial activity and degradation of organic contaminants (Nagel, et al., 1982). Hydrogen peroxide could be injected into the contaminated soil above the water table in conjunction with nutrients where it would decompose naturally or by enzymatic action to increase the dissolved oxygen content (Texas Research Institute, Inc., 1982).

It was found that air sparging was able to maintain dissolved oxygen levels of only 1 to 2 ppm in a spill area (Nagel, et al., 1982). However, addition of microbial nutrient (a specially formulated, hydrogen peroxide-based nutrient solution; FMC Aquifer Remediation Systems, Princeton, NJ) raised DO levels to over 15 ppm, thus, establishing the efficiency of hydrogen peroxide-based solutions for supplying increased oxygen levels and, thereby, enhancing the bioreclamation process. Hydrogen peroxide was selected as the source of oxygen for biodegration at the Kelly Air Force Base, Texas, because it can provide about five times more oxygen to the subsurface than aeration techniques (Wetzel, Davidson, Durst, and Sarno, 1986). The increase in microbial densities in stimulated underground spill sites is probably due to the increased oxygen from the hydrogen peroxide (Wilson, Leach, Henson, and Jones, 1986).

Hydrogen peroxide is a strong oxidant and, therefore, is nonselective (Sims and Bass, 1984). It will act with any oxidizable material present in the soil. This could be a problem because the concentration of natural organic material in the soil would be lowered, causing a reduced sorption capacity for some organics. Thus, the effectiveness of peroxide may be inhibited because it simultaneously increases mobility and decreases possible sorption sites.

Hydrogen peroxide is effective for oxidizing cyanide, aldehydes, dialkyl sulfides, dithionate, nitrogen compounds, phenols, and sulfur compounds (FMC Corporation, 1979). The following chemical groups have incompatible reactions with peroxides (i.e., the reaction products are more mobile) (Sims and Bass, 1984):

Acid chlorides and anhydrides
Acids, mineral, nonoxidizing
Acids, mineral, oxidizing
Acids, organics
Alcohols and glycols
Alkyl halides
Azo, diazo compounds, hydrazine
Cyanides
Dithio carbamates
Aldehydes
Metals and metal compounds
Phenols and cresols
Sulfides, inorganic
Chlorinated aromatics/alicycles

Hydrogen peroxide is more soluble in water than molecular oxygen and may provide more oxygen at specific sites of application (Britton, 1985). The enzymatic decomposition reactions are:

$$2H_2O_2 \dashrightarrow 2H_2O_2 + O_2$$

$$H_2O_2 + XH_2 \dashrightarrow 2H_2O + X$$

where X can be NADH, glutathione, or other biological reductants.

Hydrogen peroxide and ozone have been used in combination to degrade compounds that are refractory to either material individually (Nakayama, et al., 1979). There is an ongoing debate as to what oxidant is the best (Rich, Bluestone, and Cannon, 1986). Some believe that hydrogen peroxide is the most efficient way to move oxygen through a formation. Others find that air is the most cost-effective oxidizing agent. On the other hand, proponents of both hydrogen peroxide and ozone use aeration in their bioreclamation systems.

Hydrogen peroxide is reasonably inexpensive, is nonpersistent, and is not likely to represent a serious health hazard, if used properly (Texas Research Institute, Inc., 1982; Britton, 1985). However, it is cytotoxic (3 percent is commonly used as a general antiseptic) and may decompose (by enzymatic catalysis or nonenzymatically by in situ physicochemical processes) before reaching its targeted spill location.

This chemical can be toxic to the microorganisms that it is intended to stimulate (Rich, Bluestone, and Cannon, 1986). However, the growth rate of hydrocarbon-utilizing bacteria is not necessarily inhibited by high hydrogen peroxide concentrations (Texas Research Institute, Inc., 1982). Even growth enhancement is sometimes observed. Whether or not a given hydrogen peroxide concentration will be toxic to bacteria depends upon the concentration of the organisms when the H_2O_2 is added. Large populations are more successful at surviving high hydrogen peroxide concentrations than small populations.

Bacteria produce hydrogen peroxide from respiratory processes and have enzymes (hydroperoxidases) to protect against the compound's toxicity (Texas Research Institute, Inc., 1982). Hydrogen peroxide has been shown to be toxic to fresh bacterial cultures at levels greater than 100 ppm, although mature cultures suffered less and could function at levels as high as 10,000 ppm (Texas Research Institute, Inc., 1982). Subsequent experimentation with sand columns inoculated with gasoline and gasoline-degrading bacteria showed that 1.0, 0.5 and 0.25 percent hydrogen peroxide solutions were toxic to the bacteria (Texas Research

Institute, Inc., 1983).   In another study, using a mixed culture of gasoline degraders, the maximum concentration of H_2O_2 that could be tolerated was 0.05 percent (500 ppm), although by increasing the concentration gradually, the level of tolerance could be raised to 0.2 percent (2000 ppm).   Hydrogen peroxide could not be used in batch liquid cultures as in sand columns (Wilson, Leach, Henson, and Jones, 1986).   The liquid cultures were extremely sensitive to the compound. This was probably due to the nature of the growth in the two environments: in sand, the organisms would grow as a film with multicellular depth; in liquid, they would be unicellular, unprotected by adjacent cells.

There appears to be a critical H_2O_2:organism ratio, above which the catalase-utilizing protective mechanisms of the organisms are overwhelmed.   This ratio may be on the order of 2×10^{10}:1 in a given volume of solution, or it may be expressed as 1 ppm H_2O_2:8.9×10^5 bacteria.   Flow of hydrogen peroxide through sand-packed columns inoculated with hydrocarbon utilizers is blocked by bubble formation caused by the decomposition of the hydrogen peroxide, at as low as 0.01 percent (100 ppm) H_2O_2.

Another concern is that the hydrogen peroxide will decompose before it reaches the groundwater and cause the precipitation of iron and manganese oxides and hydroxides (Environmental Protection Agency, 1985b).   Much of the decomposition of hydrogen peroxide in soil and groundwater will be due to reactions with iron salts (Haber and Weiss, 1934).   Nonenzymatic decomposition can occur in a variety of reactions, including those in the presence of iron salts, known as Fenton chemistry (Fenton, 1894).   The following reactions show how different iron salts affect the decomposition of H_2O_2 (Haber and Weiss, 1934).

$$Fe^{++} + H_2O_2 \text{ ----- } Fe^{+++} + OH^- + OH^· \text{ (hydroxyl radical)}$$

$$OH^· + H_2O_2 \text{ ----- } H_2O + H^+ + O_2^{-·} \text{ (superoxide radical)}$$

$$O^{-·} + H_2O_2 \text{ ----- } O_2 + OH^- + OH^·$$

$$Fe^{+++} + H_2O_2 \text{ ----- } Fe^{++} + 2H^+ + O_2^{-·}$$

$$Fe^{+++} + O_2^{-·} \text{ ----- } Fe^{++} + O_2$$

Most decreases of hydrogen peroxide occur rapidly in the top 5.5 cm of a sand column, with only slight decreases thereafter, which may be due to iron stimulation (Wilson, Leach, Henson, and Jones, 1986).   On one hand, the molecular oxygen produced from these reactions would help enhance gasoline biodegradation. On the other hand, iron can cause the hydrogen peroxide to decompose before it reaches the intended site. The decomposition rate of peroxide is also greatly accelerated in the presence of another heavy metal, Cu^{++} (Bambrick, 1985).

There are a number of ways to prevent the hydrogen peroxide from decomposing. Standard practice is to add enough phosphate to the recirculated water to precipitate the iron (Wilson, Leach, Henson, and Jones, 19864). High concentrations of phosphates (e.g., 10 mg/l) can stabilize peroxide for prolonged periods of time in the presence of ferric chloride, an aggressive catalyst (Environmental Protection Agency, 1985b). The stabilizing effect of phosphate is fortuitous, since it is a major nutrient for enhancement of underground biodegradation of gasoline. However, there are problems associated with adding high phosphate concentrations, such as precipitation. Some suppliers add an organic inhibitor that will stabilize the peroxide at a rate appropriate to the rate of infiltration, so the oxygen demand of the bacteria attached to the solids is balanced by the oxygen supplied by decomposing peroxide in the recirculated water (Wilson, Leach, Henson, and Jones, 1986). Dworkin Foster, or a similar medium containing the mineral components for growth (except for a source of carbon and energy), can also stabilize hydrogen peroxide and would be a suitable solution for pumping the material underground without premature decomposition (Britton, 1985).

Enhancement of the microbial population has also been reportedly used to reduce levels of iron and manganese in the groundwater (Hallberg and Martinelli, 1976). The process, known as the Vyrodex method, was developed in Finland and has been used in Sweden and other areas where high levels of the two elements are found in the groundwater. Iron bacteria and manganese bacteria oxidize the soluble forms of iron and manganese to insoluble forms; the bacteria use the electrons adsorbed from the oxidation process as sources of energy. Dissolved oxygen is added to the groundwater to stimulate the bacteria to first remove the iron and then later the manganese. As the iron bacteria population builds up and begins to die, it supplies the organic carbon necessary for the manganese bacteria. The efficiency of the process increases with the number of aerations.

Successful use of hydrogen peroxide requires careful control of the geochemistry and hydrology of the site (Wilson, Leach, Henson, and Jones, 1986). Hydrogen peroxide can mobilize metals, such as lead and antimony; and, if the water is hard, magnesium and calcium phosphates can precipitate and plug the injection well or infiltration gallery.

Heavy metal control procedures involve techniques that effectively prevent contact between the metals and the peroxide (Bambrick, 1985). This is accomplished by using a chelating agent and silicate. The most effective of the commercially available chelating agents is the pentasodium salt of diethylenetriaminepentaacetic acid (Na_5DTPA). This is a negatively charged compound that can form a ringed structure that alters the reactivity of a positively charged ion. The heavy metal ion is bound by covalent bonds off the nitrogens and ionic bonds off the acetate group and, thus, inhibited from entering into undesirable reactions, e.g., $Na_3MnDTPA$. The breakdown of peroxide can be decreased substantially using the chelate Na_5DTPA in combination with sodium silicate and $MgSO_4$. The real value of DTPA, even when the metal level is low, is in stabilizing the peroxide liquor solution. In the lab this combination reduces the amount of peroxide decomposed to 55 percent after 2 hr. Without DTPA, 95 percent of the peroxide is useless after 1 hr. These results have also been verified in field tests.

The pH does not strongly influence the rate of hydrogen peroxide decomposition by iron salts in aqueous media (Wilson, Leach, Henson, and Jones, 1986). The pathway of its decomposition depends upon the valence of the iron. Liberation of oxygen with resulting bubble formation should occur in groundwater and soil with high concentrations of ferric iron. Addition of $FeCl_3$ to a pumping solution would be a way to form pockets of oxygen bubbles in a short time for bioreclamation of an underground gasoline spill.

An inexpensive hydrogen peroxide can be produced from a coproduct process, such as converting glucose to gluconic acid with glucose-1-oxidase (Hou, 1982).

5.1.2.4.3 Hypochlorite. Another potential oxidant is hypochlorite (Amdurer, Fellman, and Abdelhamid, 1985). It is generally available as potassium, calcium, or sodium hypochlorite (bleach) and is used in the treatment of drinking water, municipal wastewater, and industrial waste (Environmental Protection Agency, 1985b). It reacts with organic compounds as both a chlorinating agent and an oxidizing agent. Hypochlorite additions may lead to production of undesirable chlorinated by-products (e.g., chloroform) rather than oxidative degradation products. Therefore, the use of hypochlorite for in situ treatment of organic wastes is not recommended.

5.1.2.4.4 Other Electron Acceptors. Oxygen can be supplemented with other electron acceptors, such as nitrate (Wilson, Leach, Henson, and Jones, 1986). Nitrate can support the degradation of xylenes in subsurface material (Kuhn, Colberg, Schnoor, Wanner, Zehnder, and Schwarzenbach, 1985). This approach is still experimental but offers considerable promise because nitrate is inexpensive, is very soluble, and is nontoxic to microorganisms, although it is of human health concern. Nitrate itself is a pollutant limited to 10 mg/l in drinking water (Environmental Protection Agency, 1985b). Another study found that neither nitrate nor sulfate as terminal electron acceptors in an anaerobic process is effective on the types of saturated hydrocarbons found in petroleum (Texas Research Institute, Inc., 1982).

Selection of the appropriate oxidizing agent depends, in part, upon the substance to be detoxified and also upon the feasibility of delivery and environmental safety (Environmental Protection Agency, 1985b). Although there are some compounds that will not react with hydrogen peroxide but will react with ozone or hypochlorite, hydrogen peroxide appears to be the most feasible for in situ treatment.

5.1.2.4.5 Possible Approaches for Supplying Oxygen to the Subsurface (Texas Research Institute, Inc., 1982).

1. Injection of Liquified Gases

The injection of liquid oxygen or liquid air in the soil would utilize existing technology. Intermittent injection of liquid oxygen would produce a high concentration of oxygen, which would slowly diffuse into the surrounding strata. Since oxygen is 10 times more soluble in hydrocarbons than it is in water, the hydrocarbon phase could actually act as an oxygen reservoir to replace the oxygen being consumed in the aqueous phase (Faust and Hunter, 1971). Repeated injections would create a flow through the system, preventing build-up of carbon dioxide. Another technique would have to be used to add additional nutrients. This method would be best in an area where the soil contained abundant nutrients.

2. Injection of Oxygen-Releasing Compounds (With Nutrients)

The best material for implementing this approach is hydrogen peroxide (Texas Research Institute, Inc., 1982). Injection should ideally be made over the entire contaminated area, both into the water table and at points just above the water table into the gasoline-bearing soil. Recovery wells toward the center of the spill would help contain it and would aid in the recovery of gasoline as it is released from the soil. It may be possible to set up a recirculating system, whereby produced water is cleaned, fertilized and oxygenated, and reinjected into the water table. A variation would be to use a physical oxygenation technique on the injection water instead of a chemical additive.

The injection of a hydrogen peroxide solution (or highly oxygenated water) would appear to have the greatest impact on the total system, along with an appropriate nutrient solution at different depths in the soil strata and into the water table (Texas Research Institute, Inc., 1982). Injection into the contaminated soil above the water table would continuously bathe the gasoline-contaminated region with oxygenated, nutrient-filled water. A large amount of the residual gasoline would be consumed by the bacteria, and it is possible that the emulsifiers they produce would aid in mobilizing the gasoline into the water table where it could be collected by physical means at the producing well(s). Nutrients and oxygen not utilized in the soil would eventually find their way into the water table, where they could be used for cleanup on that system.

Figure A.5-2 shows a schematic of a spill and cleanup operation (Texas Research Institute, Inc., 1982).

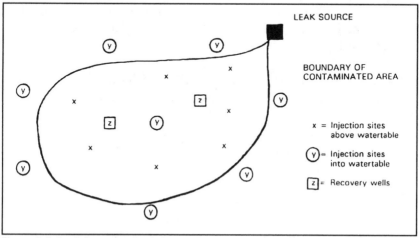

Figure A.5-2. Schematic of a Spill and Cleanup Operation (Texas Research Institute, Inc., 1982)

Since it is expensive to drill a well or injection site, it may be acceptable to have relatively shallow soil injection sites that are just deep enough to get oxygen and nutrients past the plant growth zone so they are not utilized before reaching their intended target (Texas Research Institute, Inc., 1982). If a combination recovery-cleanup operation is employed with the produced water, it could be recirculated throughout the system. Cost effectiveness would have to be determined for each situation, taking into consideration the cost of transportation to a sewage treatment plant and the cost of bringing in fresh water.

3. Venting

A venting approach would be applicable in an area with porous soil (Texas Research Institute, Inc., 1982). Blowers would be utilized to provide either suction or pressure to create a flow of air through the soil strata. Gaseous ammonia could be added to the input air, supplying nitrogen, but other methods would have to be used for the addition of other nutrients. Also, the effect on groundwater decontamination would be minimal.

In 1981, a technique for soil venting was used in Mont Belvieu, TX, to flush leaked propane and ethane out of the ground (Austin American Statesman, 1980). Liquid nitrogen was pumped underground, and the large volumes of nitrogen gas generated swept the gases through the soil. Possibly, liquid oxygen or air could be utilized in this fashion to supply oxygen to the soil strata. It is not known what effect a stray spark or flame might have on a system such as this. The potential for an underground fire exists.

5.1.2.4.6 Commercial Approaches. Enhancing the available oxygen in the soil for microbes to consume hydrocarbon contaminants has improved the rate of degradation (Chowdhury, Parkinson, and Rhein, 1986). Aquifer Remediation System's Bio XL process employs stabilized solutions of hydrogen peroxide (tradename Restore) to increase the amount of oxygen in the soil by more than 25 times, in comparison with air sparging, an earlier method. Another of its products (Restore Microbial Nutrient) prevents precipitation of chemical nutrients. Bioreclamation can now be used in low-permeability formations, where the pumping rate from recovery wells is as low as 5 gal/min.

Groundwater Technology has a similar in situ process, called END (Enhanced Natural Degradation). It is planning to introduce a new system that could cut the amount of hydrogen peroxide consumption by

75 to 90 percent by modifying the oxygen delivery system into a closed loop.

There can be severe oxygen limitation to degradation within inches of the surface of soil (Zitrides, 1983). Polybac Corporation employs tilling of the soil to provide additional oxygen, as well as to better mix a microbial inoculum with the contaminant. Otherwise, the organisms will adhere to the top layers of soil and percolate only slowly to greater depths.

A commercial product (Biostim) produced by Biosystems, Inc., can circulate 500 ppm oxygen in the soil, as opposed to the 10 ppm when air is used as the source of oxygen (Biosystems, Inc., 1986). This is achieved by using "Tysul" WW hydrogen peroxide from Du Pont. It is environmentally safe and is a good source of dissolved oxygen, since the microbes can break down the peroxide into oxygen and water.

5.1.2.4.7 Modifying Soil Oxygen Content for Aerobic Biodegradation. Although some xenobiotic organic compounds appear to require the slow anaerobic metabolism for decomposition, most of these compounds are susceptible to attack by aerobic organisms (Alexander, 1977; Brunner and Focht, 1983). Therefore, assuring the aerobiosis of the soil will enhance the rate of biological decomposition for some compounds. Active microflora have been observed in the top 15 cm of soil, and tilling is suggested as an effective means of promoting aeration (Raymond, Hudson, and Jamison, 1976). Tilling the soil for aeration is common practice in agriculture and has been recommended for hazardous waste-contaminated soil reclamation by practitioners and researchers (Thibault and Elliott, 1979; Arthur D. Little, Inc., 1976). The soil must be tilled at periodic intervals to assure adequate aeration. However, tillage will increase the susceptibility of the site to erosion. Spills often contaminate just the upper layer of soil. Even in these situations, an oxygen limitation on degradation rate can occur, which can be easily solved by tilling (Thibault and Elliott, 1979).

Tilling equipment can aerate surface soils and mix wastes or reagents into the soil (Sims and Bass, 1984). Oxygen must first be dissolved in the interstitial water of the soil, since the viable microorganisms are present in the aqueous environment (Thibault and Elliott, 1979). Therefore, it is important that the contaminated soil be kept moist during the cleanup. It has been concluded that soil moisture is, in itself, a simple and low-cost method of supplying some aeration (Thibault and Elliott, 1980). Oil tilled into the soil immediately and thoroughly after application also significantly reduces the possibility of it being moved hydrologically out of the area (Raymond, Hudson, and Jamison, 1976).

Aeration of subsurface soils not accessible to tillage equipment can be accomplished using construction equipment, such as a backhoe, or a well point injection system, for soils deeper than about 2 ft. Diffusers attached to paint sprayer-type compressors have been used to inject air into a series of 10 wells (Raymond, Jamison, and Hudson, 1976). They deliver about 2.5 cfm to enhance microbial degradation. Various nutrients are added simultaneously. The diffusers are positioned 5 ft from the bottom of the well and below the water table. Aeration through well points has been primarily used for saturated soils and has been shown to be effective for delivering oxygen to the subsurface. It is uncertain whether the technique would also work for unsaturated soils.

If the site is too wet, a drainage system should be installed (see Appendix A, Section 5.1.2.1). Soils with high water tables can be drained using common agricultural techniques. Controls should be set up to prevent run-on and run-off of precipitation (Thibault and Elliott, 1979).

Site geology, in most cases will determine the methods of aeration to be used in a given situation (Raymond, Jamison, and Hudson, 1976). For example, in a fractured dolomite and clay formation, lack of homogeneity makes well injection and distribution of oxygen difficult.

5.1.2.4.8 Modifying Soil Oxygen Content for Anaerobic Biodegradation. Oxygen levels can be decreased by compacting the soil or by saturating the soil with water (Sims and Bass, 1984). By reducing pore sizes and restricting aeration, anaerobic microsite frequency in the soil will increase. Compaction helps draw moisture to the soil surface. This lessens the problems of leaching that may occur, if anaerobiosis is achieved by water addition. Volatilization may also be suppressed by surface soil compaction. Water may still have to be added to achieve the required degree of anaerobiosis; however, it would be less than for an uncompacted soil, also minimizing the leaching potential. After the available oxygen in the water is depleted, the saturated conditions would prevent percolation of additional oxygen to the submerged area. Diking is a common agricultural practice that may be applicable to decreasing the soil oxygen content (Arthur D. Little, Inc., 1976). This would establish and maintain anaerobic conditions as long as the land is kept under water. Another possible method of rendering the site anaerobic would be to add excessive amounts of easily biodegradable organics so the oxygen would be depleted (Environmental Protection Agency, 1985b).

5.1.2.5 Nutrients

Addition of nutrients to hydrocarbon-contaminated surface soil has proved useful in increasing microbial degradation. The total numbers of microbes increase greatly after a petroleum spill. An increase was noted from 10^6 to 10^8 organisms/g after an oil well blowout (Odu, 1972). Application of fertilizer stimulated greater microbial growth and utilization of some components of oil (Westlake, Jobson, and Cook, 1978). Low concentrations of readily metabolized organic compounds (peptone, calcium lactate, yeast extract, nicotinamide, riboflavin, pyridoxine, thiamine, ascorbic acid) often promote the growth of the oxidizer, but high concentrations will retard the degradation of the hydrocarbons (Zobell, 1946; Morozov and Nikolayov, 1978).

Nitrogen and phosphate are the nutrients most frequently present in limiting concentrations in soils (Environmental Protection Agency, 1985b). Other nutrients required for microbial metabolism include potassium, magnesium, calcium, sulfur, sodium, manganese, iron, and trace metals. The elements essential for biological growth and sources for them are presented in Table A.5-7 (Mattingly, 1975).

The nitrogen requirement value (NRV) is the amount of nitrogen required by microorganisms to decompose/degrade a particular organic chemical waste (Parr, Sikora, and Burge, 1983). It depends mainly upon two factors: the chemical composition of the waste and the rate of decomposition. This value is also affected by the other soil factors.

After contact with an oily waste, microbial activity initially decreases (Hornick, Fisher, and Paolini, 1983). This may be due to the same initial decrease in mineral nitrogen resulting from nitrogen immobilization by hydrocarbon-metabolizing microbes using up all available nitrogen. In time, the microorganisms will adapt to the high C:N ratio and increase the total microbial population (Overcash and Pal, 1979).

Phosphorus concentrations in the soil solution are usually low, ranging from 0.1 to 1 ppm, since this element is mostly associated with the solid phase in soils (Hornick, 1983). In acid soils, phosphorus reacts with iron and aluminum hydroxides to produce adsorbed forms of phosphorus that are in equilibrium with the soil solution or are precipitated and, thus, occluded by the minerals. At low concentrations

Table A.5-7.   Essential Elements for Biological Growth (Based on Requirements for Plant Growth) (Mattingly, 1975)

Elements	Source
Major nutrients	
Carbon	Air and water
Hydrogen	
Oxygen	
Nitrogen	Soil, inorganic fertilizers,
Phosphorus	or in waste
Potassium	
Sulfur	
Calcium	Soil liming materials,
Magnesium	or in waste
Minor nutrients	
Iron	Soil, soil amendments,
Manganese	or in waste
Boron	
Molybdenum	
Copper	
Zinc	
Chlorine	
Sodium	
Cobalt	
Vanadium	
Silicon	

in calcareous soils, phosphorus is adsorbed onto calcium carbonate; at high concentrations, calcium phosphate minerals are formed (Mattingly, 1975). The common forms of phosphorus in soil are $H_2PO_4^{-2}$ in basic soil solutions.

The quantity of nitrogen and phosphorus required to convert 100 percent of the petroleum carbon to biomass may be calculated from the C:N and C:P ratios found in cellular material (Dibble and Bartha, 1979a). Accepted values for a mixed microbial population in the soil are: C:N, 10:1 (Waksman, 1924); and C:P, 100:1 (Thompson, Black, and Zoellner, 1954). In reality, a complete assimilation of petroleum carbon into biomass is not achievable under natural conditions. Some of the petroleum compounds are recalcitrant or are metabolized slowly over long periods. From petroleum compounds that are readily metabolized, some carbon will be mineralized to carbon dioxide. Thus, efficiency of conversion of substrate (petroleum) carbon to cellular material is less than 100 percent. The optimal C:N and C:P ratios are expected to be wider than the theoretical values.

Under most growth conditions, about half of the carbon available from growth hydrocarbons eventually becomes cellular biomass (Texas Research Institute, Inc., 1982). This consists primarily of proteins, nucleic acids, amino acid, purines, pyrimidines, lipids, and polysaccharides.

Measurement of soil organic carbon, organic nitrogen, and organic phosphorus allows the determination of its C:N:P ratio and an evaluation of nutrient availability (Sims and Bass, 1984). If the ratio of organic C:N:P is wider than about 300:15:1 (100:15:13, Thibault and Elliott, 1980; 100:15:3, Thibault and Elliott, 1979), and available (extractable) inorganic forms of nitrogen and phosphorus do not narrow the ratio to within these limits, supplemental nitrogen and/or phosphorus should be added, such as by addition of commercial fertilizers (Kowalenko, 1978). One such product is POLYBAC[R] N Biodegradable Nutrients, which contains the proper balance of nitrogen and phosphorus required in a form readily available for microbial uptake (Thibault and Elliott, 1979).

The ratio depends upon the rate and extent of degradation of the chemicals involved (Josephson, 1983) and may vary according to the particular contaminant present. Biodegradation of complex oily sludges in soil occurs most rapidly when nitrogen is added to reduce the carbon to nitrogen (C:N) ratio to 9:1 (Brown, Donnelly, and Deuel, 1983). That of petrochemical sludge is most rapid when nitrogen, phosphorus, and potassium are added at a rate of 124:1, C:NPK.

During experiments on landfarming waste oil, it was determined that carbon-to-nitrogen and carbon-to-phosphorus ratios of 60:1 and 800:1, respectively, were optimal under the conditions used (Dibble and Bartha, 1979a). Addition of yeast extract or domestic sewage did not prove beneficial. Urea formaldehyde was found to be the most satisfactory nitrogen source tested, since it effectively stimulated biodegradation and did not leach nitrogen, which could contaminate the groundwater (Dibble and Bartha, 1979b). A problem with this technology is that run-off water from the site could contain high amounts of oil and fertilizer (Kincannon, 1972).

In one situation, the addition of sodium nitrate enhanced the oxidation of hydrocarbons and the ultimate decay of the resulting organic carbon compounds to inorganic carbon compounds (Dietz, 1980). Addition of potassium orthophosphates, KH_2PO_4 and K_2HPO_4, had no effect on biodegradation in this application. The phosphate precipitated in a very early stage, due to the presence of calcium. Calcium promotes flocculation, the clumping of tiny soil particulates, and may prevent thorough incorporation of phosphate into the soil (Brady, 1974).

Microorganisms may be limited by phosphorus but not nitrogen (Thorn and Ventullo, 1986). Neither nitrogen nor phosphorus enrichment alone stimulated the biodegradation of phenol in topsoil (Atlas and Bartha, 1973). However, in two different types of subsurface soils, addition of these nutrients significantly stimulated mineralization. Phosphorus enrichment had the greatest effects, and the effects of simultaneous nitrogen and phosphorus amendments were similar to those observed with phosphorus alone. Phosphorus limitation may be widespread in subsurface soils. If the input of phosphorus into the subsurface is disproportionate to that of organic compounds, phosphorus limitation could greatly reduce the ability of microbes in the lower soil profile to attenuate pollutants during their transit to the underlying groundwater, assuming oxygen or another electron acceptor is not limiting. It should be noted that addition of phosphates can result in the precipitation of calcium and iron phosphates (Environmental Protection Agency, 1985b). If calcium concentrations are high, the added phosphate can be tied up by the calcium, and would, therefore, not be available to the microorganisms.

If calcium is present at 200 mg/l, it is likely that calcium supplementation is unnecessary (Environmental Protection Agency, 1985b). Calcium deficiencies usually occur only in acid soils and can be corrected by liming (JRB and Associates, Inc., 1984b). If the soil is deficient in magnesium, the use of dolomitic lime is advised. It is desirable to have a high level of exchangeable bases (calcium, magnesium, sodium, and potassium) on the surface exchange sites of the soil for good microbial activity and for preventing excessively acid conditions. Sulfur levels in soils are usually sufficient; however, sulfur is also a constituent of most inorganic fertilizers. Micronutrients are also present in adequate amounts in most soils.

Key trace elements are also essential to the stimulation of bacterial growth (Kincannon, 1972). These are required in such small doses that most are already abundant in the soil. They include sulfur, sodium, calcium, magnesium, and iron. Copper, zinc, and lead are normally considered to exhibit harmful effects on biological growth. Addition of yeast cells can serve as a nutrient source (Lehtomakei and Niemela, 1975). Organic and inorganic nutrients in natural waters affect the rate of mineralization of organic compounds in trace concentrations (Kaufman, 1983). Inorganic nutrients, arginine, or yeast extract often enhanced, but glucose reduced, the rate of mineralization.

The concentration of nutrients and organics should be kept as uniform as possible to protect against shock loading (Environmental Protection Agency, 1985b). Nitrogen must be applied with caution to avoid excessive application (Saxena and Bartha, 1983). Nitrate or other forms of nitrogen oxidized to nitrate in the soil may be leached to the groundwater (nitrate is itself a pollutant limited to 10 mg/l in drinking water) (Environmental Protection Agency, 1985b). Some nitrogen fertilizers may also tend to lower the soil pH, necessitating a liming program to maintain the optimal pH for biological activity. Low concentrations of readily metabolized organic compounds (peptone, calcium lactate, yeast extract, nicotinamide, riboflavin, pyridoxine, thiamine, ascorbic acid) often promote the growth of the oxidizer, but high concentrations will retard the degradation of the hydrocarbons (Zobell, 1946; Morozov and Nikolayov, 1978). The quantity of organic material to add must be determined in treatability studies (Sims and Bass, 1984). Nutrient formulations should be devised with the help of experienced geochemists to minimize problems with precipitation and dispersion of clays (Environmental Protection Agency, 1985b). Special soil preconditioners and nutrient formulations to reduce these problems and maximize nutrient mobility and solubility are being investigated.

Population turnover allows for the recycling of nutrients (Dibble and Bartha, 1979a). However, it is expected that fertilizer in the optimal ratios will have to be reapplied, as necessary. The best fertilizers for soil application are in a form of readily usable nitrogen and phosphorus and also in a slow-release form to provide a continuous supply of nutrients, which is beneficial in terms of fertilizer savings and minimized leaching from the oil-soil interface (Atlas, 1977).

A liquid fertilizer containing 3340 lb ammonium sulfate, 920 lb disodium phosphate, and 740 lb monosodium phosphate was injected into wells at a contaminated site at in Marcus Hook, PA (Raymond, Jamison, and Hudson, 1976). Addition of nutrients in this form accelerated the removal of contaminating gasoline.

There are many substances that would be suitable as fertilizers, and their compositions and origins differ considerably (Sims and Bass, 1984). The choice of an appropriate fertilizer can be complicated, and an agronomist should be consulted to develop a fertilization plan at a hazardous waste site. A plan may include types and amounts of nutrients, timing and frequency of application, and method of application. The nutrient status of the soil and the nutrient content of the wastes must be determined to formulate an appropriate fertilization plan.

An optimum fertilization program has been proposed (Kincannon, 1972). Chemicals are added so as to attain a slight excess of nitrogen, phosphorus, and potassium in the contaminated area. In addition, soil testing for ammonia and nitrates is conducted at regular monthly intervals. Small doses of ammonium nitrate are added, as needed, to maintain the ammonium or nitrate surplus. Urea is used as a nitrogen source to avoid the initial increase in soil salts, which may result from additions of other fertilizer stocks. Ammonium nitrate is subsequently applied, once urea is deemed no longer necessary. Potash is added as a potassium source.

An application method must also be selected. In agricultural application, fertilizers are either applied evenly over an area or concentrated at given points, such as banded along roots. However, at a hazardous waste site, fertilizer will likely be applied evenly over the whole contaminated area and incorporated by tilling, if necessary. Nutrients can also be injected through well points below the plow layer.

With broadcast fertilization, the fertilizer can be left on the surface or incorporated with a harrow (2 to 3 cm deep), a cultivator (4 to 6 cm deep), or a plow (a layer at bottom of furrow; e.g., 15 cm deep). The depth depends upon the solubility of the fertilizer and the desired point of contact in the soil. In general, nitrate fertilizers move freely, while ammonia nitrogen is adsorbed by soil colloids and moves little until converted to nitrate. Potassium is also adsorbed and moves little except in sandy soils. Phosphorus does not move in most soils. Therefore, potassium and phosphorus need to be applied or incorporated to the desired point of use.

Site geology, in most cases, will determine the methods of fertilization to be used in a given situation (Raymond, Jamison, and Hudson, 1976). For example, in a fractured dolomite and clay formation, lack of homogeneity makes well injection and distribution of nutrients difficult. Use of diammonium phosphate could result in excessive precipitation, and nutrient solution containing sodium could cause dispersion of the clays, thereby, reducing permeability (Environmental Protection Agency, 1985b). High calcium could cause precipitation of added phosphate, rendering it unavailable to microbial metabolism. If a site is likely to encounter problems with precipitation, iron and manganese addition may not be desirable. If the total dissolved solids content in the water is extremely high, it may be desirable to add as little extra salts as possible.

Earthworms also contribute to the degradation of organic materials in soil (Hornick, 1983). As a result of their movement, they carry nutrients to deeper soils and improve soil aeration. Macrofauna, in general, play an essential role in the decomposition of wastes, and the addition of materials to the soil that are toxic to these organisms can alter the rate of decomposition.

Results of oil biodegradation in Marcus Hook, PA, and Corpus Christi, TX, indicated that fertilizer was not a factor in biodegradation until approximately 50 percent of the oil had been degraded (Raymond, Hudson, and Jamison, 1976). However, other environmental factors may have affected these studies, and this cannot be regarded as conclusive.

A large kerosene spill (1.9 million l) in New Jersey was cleaned up with enhancement (Dibble and Bartha, 1979c). Much of the kerosene was recovered by physical means and by removing 200 m^3 of contaminated soil. Following stimulation of microbial degradation by liming, fertilization, and tillage, phytotoxicity was reduced.

Addition of nitrogen and phosphorus fertilizer at another site resulted in a doubling of the oil biodegradation rate of 70 bbl/acre/month to as much as 1.0 lb/ft^3/month (Kincannon, 1972). It is recommended that monthly determinations of nitrogen and phosphorus levels in the soil and periodic fertilizer application, when necessary, will optimize the fertilization process. The cost of soil disposal of oily wastes was estimated at $3.00/bbl. Degradation rates of up to 100 bbl/acre/month were reported, when the oil was applied to fertilized soils (Francke and Clark, 1974).

5.1.2.6 Organic Matter

Organic matter is generally an amorphous organic residual in soils, which, when present in sufficient amounts, has a beneficial effect on the physical and chemical properties of the soil because of its high cation exchange capacity (the total amount of cations held exchangeably by a unit mass or weight of a soil), high specific or reactive surface, and large amounts of exchangeable bases (Hornick, 1983). Humic substances (humic acids, fulvic acids, and humins; National Academy of Sciences, 1977) constitute 60 to 80 percent of the total organic content of most groundwaters and sediments (Khan, 1980). These organics tend to be recalcitrant to degradation. The crude humin consists of humic acid and hymatomelanic acids containing functional carboxyl and phenolic hydroxyl groups responsible for exchange and adsorption reactions (Hornick, 1983). Both humates and fulvates show a high degree of reactivity due to their acidic functional groups. The reaction of these materials with cations in the soil solution is strongly pH dependent.

Other organic materials involved in metal reactions and complexation in soils are plant root exudates and various degradation products, which can serve as the base for the humic fraction of the soil (Hornick, 1983). Easily decomposed organic contaminants can become part of an important soil process and result in a substantial increase in beneficial organic materials. It is likely that maintaining a supply of biodegradable organic matter in site soils would allow a higher population of diverse microbes capable of degrading many kinds of toxic organic compounds (Kaufman, 1983).

Humus increases the water-holding capacity of soil by swelling when wet to absorb two to three times its weight in water (Hornick, 1983). Because of its surface area, surface properties, and functional groups, humified soil can serve as a buffer, an ion exchanger, a surfactant, a chelating agent, and a general sorbent to help in the attenuation of hazardous compounds in soils (Ahlrichs, 1972). Enzyme activities of soil organisms can be responsible for coupling xenobiotic compounds and their breakdown products to soil humic materials (Bollag, 1983). Bound hazardous organic compounds, including toxic metabolites, should be monitored. Humus-bound xenobiotic compounds may be slow to mineralize or be transformed to innocuous forms (Khan, 1982), and they may become an integral part of the soil matrix. In these cases, the humic content of the soil should probably not be increased. Hazardous materials bound to humus might be released by microbial action and be subject to leaching, volatilization, or reattachment to soil organic matter. This suggests that treatment is not complete until it can be demonstrated that these compounds are absent or at a safe level in the soil (Morozov and Nikolayov, 1978).

Natural organic matter can be added to the soil, such as in the form of synthetic commercial organics, cattle manure, sewage sludge, or crop residues (Arthur D. Little, Inc., 1976). Commercial synthetic organics are expensive and their suitability for microbial growth is uncertain. Sewage sludge and cattle manure are the least expensive supplements; however, their use is limited since they contain variable quantities of trace elements that may disturb the expected soil mechanisms for degradation. They also contain populations of organisms, which, although they are usually enteric and do not survive long in the soil, may represent enough competition to slow the build-up of the desired soil microorganisms. Eight tons/acre of alfalfa meal has been shown to be as effective in stimulating microorganisms as 80 tons/acre of cattle manure. Considerable energy source is removed in the digestive tract of the cattle. The addition of organic wastes, such as animal manure and sewage sludge compost, decreases soil bulk density and increases infiltration and permeability, since organic wastes tend to increase soil aggregation and porosity (Hornick, Murray, and Chaney, 1979).

Mixed results have been obtained by different researchers with using manure amendments to increase the rate of degradation of organic chemicals. While some workers reported that manure increased the rate of degradation of ten organic chemicals tested, Doyle (1979) found that manure did not significantly reduce the degradation of any chemical examined. The breakdown of several compounds was positively correlated with the increased total microbial activity of manure-amended soil. Sewage sludge, however, enhanced the breakdown of only two compounds, while decreasing the rate of degradation of nine others.

Some advantages of using municipal sludges in organic waste treatment are that they contain active indigenous populations of microorganisms with degradative potential and they provide necessary nutrients for biodegradation (Sims and Bass, 1984). However, high levels of heavy metal will adversely affect this population.

Nonspecific, readily biodegradable organic matter should be added and mixed into the soil as dry materials or as slurries (Sims and Bass, 1984). Straw has been added to soils to increase adsorption of s-triazine herbicides (Walker and Crawford, 1968). Fungal mycelium and baker's yeast also improved soil sorption, with nonliving cells exhibiting greater sorption capacity than living cells (Shin, Chodan, and Wolcott, 1970; Voerman and Tammes, 1969). The soil moisture level should be optimized when adding organic matter, and frequent mixing is required to maintain aerobic conditions (Sims and Bass, 1984). Controls to manage the run-on and run-off from the site, as a result of tillage, are necessary to prevent drainage and erosion problems (Kowalenko, 1978). Retreatment may be necessary at intervals as nutrients are used up. The potential achievable level of treatment ranges from low to high, depending upon the solubility, sorption, and biodegradability of the organic constituents.

Hazardous constituents may be initially bound to organic materials, but later released as organic materials decompose (Sims and Bass, 1984). The formation of organo-metal complexes through the organic matter chelation of metals is an important factor governing metal availability (Schnitzer and Khan, 1978). In waste-amended soils, the addition of high amounts of organic matter ensures a predominance of organic matter reactions. The mobility of heavy metals added by wastes is related to the organic matter content of soils, pH, hydrous oxide reactions, and the oxidation-reduction or redox potential of a soil.

Addition of organic material and maintenance of aerobic conditions can result in the oxidation of arsenite to arsenate (Sims and Bass, 1984). Further treatment with ferrous sulfate will form highly insoluble $FeAsSO_4$. Anaerobic conditions must be avoided with this technology to prevent the reduction and methylation of arsenic to volatile forms, although anaerobic microsites can probably not be completely avoided even in carefully managed soils.

It is generally accepted that subsurface microbes are oligotrophic (Wilson, McNabb, Balkwill, and Ghiorse, 1983); however, in one study, carbon (cellulose) enrichment had little effect on mineralization of phenol in any soil examined (Thornton-Manning, Jones, and Federle, 1987). This response could have been due to the recalcitrance of the added carbon source or inorganic nutrient limitation. The most extensive mineralization occurred in a surface soil, which had the lowest content of organic matter.

It may be necessary to conduct laboratory experiments to determine the biochemical fate of given hazardous compounds in organically enriched soil or compost, (Kaplan and Kaplan, 1982).

Terrestrial oil spillages will probably result in the death of plants, releasing large amounts of nonhydrocarbon organic matter into soil, which might serve as an alternate source of carbon for heterotrophic microorganisms, thereby, interfering with the degradation of the contaminants (Atlas, 1977).

5.1.2.7 Oxidation-Reduction Potential

Table A.5-8 shows a succession of events in development of anaerobic conditions, which can occur in water-logged soils or poorly drained soils receiving excessive loadings of organic chemical wastes or crop residues.

Table A.5-8.   Succession of Events Related to the Redox Potential (JRB and Associates, Inc., 1984b; Takai and Kamura, 1966)

Period of Incubation	System	Redox Potential (mv)	Nature of Microbial Metabolism	Formation of Organic Acids
Early	Disappearance of O_2	+ 600 to + 400	Aerobes	None
	Disappearance of NO^{-3}	+ 500 to + 300	Facultative	Some accumulation
	Formation of Mn^{+2}	+ 400 to + 200	anaerobes	after addition of
	Formation of Fe^{+2}	+ 300 to + 100		organic matter

Table A.5-8.   continued

Period of Incubation	System	Redox Potential (mv)	Nature of Microbial Metabolism	Formation of Organic Acids
Later	Formation of S^{-2}	0 to -150	Obligate anaerobes	Rapid accumulation
	Formation of H_2	-150 to -220		Rapid decrease
	Formation of CH_4	-150 to -220		

Oxygen levels in aquatic surface and subsurface environments can also be expressed in terms of the logarithm of the electron concentration pe (Bitton and Gerba, 1985). Values for Eh and pe for various microbiological processes are (at 25°C and pH 7):

Process	pe	Eh (in mV)
Aerobic respiration	+13.75	+810
Denitrification	+12.65	+750
Sulfate reduction	-3.75	-220
Methane formation	-4.13	-240

The following classification of oxygen levels in soils, based upon their redox potential at pH 7, has been proposed (Patrick and Mahapatra, 1968):

Soil Type	Redox Potential
Oxidized soil	>400 mV
Moderately reduced soil	100 to 400 mV
Reduced soils	-100 to 100 mV
Highly reduced soils	-300 to -100 mV

It may be feasible in some cases to enhance reducing conditions intentionally in the subsurface, thereby, lowering the redox potential (Environmental Protection Agency, 1985b). The pH can be adjusted with the addition of dilute acids or bases.

5.1.2.8 Attenuation

The mixing of indigenous soil layers, or the addition of exogenous soil to contaminated soil, is a means of increasing the extent of immobilization of chemical contaminants at hazardous waste sites (Sims and Bass, 1984).   This may also aid in decreasing toxicity of the contaminated soil to soil microorganisms due to high concentrations of constituents, to bring the concentrations to levels that can be successfully biodegraded.

This treatment is applicable to all organic wastes. However, organics that are very soluble in water may be more effectively treated by other methods, since large amounts of soil may be required to reduce the mobility of the compound. If very toxic components are present in the waste, destructive treatment would be the preferable treatment alternative.

The indigenous soil profile is tilled to mix uncontaminated soil with the contaminated layers, importing soil or clay, if necessary (Sims and Bass, 1984). The ease of use of this method depends upon site/soil trafficability considerations for tillage and incorporation of added material. Tillage may cause erosion. The level of attenuation achievable is potentially high with suitable size, soil, and waste characteristics. The mixing of new material may alter the properties of the natural soil; thus, the effectiveness of this may vary for different compounds and may not be as expected. However, this method should be reliable under most conditions. There is limited field experience in this technology.

5.1.2.9 Texture and Structure

5.1.2.9.1 Soil Texture. Soil composition influences infiltration rate and permeability, water holding capacity, and adsorption capacity for various waste components (Hornick, 1983). Clay soils have a greater capacity for physicochemical attenuation of contaminants than coarse sands or fissured rocks (Pye and Patrick, 1983). A predominance of clay and silt particles in finer textured soils results in a very small pore size, with a slow infiltration rate of water (Hornick, 1983). The presence of montmorillonite, with high shrink-swell tendencies, would cause swelling of the soil with added moisture or water and block any further water movement.   Run-off or flooding could then occur, and anaerobic conditions would be induced. Coarse soils of sand and gravel have large interconnecting pores and allow rapid water movement. However, if such a site is excessively drained, nutrients in added material will move too rapidly to be sufficiently adsorbed on the soil. The groundwater can be contaminated, if there is no restrictive layer between the coarse layer and the water table.

5.1.2.9.2 Bulk Density. This is a measure of dry soil weight per unit volume and determines pore space through which water can move (Hornick, 1983). The frequent use of heavy machinery either to work the soil or apply wastes compacts the soil and, thus, increases the bulk density.

5.1.2.9.3 Water-Holding Capacity. This capacity is directly related to the soil's bulk density and texture (Hornick, 1983). Soils with very fine or very coarse textures or high bulk densities can not maintain an adequate supply of water: the water content determines available oxygen, redox potential, and microbial activity of a soil system.

5.1.3 Alteration of Organic Contaminants

5.1.3.1 Addition of Surfactants

Spontaneous dissolution rates are important factors affecting the rates of biodegradation (Thomas, Yordy, Amador, and Alexander, 1986). Growth of pure cultures of bacteria on naphthalene, phenanthrene, and anthracene is fastest on the solid substrates with the highest water solubilities (Wodzinski and Johnson, 1968). The rate of dissolution of compounds, such as naphthalene, is directly related to their surface areas (Thomas, Yordy, Amador, and Alexander, 1986), and increasing the surface area of hexadecane increases its microbial destruction (Fogel, Lancione, Sewall, and Boethling, 1985). Degradation of polychlorinated biphenyls by a Pseudomonas sp. is enhanced when the surface area of the substrate is increased by emulsification (Liu, 1980). In fact, the first step in hydrocarbon assimilation by Candida lipolytica is the microbial enhancement of the solubilization of the substrate (Goma, Pareilleux, and Durand, 1974).

This principle can be utilized to improve the degradation of organic compounds in contaminated sites. If it is inconvenient, expensive, or too time consuming to supply oxygen or other electron acceptors to organisms metabolizing oily contaminants in situ, it might be possible for the subsurface organisms to emulsify the hydrocarbons, if conditions are suitable (Wilson, Leach, Henson, and Jones, 1986). Use of bioemulsification of oils for microbial enhancement of oil recovery from petroleum reservoirs (Cooper, 1982) should be directly applicable to petroleum product spills. Bacteria from a well contaminated by a spill of JP-5 jet fuel could emulsify the fuel, if the well water was supplemented with phosphate and nitrate (Ehrlich, Schroeder, and

Martin, 1985). The surfactants not only emulsify the hydrocarbons, but also aid in mobilizing them through soil and water (Vanloocke, Verlinde, Verstraete, and DeBurger, 1979). In favorable geological situations, the mobile emulsions could be removed by pumping for treatment on the surface (Ehrlich, Schroeder, and Martin, 1985).

Detergents are able to increase microbial membrane permeability (Gloxhuber, 1974), and substances from humic acids may have the same effect (Visser, 1982). The addition of humic products to a culture medium resulted in a 2000-fold increase in growth (Visser, 1985). These substances appear to induce a change in metabolism, allowing the organisms to proliferate on substrates they could not previously utilize. Tween 20-80 and Brij 35 increased microbial ATP levels, possibly as a result of an increased metabolic rate with the greater amount of nutrients; however, the mechanisms involved with the humic material have not yet been elucidated. It has been recommended that humic products be incorporated in media for determination of microbial activities in terrestrial and aquatic environments. Humic substances constitute the major part of the natural organic constituents of most waters and sediments, typically forming 60 to 80 percent of the total organic content (Khan, 1980).

The removal of petroleum hydrocarbons from soil was improved by orders of magnitude by use of a 2 percent aqueous solution of Adsee 799 (Witco Chemical) and Hyonic NP-90 (Diamond Shamrock) rather than just water washing (Ellis, Payne, and McNabb, 1985). This combination has adequate solubility in water, minimal mobilization of clay-sized soil fines (to maintain soil permeability), good oil dispersion, and adequate biodegradability. It is potentially useful for in situ cleanup of hydrophobic and slightly hydrophilic organic contaminants in soil. Removal efficiency of the latter would be significantly improved by use of aqueous surfactants in soils with high TOC (total organic carbon) values.

It may even be possible to add or select for organisms that produce emulsifiers. Surface active agents are excreted into the aqueous medium when certain organisms are grown on liquid hydrocarbons, particularly n-alkanes (Zajic and Gerson, 1977). When exposed to hydrocarbons, the lipid content of the cell wall increases significantly, which increases the affinity of the microbe for the hydrocarbon (Blanch and Einsele, 1973). The surface active agents are mainly lipids, lipoproteins, or sugar-lipid complexes, which reduce the interfacial tension between the hydrocarbon and the aqueous medium. Emulsions formed are stabilized by the polysaccharide polymers secreted extracellularly by the microbes. These can be then absorbed through the lipophilic cell wall to be utilized by the microorganism as a carbon and energy source.

A strain of <u>Corynebacterium</u> sp., isolated from sewage sludge, was grown in mineral salts medium with hexadecane (3.0 percent v/v) as a carbon and energy source (Panchal, Zajic, and Gerson, 1979). Both hydrophobic and hydrophilic emulsifiers were isolated from the same culture broth. The lipid extract was a very potent emulsifying agent, while the polysaccharide was very weak, unless used at a high concentration and in combination with Tween 20.

5.1.3.2 Photolysis

The major photoreaction taking place with pesticides in the atmosphere is oxidation (Crosby, 1971) involving the OH° radical or ozone, of which the OH° radical is the species of greatest reactivity (Lemaire, Campbell, Hulpke, Guth, Merz, Philop, and Von Waldow, 1982). Based on a first order rate of reaction of vapor phase reactions with the OH° radical, the half-life of a specific chemical species can be estimated, if its OH° radical reaction rate constant is known using:

$$t_{1/2} = 0.693/k_{OH}°[OH°])$$

where:

$t_{1/2}$ = time to decrease component concentration by 50%

$k_{OH}°$ = OH radical reaction rate constant (cm^3/molecule)

$[OH°]$ = atmospheric OH radical concentration
(4×10^5 molecules/cm^3)

= 6645×10^{-19} moles /cm^3)

Table A.5-9 presents the OH° radical reaction rate constants for various organic compounds (Sims and Bass, 1984). The higher the number, the faster the oxidation of the compounds.

In order to assess the potential for use of photodegradation, the specific compound's atmospheric reaction rate (log $K_{OH}°$) and the anticipated reaction products must be known (Crosby, 1971). If a compound is poorly photoreactive (e.g., a $t_{1/2}$ in the atmosphere greater than 1 day) volatilization suppression may be required to maintain safe ambient air concentrations at the site.

Table A.5-9.   Rate Constants for the Hydroxide Radical Reaction in Air with Various
Organic Substances (Sims and Bass, 1984)

$k_{OH}{}^{o}$ in Units of (Mole-sec)$^{-1}$

Substance	$\log_{air} k_{OH}{}^{o}$
Acetaldehyde	9.98
Acrolein	10.42
Acrylonitrile	9.08
Allyl chloride	10.23
Benzene	8.95
Benzyl chloride	9.26
Bis(chloromethyl)ether	9.38
Carbon tetrachloride	<5.78
Chlorobenzene	8.38
Chloroform	7.78
Chloromethyl methyl ether	9.26
Chloroprene	10.44
o-,m-,p-cresol	10.52
p-cresol	10.49
Dichlorobromobenzene	8.26
Diethyl ether	9.73
Dimethyl nitrosamine	10.37
Dioxane	9.26
Epichlorohydrin	9.08
1,2-Epoxybutane	9.16
Epoxypropane	8.89
Ethanol	9.28
Ethyl acetate	9.06
Ethyl propionate	9.03
Ethylene dibromide	8.18
Ethylene dichloride	8.12
Ethylene oxide	9.08
Formaldehyde	9.78
Hexachlorocyclopentadiene	10.55
Maleic anhydride	10.56
Methanol	8.78
Methyl acetate	8.04
Methyl chloroform	6.86
Methyl ethyl ketone	9.32
Methylene chloride	7.93
Methyl propionate	8.23
Nitrobenzene	7.56
Nitromethane	8.81
2-Nitropropane	10.52
n-Nitrosodiethylamine	10.19
Nitrosoethylurea	9.89
n-Propylacetate	9.41
Perchloroethylene	8.01
Phenol	10.01
Phosgene	nonreactive
Polychlorinated biphenyls	<8.78

Table A.5-9.   continued

k_{OHo} in Units of (Mole-sec)$^{-1}$	
Substance	$\log_{air} k_{OHo}$
Propanol	9.51
Propylene oxide	8.89
Tetrahydrofuran	9.95
Toluene	9.52, 9.56
Trichloroethylene	9.12
Vinylidene chloride	9.38
o-,m-,p-xylene	9.98

Source: Adopted from (Lemaire, Campbell, Hulpke, Guth, Merz, Philop, and Von Waldow, 1982 and Cupitt, 1980)

Groups that typically do not undergo direct photolysis include saturated aliphatics, alcohols, ethers, and amines (Sims and Bass, 1984). Photodegradable organic wastes generally include compounds with moderate to strong absorption in the >290-nm wavelength range. Such compounds generally have an extended conjugated hydrocarbon system or a group with an unsaturated hetero atom (e.g., carbonyl, azo, nitro). Enhanced photodegradation of soil contaminants may be accomplished through the addition of various proton donor materials to the contaminated soils.

Table A.5-10 presents additional constants, with an estimation of the likelihood of a photolysis reaction occurring within the ambient atmosphere (Sims and Bass, 1984).

5.1.3.3 Supplementing Threshold Concentrations of Contaminants

Table A.5-10. Atmospheric Reaction Rates and Residence Times of Selected Organic Chemicals (Sims and Bass, 1984)

Compound	$^k OH \times 10^{12}$ (cm^3/molecule/ sec^{-1})	Direct Photolysis Probability	Physical Removal Probability	Residence Time (days)	Anticipated Photoproducts
Acetaldehyde	16	Probable	Unlikely	0.03-0.7	H_2CO, CO_2
Acrolein	44	Probable	Unlikely	0.2	$OCH-CHO, H_2CO$ $HCOOH, CO_2$

Table A.5-10. continued

Compound	$^kOH \times 10^{12}$ (cm^3/molecule/ sec^{-1})	Direct Photolysis Probability	Physical Removal Probability	Residence Time (days)	Anticipated Photoproducts
Acrylonitrile	2	-	Unlikely	5.6	H_2CO, CN^O, $HC(O)CN, HCOOH$
Carbon Tetra-chloride	<0.001	-	Unlikely	>11,000	Cl_2CO, Cl^O
o-,m-,p-cresol	55	-	Unlikely	0.2	Hydroxynitro-toluenes, ring cleavage products
Formaldehyde	10	Probable	Unlikely	0.1-1.2	CO, CO_2
Nitrobenzene	0.06	Possible	Unlikely	190	Nitrophenols, ring cleavage products
2-Nitropropane	55	Possible	Unlikely	0.2	H_2CO, CH_3CHO
Phenol	17	-	Possible	0.6	Dihydroxyben-zenes, nitro-phenols, ring cleavage products
Toluene	6	-	Unlikely	1.9	Benzaldehyde, resols, ring cleavage products, nitro compounds
o-,m-,p-xylene	16	-	Unlikely	0.7	Substituted benzaldehydes, hydroxy xylenes, ring cleavage products, nitro compounds

Source: (Cupitt, 1980)

5.2 OPTIMIZATION OF GROUNDWATER BIODEGRADATION

Some of the information in this section may duplicate material covered in Section 5.1, Optimization of Soil Biodegradation, and Section

5.3, Optimization of Freshwater, Estaurine, and Marine Biodegradation; however, it is presented here under a separate heading, with other related information, to accommodate those readers who may specifically wish to address treatment of groundwater contamination only.

5.2.1 Biological Enhancement

5.2.1.1 Seeding of Microorganisms

5.2.1.2 Acclimation

Naphthalene could be degraded rapidly in aerobic groundwaters contaminated by PAHs (Lee and Ward, 1984b) or in groundwater near oil and gas beds (Slavnia, 1965). It was biodegraded in an aquifer recharged with reclaimed water from wastewater treatment after an initial lag (Roberts, McCarty, Reinhard, and Schreiner, 1980). These results demonstrate the importance of acclimation of the organisms to the contaminant.

It is possible that adaptation requirements may be different for mineralization and degradation (Swindoll, Aelion, Dobbins, Jiang, Long, and Pfaender, 1988). A microbial community isolated from fine sand aquifer subsoil about 5 ft below the surface was capable of biodegrading xenobiotics. The community appeared to be preadapted to utilization of some of the chemicals. Phenol was initially mineralized at a rapid rate, which leveled off such that a maximum percent respiration of the compound was reached, usually within weeks. While m-cresol, m-aminophenol, and aniline exhibited an initially slow degradation rate, which was maintained throughout the incubation period (7 months), a linear increase in percent mineralized occurred with time. Only p-nitrophenol was found to require an adaptation period, after which the rate of degradation was very rapid.

5.2.2 Optimization of Groundwater Factors

5.2.2.1 Temperature

Figure A.5-3 shows typical groundwater temperatures throughout the United States (Environmental Protection Agency, 1985b).

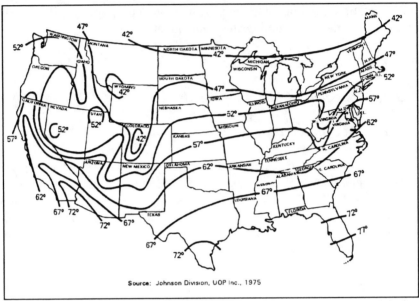

Figure A.5-3. Typical Groundwater Temperatures (° at 100-ft. Depth) in the United States (Environmental Protection Agency, 1985b)

5.2.2.2 Oxygen Supply

Oxygen may be especially limiting in areas of high oil concentration or when oil sinks into sediments (ZoBell, 1969). Oxygen concentration was 3 to 5 mg/l in samples from wells outside the area of jet-fuel contamination, but was zero in wells containing jet fuel (Ehrlich, Schroeder, and Martin, 1985). Evidently, aerobic bacteria using jet fuel as an energy source rapidly consume all the available oxygen. The absence of nitrate and oxygen in the groundwater contaminated by jet fuel suggests that <u>in situ</u> degradation might be enhanced, if additional oxygen and inorganic nitrogen are available. In fact, degradation of a plume of contaminated groundwater consisting of naphthalene, 1-methylnaphthalene, 2-methylnaphthalene, dibenzofuran, and fluorene (1000 to 100 mg/l) was controlled, not by the rate of utilization of the pollutants by the microorganisms, but by the extent of utilization allowed by the supply of oxygen (Wilson, McNabb, Cochran, Wang, Tomson, and Bedient, 1985).

The solubility of benzene (1780 mg/l) is much greater than the capacity for its aerobic degradation in groundwater (Wilson, Leach, Henson, and Jones, 1986). The concentration of the contaminants is an important factor. Concentrated plumes cannot be degraded aerobically until dispersion or other processes dilute the plume with oxygenated water. High concentrations of organic contaminants in the groundwater will deplete the oxygen and aerobic metabolism will stop. Anaerobic biotransformations may take over, but the hydrocarbon degradation would be very slow.

Air can be added to extracted groundwater before reinjection, or it can be injected directly into the aquifer (Environmental Protection Agency, 1985b). The first method, in-line aeration, involves adding air into the pipeline and mixing it with a static mixer to provide a maximum of 10 mg/l oxygen. This concentration will degrade about 5 mg/l hydrocarbons and would, therefore, provide an inadequate oxygen supply. A pressurized line can increase oxygen concentrations, as can the use of pure oxygen.

The equilibrium oxygen concentration in water increases with increased air pressure according to Henry's Law:

where:

$$C_L = PH_k$$

C_L = concentration of oxygen in liquid (mg/l)

= volume fraction (0.21 for O_2 in air)

P = air pressure (atm)

H_k = Henry's Law Constant for oxygen

The value of Henry's Law Constant is 43.8 mg/l/atm at 68°F (20°C). Pressure increases with groundwater depth at the rate of 0.0294 atm/ft^2.

The use of in situ aeration wells is a more suitable method for injecting air into contaminated leachate plumes (Environmental Protection Agency, 1985b). In-line or in situ oxygenation systems can achieve higher oxygen solubilities and more efficient oxygen transfer to the microorganisms than conventional aeration. A bank of aeration wells can be installed to provide a zone of continuous aeration through which the contaminated groundwater would flow. Oxygen saturation conditions can be maintained for degrading organics during the residence time of groundwater flow through the aerated zone. The required time for aeration can be derived from bench-scale studies. Residence time (t_r through the aerated zone can be calculated from Darcy's equation using groundwater elevations and hydraulic conductivity, as follows:

$$t_r = (L_a)^2/K(h_1-h_2)$$

where:

t_r = residence time (sec)

K = hydraulic conductivity (ft/sec)

L_a = length of aerated zone (ft)

h_1 = groundwater elevation at beginning of aerated zone (ft)

h_2 = groundwater elevation at end of aerated zone (ft)

Solubilities of oxygen in various liquids are four to five times higher under pure oxygen systems than with conventional aeration (Environmental Protection Agency, 1985b). In-line injection of pure oxygen can impart 40 to 50 ppm of dissolved oxygen to water (Brown, Norris, and Raymond, 1984), which will provide sufficient dissolved oxygen to degrade 20 to 30 mg/l of organic material, assuming 50 percent cell conversion. This oxygen will not be consumed immediately, as the oxygen is from aeration. However, pure oxygen is expensive to use and the oxygen is likely to bubble out of solution (degas) before the microbes can utilize it.

There are various methods for injecting air or pure oxygen (Environmental Protection Agency, 1985b). These include use of pumps, propellers, stirrers, spargers, sprayers, and cascades (Texas Research Institute, Inc., 1982). Air can be sparged into wells using diffusers (e.g., diffusers attached to paint sprayer-type compressors that can deliver about 2.5 ft³/min) (Environmental Protection Agency, 1985b). A blower can also be used to provide the flow rate and pressure for aeration, such as 5 psi pressure in a 10-ft aeration well, with an air flow of 5 ft³/min. The solubility of oxygen is very low, approximately 8 mg/l at groundwater temperatures (Wilson and Rees, 1985). Diffusers that sparge compressed air into the groundwater cannot exceed the solubility of oxygen in water (Lee and Ward, 1986). A newly developed method that holds great promise for introducing oxygen to the subsurface is microdispersion of air in water using colloidal gas aprons (CGA), which creates bubbles 25 to 50 um in diameter (Environmental Protection

Agency, 1985b). With selected surfactants, dispersions of CGA's can be generated containing 65 percent air by volume.

An in situ aeration well zone must be wide enough to allow the total plume to pass through, and there must be sufficient air to generate a substantial radius of aeration without causing an air barrier to the flow of groundwater (Environmental Protection Agency, 1985b).

Hydrogen peroxide, which decomposes to form water and oxygen, can supply much greater oxygen levels (Lee and Ward, 1986) and has been used successfully to clean up several spill sites (Environmental Protection Agency, 1985b). Advantages of hydrogen peroxide include:

1. Greater oxygen concentrations can be delivered to the subsurface. Only 100 mg/l H_2O_2 provides 50 mg/l oxygen.
2. Less equipment is required to oxygenate the subsurface. Hydrogen peroxide can be added in-line along with the nutrient solution. Aeration wells are not necessary.
3. Hydrogen peroxide keeps the well free of heavy biogrowth. Such growth and clogging can be a problem in air injection systems.

Although hydrogen peroxide may be cytotoxic at levels as low as 200 ppm, it can be added to acclimated cultures at up to 1000 ppm (Environmental Protection Agency, 1985b). The remediation at Grange, IN, involved adding an initial concentration of 100 ppm, and increasing it to 500 ppm over the course of the treatment. Although the compound may degas above 100 ppm and form air bubbles that block the pores in the aquifer, it may be possible to overcome this limitation by stabilizing the hydrogen peroxide solution (Lee and Ward, 1986).

There is field evidence for enhanced degradation with the use of hydrogen peroxide (Yaniga and Smith, 1984). During air sparging with 100 ppm hydrogen peroxide, the dissolved oxygen concentrations in monitoring wells at a site contaminated by gasoline increased from 4 to 10 ppm. This was accompanied by an increase in the numbers of gasoline-utilizing organisms and a reduction in the size of the gasoline plume and a decrease from 4 to 2.5 ppm hydrocarbon. However, other restoration measures were concurrently being employed.

Ozone is also used for disinfection and chemical oxidation of organics in water and wastewater treatment (Environmental Protection Agency, 1985b), but ozone has the same limitations as hydrogen peroxide (Lee and Ward, 1985). In commercially available ozone-from-air generators, ozone is produced at a concentration of 1 to 2 percent in air (Environmental Protection Agency, 1985b). In bioreclamation, this ozone-in-air mixture could be contacted with pumped leachate using in-line injection and static mixing or using a bubble contact tank. A dosage of 1 to 3 mg/l of ozone can be used to attain chemical oxidation. However, the dosage should not be greater than 1 mg/l of ozone per mg/l total organic carbon; higher concentrations may be deleterious to the microorganisms. At many sites, this may limit the use of ozone as a pretreatment method to oxidize refractory organics, making them more amenable to biological oxidation.

In a hydrocarbon-contaminated area in West Germany, the water was pumped out, treated with ozone, and recirculated to the aquifer via injection wells (Lee and Ward, 1985; Environmental Protection Agency, 1985b). About 1 g of ozone per gram of dissolved organic carbon was added to the groundwater, with a contact time of 4 min in the above ground reactor. This increased the dissolved oxygen levels to 9 mg/l, with a residual of 0.1 to 0.2 g of ozone per cubic meter in the treated water. The microbial counts subsequently increased in the wells, with a decrease in dissolved organic carbon and mineral oil hydrocarbons. The ozone may have also reacted with the hydrocarbon for partial destruction of the organics. Oxidizing the subsurface could result in the precipitation of iron and manganese oxides and hydroxides. If this is extensive, the delivery system and possibly even the aquifer could become clogged.

Section 5.1.2.4 of Appendix A provides an in-depth review of the use of ozone and hydrogen peroxide for subsurface aeration. It describes problems associated with using these oxidants and suggests ways to deal with the problems. One such process, called the Vyrodex method, has been developed specifically to reduce the levels of iron and manganese in groundwater. This is accomplished by the addition of iron and manganese bacteria, and stimulation with dissolved oxygen, to prevent the heavy metals from decomposing the hydrogen peroxide before it can reach its intended site (Knox, Canter, Kincannon, Stover, and Ward, 1968).

In cases where the extent of the pollution is large or the water table extends to a depth where physical removal of contaminated material is totally impractical, alternative methods are used (Wilson, Leach, Henson, and Jones, 1986). One of these methods is construction of one or a series of surface infiltration galleries. These galleries take water that has been treated and recirculate it back through the contaminated unsaturated zone. Oxygen is generally added to the infiltrated water during an in-line stripping process for volatile organic contaminants or through aeration devices placed in the infiltration galleries. Recirculation of the water also facilitates movement of contaminants to the recovery well.

The dislodged or solubilized contaminant can be treated in a surface treatment system before the water is reinjected. Controlling the rate of groundwater flow is critical to moving oxygen and nutrients to the contaminated zone and optimizing the degradation process. Silty or shaley materials accept water very slowly. Aquifer flow rates should be sufficiently high so that the aquifer is flushed several times over the period of operation (e.g., twice a year over a three-year treatment period) (Environmental Protection Agency, 1985b). The operating period will depend upon the biodegradation rate of the contaminants in the plume and the amount of recycle. If the period of operation is excessively long, for example, more than five years, the operating costs of bioreclamation may outweigh the capital costs of another remedial alternative.

Air sparging in injection wells, in conjunction with nutrient addition, appeared to be a contributing factor in removing free gasoline from the groundwater in a cleanup operation in Millville, NJ (Raymond, Jamison, Hudson, Mitchell, and Farmer, 1978). However, dissolved oxygen measurements from surrounding wells did not reflect the additional oxygen input. This approach attacks the oxygenation problem only from the groundwater perspective. Residual gasoline in the soil above the groundwater will continue to leach into the groundwater with time.

5.2.2.3 Nutrients

5.2.2.4 Oxidation-Reduction Potential

Nitrate respiration may be a feasible approach to decontaminating an aquifer (Environmental Protection Agency, 1985b). Denitrification (the reduction of NO_3 to NH_3 or N_2) has been demonstrated to occur in contaminated aquifers. Nitrate respiration was used successfully in the treatment of an aquifer contaminated with aromatic and aliphatic hydrocarbons. Nitrate can be added in-line along with other nutrients and intimate mixing with groundwater can occur. The cost is moderate; only the nutrient feed system and an in-line mixer are required.

Nitrate, however, is a pollutant, limited to 10 ppm in drinking water. Consequently, it may be more difficult to obtain permits for use of nitrate than for oxygen or hydrogen peroxide. Also, degradation rates under anaerobic conditions are not as rapid, and the substrate range is more limited. There is no reason why nitrate respiration would be a better treatment approach, given the amount of success that has been already demonstrated with aerobic processes.

Many compounds can be transformed under anaerobic conditions, but not aerobically; examples are chloroform, bromodichloromethane, dibromochloro-methane, bromoform, and 1,1,1-trichloroethane (McCarty, Rittmann, and Bouwer, 1984). Different redox conditions may also affect the transformation of a compound. For example, chloroform and 1,1,1-trichloroethylene can be degraded under methanogenic conditions but not under denitrification conditions (Bouwer and McCarty, 1983b; Bouwer and McCarty, 1983a).

5.2.3 Alteration of Organic Contaminants

5.2.3.1 Addition of Surfactants

5.3 OPTIMIZATION OF FRESHWATER, ESTAURINE, AND MARINE BIODEGRADATION

Some of the information in this section may duplicate material covered in Section 5.1, Optimization of Soil Biodegradation, and Section 5.2, Optimization of Groundwater Biodegradation; however, it is presented here under a separate heading, with other related information, to accommodate those readers who may specifically wish to address treatment of freshwater, estaurine, and marine contamination only.

5.3.1 Biological Enhancement

5.3.1.1 Seeding of Microorganisms

Mixed marine enrichment cultures have been investigated for treating spilled oil (Atlas and Bartha, 1972c). Mixed enrichments can no doubt degrade a highly complex substrate, such as petroleum, more effectively than any single microorganisms, but the practical use of an enrichment of unknown composition is likely to encounter licensing difficulties because of its potential side effects on marine life. It may be necessary to construct an effective mixed culture from known microorganisms with wide and complementing substrate ranges.

In aqueous environments, added seed organisms are likely to dissipate from the oiled area unless they are added in high enough concentrations and in a metabolic state so as to allow for immediate colonization of the oil. It has also been suggested that seed bacteria could be encapsulated to ensure that they adhere to and remain with the oil in aqueous environments (Azarowicz, 1973).

Several commercial mixtures of microorganisms for seeding have been tested and found to be ineffective for degradation of petroleum in the marine environment, although the literature supplied with the mixtures included claims of effectiveness for cleaning up ocean oil spills (Azarowicz, 1973). Even under optimal temperature, aeration, and nutrient conditions, such treatment of oil on seawater failed to stimulate petroleum biodegradation beyond the rate and extent carried out by indigenous microbial populations. The possibility of seeding a hydrocarbonoclastic bacterium into Arctic saline and freshwater ponds has been investigated, using an organism isolated from an estuarine environment (Spain, Milhous, and Bourquin, 1981). This organism stimulated biodegradation in a saline pond but not in a freshwater pond.

Seeding the salt water pond with an oil-degrading Pseudomonas sp., at the same time as adding fertilizer, resulted in the greatest breakdown of the material (Atlas and Busdosh, 1976). However, seeding of the freshwater pond did not increase degradation above that observed with the fertilizer alone. In both ponds, the oil-degrading population decreased one week after seeding. In the saline pond, the counts recovered and increased greatly over a three-week period. In the freshwater pond, the organism had disappeared after two weeks. Apparently, the seed organism was unable to survive in the freshwater ecosystem. This suggests that it is necessary to have a variety of seed organisms available for biodegradation in different environments.

The normal nutrient content of 100 ml of seawater will support less than 0.1 mg (dry wt) microbial biomass (Atlas and Bartha, 1972b). Any seeding operation should be connected with application of an appropriate fertilizer. In a nutritionally unfavorable environment, there would be little, if any, benefit from relying on inoculation alone. A hydrophobic binder may achieve the selective retention of the mineral supplements within the oil slick (see Section 5.3.2.3).

Since there are low concentrations of microorganisms in the ocean, seed organisms would have less competition with the indigenous population (Atlas, 1977). Predation is a limiting factor in this environment, to which the seed would also be subjected. Ciliates have been observed to strip yeast and bacterial cells from the surface of oil globules during a microbial bloom following an oil spill. Use of yeast strains that can grow within the oil globule offers additional protection from predators. Many marine isolates of Trichosporon sp. show this capability.

A mixture of hydrocarbonoclastic yeasts was seeded into an estuarine environment (Cook, Massey, and Ahearn, 1973). Only two of these yeasts were able to persist for a long period of time: Candida lipolytica lasted three to five months and C. subtropicalis lasted over one year in freshwater and seven months in estuarine environments. The Candida species that persisted were capable of degrading a wide range of alkanes and alkenes with no adverse ecological side effects and were not pathogenic.

A strain of Arthrobacter has been used for cleaning out oil residues from tanker holds (Rosenberg, Englander, Horowitz, and Gutnick, 1975). The dispersing agent produced by the bacterium was not toxic, but the resulting emulsion was. The toxicity (from polar metabolites from the hydrocarbons) could be reduced considerably by dialysis against seawater.

Encapsulation of seed bacteria may be used to ensure the organisms adhere to and remain with the oil in aqueous environments (Gholson, Guire, and Friede, 1972).

5.3.1.2 Acclimation

Natural microbial populations in water/sediment cores from all freshwater sites tested were able to adapt to more rapidly degrade organic substrates supplied at low concentrations (Spain, Milhous, and Bourquin, 1981). However, none of the estuarine or marine populations adapted. The extent of adaptation depended upon preexposure concentrations, but the relationship was not linear. Adaptation was maximal at 15 days after initial exposure and declined gradually until no longer detectable after 50 days. Adaptation periods were found to vary with substrate concentration and were inversely proportional to the concentration (Steen, Paris, and Latimer, 1981).

5.3.2 Optimization of Aquatic Factors

5.3.2.1 Temperature

The effect of temperature on the rate of degradation depends partly upon the compound. For instance, rates of napthalene and anthracene biodegradation in intertidal sediments increase with increasing temperature (Bauer and Capone, 1985); however, the rate for cresol mineralization in estuarine water does not depend upon temperature (Bartholomew and Pfaender, 1983). In studies on the degradation by a

Nocardia sp. of Bunker C, hexadecane, and a hydrocarbon mixture at $5°$ and $15°C$, a $10°C$ decrease in temperature resulted in a 2.2-fold decrease in generation time of the bacteria and a slower degradation rate (Mulkins-Phillips and Stewart, 1974b). The rate of natural biodegradation of oil in marine temperate-to-polar zones is probably limited by low temperatures and phosphorus concentrations.

5.3.2.2 Oxygen Supply

CFU formation of organisms from a lake in Antarctica with perpetual high dissolved oxygen (HDO) and that from a lake in Virginia with perpetual saturated atmospheric dissolved oxygen (ADO) were sensitive to HDO toxicity (Mikell, Parker, and Simmons, 1984). The former might have been due to inadequate nutrient supply, while the latter had optimum nutrients available. High oxygen uptake may involve metabolically produced intracellular toxic oxygen by-products, such as peroxide and superoxide. Isolates from the first lake were catalase positive, with inducible superoxide dismutase, and all were pigmented.

Petroleum hydrocarbon degradation markedly occurred in superficial marine sediments (0 to 1 cm) where the oxygen concentration was 8 ppm, whereas, such degradation was slower, but detectable in the system incorporating 2 to 3 ppm. Under anaerobic conditions, no degradation was detected (Bertrand, Esteves, Mulyono, and Mille, 1986).

5.3.2.3 Nutrients

Since physical removal or burning of accidentally spilled oil is seldom feasible, and dispersion or sinking may adversely affect marine life, artificially stimulated biodegradation is being considered as a possible alternative (Atlas and Bartha, 1972b). For this approach to be successful, it is essential that the biodegradation-limiting parameters in seawater should be properly identified.

Oil biodegradation in arctic coastal ponds was found to be nutrient-limited (Atlas and Busdosh, 1976). The addition of an oleophilic nitrogen and phosphorus fertilizer permitted degradation in a previously inactive freshwater pond by increasing the number of oil-degrading microorganisms. It also enhanced the degradation observed in a saltwater pond.

The phosphorous and nitrogen contents of 100 ml unsupplemented seawater would support less than 0.1 mg (dry wt) microbial biomass (Atlas and Bartha, 1972b). This appears to be a major cause for the slow rate of petroleum biodegradation in the sea, especially when vigorous circulation does not occur in the water column underneath the oil slick. Either nitrogen or phosphorous deficiency will tend to produce cells with abnormally high lipid stores and low metabolic activity and may also cause the accumulation of extracellular intermediary metabolites.

An oleophilic nitrogen and phosphorus fertilizer has been developed (Atlas and Bartha, 1973a) and tested for its ability to stimulate petroleum degradation by indigenous organisms in several environments) (Atlas and Schofield, 1975; Spain, Milhous, and Bourquin, 1981; Atlas, 1975a). The fertilizer contains paraffinized urea and octyl phosphate. (Optimal C/N and C/P ratios were 10:1 and 100:1, respectively) (Atlas and Bartha, 1973a). When it was tested in situ and in vitro in near shore areas and several ponds, it stimulated biodegradation of an oil slick by 30 to 40 percent. It did not have any harmful effects on the algae or invertebrate assay organisms.

Iron is also in low concentration in seawater (Atlas, 1977) and is potentially limiting for petroleum biodegradation in nitrogen and phosphorus-supplemented seawater (Dibble and Bartha, 1976). It can be encapsulated or supplied in oleophilic form to meet the nutritional requirements of oil-degrading microorganisms. Additional stimulation was found when oleophilic iron and ferric octoate were added along with nitrogen and phosphorus. Oleophilic iron may be especially beneficial in open oceans, where iron concentrations are low. Use of an oleophilic fertilizer that concentrates the nutrients at the water surface has been shown to increase the degradation of oil slicks (Ward and Lee, 1984) and reduces nutrient loss by diffusion (Atlas and Bartha, 1972d).

In environments without extensive oil pollution history, oleophilic fertilizers may be used to best advantage in combination with microbial inocula (Bartha and Atlas, 1977). Microencapsulation techniques may offer the best solution for a single formulation of organisms and fertilizer.

Another suggestion is to encapsulate a nitrogen-phosphorus fertilizer in a matrix that would allow it to float and be slowly released (Gholson, Guire, and Friede, 1972). A slow-release fertilizer containing paraffin-supported magnesium ammonium phosphate as the active ingredient considerably enhances petroleum biodegradation (Olivieri, Bacchin, Robertiello, Oddo, Degen, and Tonolo, 1976). This mixture is recommended for improving oil biodegradation in seawater (Kator, Miget, and Oppenheimer, 1972).

Assuming that oxygen is always in excess and the contaminating oil has a very large surface exposure to the aqueous phase, the amount of nitrogen required for the destruction of a unit quantity of oil can be calculated. It has been found to be about 4 nmol of nitrogen/ng of oil (Floodgate, 1979). The values for nitrogen demand (analogous to the concept of biochemical oxygen demand) can be converted into volumes of water containing the required amount of nitrogen (Floodgate, 1976). If the nitrogen turnover rate is known for a given body of water, the rates of degradation to be expected can be calculated for summer and winter temperatures. These were found to be 30 g/m^3/year in summer and around 11 g/m^3/year in winter for the Irish Sea. The concentration of elemental nitrogen required to bring about the disappearance of 1 mg of hexadecane by a <u>Nocardia</u> sp. is 0.5 mg (Mulkins-Phillips and Stewart, 1974b). This confirms suggestions that the rate of natural biodegradation of oil in marine environments is not limited by the concentrations of nitrogen occurring naturally, but rather by low temperatures and phosphorus concentrations.

Application of fertilizers to freshwater environments may result in undesirable eutrophication and outgrowth of algae (Atlas, 1977). In eutrophic lakes and ponds, addition of nitrogen and phosphorus may not be necessary. In Lake Mendota, which is becoming eutrophified, nitrogen and phosphorus were found to limit rates of oil biodegradation. However, in oligotrophic water bodies and in marine environments, concentrations of usable nitrogen and phosphorus compounds in surface waters are generally too low to support extensive microbial degradation of oil. If nitrogen and phosphorus are added, they should be in a form that will not allow them to dissipate from the oil-water interface, such as using an encapsulated fertilizer in a matrix that would float and release the compounds slowly.

5.3.3 Alteration of Organic Contaminants

5.3.3.1 Addition of Surfactants

The use of microorganisms as oil dispersants would cause a minimum of environmental damage with a low-toxicity surfactant (Bartha and Atlas, 1977). An <u>Arthrobacter</u> strain, designated as RAG 1, has proved to be a highly efficient emulsifier when growing on crude oil (Reisfeld, Rosenberg, and Gutnik, 1972). The low cell yield, coupled with extensive decrease of benzene-extractable hydrocarbons, suggests the production of degradation intermediates, most likely fatty acids. Fatty acids have been established as biogenic dispersants. In addition, high molecular weight extracellular polymers that are anthrone positive and are precipitated by 95 percent ethanol are produced by several pseudomonads and by Corynebacterium hydrocarbonoclastus (Zajic, Suplisson, and Volesky, 1974). The polymers were effective dispersants also in the absence of the bacteria and acted as flocculants (Knetting and Zajic, 1972).

Several dispersants were tested on marine samples to determine their effects on the degradation of oil (Atlas and Bartha, 1973e). They were found to increase the rate of mineralization, provided the seawater was amended with nitrate and phosphate, but did not increase the extent of degradation. Four dispersants were tested and only one was found that stimulated biodegradation (Mulkins-Phillips and Stewart, 1974c). Two "oil herders" were tested and determined to increase the mineralization rate, but not the extent of degradation (Atlas and Bartha, 1973e).

In another instance, mixed bacterial cultures isolated from foam on the surface of contaminated seawater were found to induce the production of surface-active agents that emulsified the hydrocarbons on which they were growing and improved degradation of these compounds. This ability was not detected with cultures from sea water or sediments.

Within terrestrial ecosystems, such as beach sands, the presence of some emulsifying agents has been found to have a neutral effect on rates of oil biodegradation (Bloom, 1970). Tarry materials occurring on soil or beaches can be physically broken up, e.g., by dicing (Atlas, 1977; Bloom, 1970; Cobet and Guard, 1973). This increases the available surface area for microbial colonization. Breaking up tar globules also increases surface area and, thus, the availability of oxygen, water, and nutrients (Gibbs, 1976).

5.4 TREATMENT TRAINS WITH CHEMICAL AND BIOLOGICAL PROCESSES

No two contamination incidents are exactly alike (Bartha and Atlas, 1977). Consequently, control responses should be flexible and tailored to the situation, rather than follow a rigid pattern. Stimulated biodegradation is not expected to replace all other control measures, but it should rather add further flexibility to integrated control programs.

Treatment trains employing one or more treatment processes may be required for complex waste streams (Lee and Ward, 1986); and bioreclamation can be preceded by, or otherwise used in combination with, other treatments that could reduce toxic concentrations to a tolerable level (Environmental Protection Agency, 1985b). These on-site or in situ treatment techniques could destroy, degrade, or by other means reduce the toxicity of contaminants. Chemical detoxification techniques include injection of neutralizing agents for acid or caustic leachates, addition of oxidizing agents to destroy organics or precipitate inorganic compounds, addition of agents that promote photodegradation or other natural degradation processes, extraction of contaminants, immobilization, or reaction in treatment beds (Lee and Ward, 1986). Some of these processes are discussed in this section. Biological on-site methods, such as treatment of withdrawn groundwater, can be used in conjunction with the in situ practices.

Biological treatment is the least expensive method of organic destruction (Environmental Protection Agency, 1985b). About 99 percent of all organic compounds can be destroyed by biological reactions. When used with other treatment technologies, essentially all the organic contaminants can be removed and destroyed.

Table A.5-11 shows a number of unit operations and the waste types for which they are effective (Ward and Lee, 1984). Table A.5-12 indicates that biological methods can attain a greater level of treatment in groundwater than either stripping or sorption (Knox, Canter, Kincannon, Stover, and Ward, 1968).

5.4.1 Supplementary Processes

Volatile organics, extractable organics, and inorganics (heavy metals) of concern in contaminated groundwater can be treated successfully by two alternative processes (Stover and Kincannon, 1983). 1) One process consists of chemical precipitation to remove metals, steam stripping, and activated carbon adsorption. 2) The alternative consists of combined physical-chemical and biological treatment. Metals treatment in the latter would be a safety measure against possibly higher concentrations than anticipated. It would also be required for removing high iron and manganese concentrations. Combining the unit processes of chemical precipitation, steam stripping, and biological treatment is the most feasible treatment alternative. With the first method, the concentrations of residual organics, measured as TOC, would still be too high.

Table A.5-11. Summary of Suitability of Treatment Processes (Ward and Lee, 1984)

Process	Volatile Organics	Nonvolatile Organics	Inorganics
Air stripping	Suitable for most cases	Not suitable	Not suitable
Steam stripping	Effective concentrated technique	Not suitable	Not suitable
Carbon adsorption	Inadequate removal	Effective removal technique	Not suitable
Biological	Effective removal technique	Effective removal technique	Not suitable metals toxic
pH adjustment precipitation	Not applicable	Not applicable	Effective removal technology
Electrodialysis	Not applicable	Not applicable	Inefficient operation-- inadequate removal
Ion exchange	Not applicable	Not applicable	Inappropriate technology-- difficult operation

5.4.1.1 Neutralization

Neutralization involves injecting dilute acids or bases into the groundwater to adjust the pH (Environmental Protection Agency, 1985b). This can serve as a pretreatment prior to _in situ_ biodegradation to optimize the pH range for the microorganisms. Adjustment of the pH may be required to make the water less corrosive and suitable for other unit processes (Stover and Kincannon, 1983).

5.4.1.2 Oxidation/Reduction

Oxidation of inorganics in soils is, for all practical purposes, limited to oxidation of arsenic and possibly some lead compounds by use of potassium permanganate. Hydrogen peroxide, ozone, and hypochlorites

Table A.5-12. Removal Mechanisms of Toxic Organics from Groundwater (Knox, Canter, Kincannon, Stover, and Ward, 1968)

Compound	Percent Treatment Achieved		
	Stripping	Sorption	Biological
Phenol	--	--	99.9
Aromatics			
Benzene	2.0	--	97.9
Toluene	5.1	0.02	94.9
Ethylbenzene	5.2	0.19	94.6
Halogenated Hydrocarbons			
Methylene Chloride	8.0	--	91.7
Polynuclear Aromatics			
Phenanthrene	--	--	98.2
Naphthalene	--	--	98.6
Other			
Ethyl Acetate	1.0	--	98.8

are the most useful oxidizing agents available. Ozone oxidizes many organic compounds that cannot be easily broken down biologically, including chlorinated hydrocarbons, alcohols, chlorinated aromatics, pesticides, and cyanides (Lee and Ward, 1986).

Chromium can be reduced from the hexavalent state to the trivalent and then precipitated with hydroxide (Lee and Ward, 1986). Reducing agents for chromium include gaseous sulfur dioxide, iron sulfate, waste pickling liquor from metal plating industries, and sodium bisulfite with sulfuric acid commonly used to reduce the pH (Ehrenfeld and Bass, 1984). Levels of less than 1 ppm chromium can be achieved.

Wet-air oxidation involves addition of air at high pressures and temperatures in a form of combustion (Lee and Ward, 1986). A catalyst promotes the oxidation process. Dilute wastes that cannot be treated with incineration can be handled by this process with greater than 99 percent destruction.

5.4.1.3 Precipitation

Precipitation of certain waste components can be accomplished by adding a chemical that reacts with the hazardous consitituent to form a sparingly soluble product or by adding a chemical or changing the temperature to reduce the solubility of the hazardous constituent (Ehrenfeld and Bass, 1984).

Chemical precipitation with carbonate, sulfides, or hydroxides has been used routinely to chemically treat wastewaters containing heavy metals and other inorganics (Knox, Canter, Kincannon, Stover, and Ward, 1984). Sulfides are probably the most effective for precipitating heavy metals; however, sulfide sludges are susceptible to oxidation to sulfate, which may release the metals.

The hydroxide system with lime or sodium hydroxide is widely used but may produce a gelatinous sludge, which is difficult to dewater (Knox, Canter, Kincannon, Stover, and Ward, 1984). Removal of metals by chemical precipitation with lime requires a pH at which a soluble form of the metal is converted to an insoluble form (Stover and Kincannon, 1983).

Soda ash is employed with the carbonate system and may be difficult to control (Knox, Canter, Kincannon, Stover, and Ward, 1984). Alum is another common agent used in chemical precipitation. The effectiveness of these chemical treatments will vary with the nature and concentration of the constituents of the waste stream (Lee and Ward, 1986). A process design for chemical precipitation will have to consider the systems for chemical addition and mixing, the optimal chemical dose, the time required for flocculation and the removal and disposal of the sludge.

5.4.1.4 Permeable Treatment Beds

Permeable treatment beds can intercept a plume and provide a reactor for either chemical treatment or precipitation (Lee and Ward, 1986). These are essentially excavated trenches placed perpendicular to groundwater flow and filled with an appropriate material to treat the plume as it flows through the material (Environmental Protection Agency, 1985b). Various materials could be used as the fill. Limestone (limestone containing little magnesium carbonate is more effective in removing ions than dolomitic limestone) or crushed shell could be used for neutralizing acidic groundwater and retaining certain metals (such as cadmium, iron, and chromium). Activated carbon could be employed for removing nonpolar organic compounds from contaminated plumes, but not polar organics or heavy metals. Glauconitic green sands have the potential for removing heavy metals, especially copper, mercury, nickel, arsenic, and cadmium. Synthetic ion exchange resins may also be used as fill to remove heavy metals, although they may have a short lifetime, high costs, and be difficult to regenerate. Permeable treatment beds may plug or exhibit channeling, which will reduce their effectiveness. These beds have the potential to reduce the quantities of contaminants present in leachate plumes and are applicable to relatively shallow groundwater tables containing a plume. They have the potential problem of the saturation of bed material, plugging of the bed with precipitates, and short life of the treatment materials.

5.4.1.5 Soil Flushing

Soil flushing is an extraction process to remove organic and inorganic contaminants from contaminated soils (Environmental Protection Agency, 1985b). Water or an aqueous solution is injected into the area of contamination, and the contaminated elutriate is pumped to the surface for removal, recirculation, or on-site treatment and reinjection. During elutriation, sorbed contaminants are mobilized into solution by reason of solubility, formation of an emulsion, or by chemical reaction with the flushing solution.

Solutions with the greatest potential for use in soil flushing would be either water, acids/bases, complexing and chelating agents, surfactants, and certain reducing agents (Environmental Protection Agency, 1985b). Water can be used to flush water-soluble or water-mobile organics and inorganics. Organics amenable to water flushing can be identified according to their soil/water partition coefficient, or estimated using the octanol/water coefficient. Inorganics that can be flushed from soil with water are soluble salts, such as the carbonates of nickel, zinc, and copper. Adjusting the pH with dilute solutions of acids or bases will enhance inorganic solubilization and removal.

Weak acids (e.g., sodium dihydrogen phosphate and acetic acid) or dilute solutions of strong acids (e.g., sulfuric) could be used if the soil contains sufficient alkalinity to neutralize it (Environmental Protection Agency, 1985b). Complexing and chelating agents may also be employed. These can mobilize metals strongly adsorbed to manganese and iron oxides in soils. Surfactants can improve the solvent property of the recharge water, emulsify nonsoluble organics, and enhance removal of hydrophobic organics sorbed onto soil particles. This is a promising in situ chemical treatment method.

It would be economically feasible to recycle the elutriate from soil flushing back through the contaminated soil (Environmental Protection Agency, 1985b). Soil flushing methods involving the use of water surfactants appear to be the most feasible and cost-effective chemical treatment for organics. They can use relatively cheap, innocuous treatment reagents, can be used to treat a broad range of waste constituents, and do not result in toxic degradation products. Although a laboratory experiment can be conducted to estimate the number of times the groundwater would have to be turned over or filtered through the contaminated soil to achieve to required level of water quality (Stover and Kincannon, 1983), this process has not proved effective.

5.4.1.6 Carbon Adsorption

Organics can be trapped on activated carbon or on resins by both physical and chemical forces and removed from the groundwater (Nielsen, 1983). The degree of sorption onto the carbon depends upon (Knox, Canter, Kincannon, Stover, and Ward, 1984):

1. The solubility of the compound, insoluble compounds being more likely to be adsorbed
2. The pH of the water, which controls the degree of ionization of the compounds; acids are adsorbed better under acidic conditions and adsorption of amine-containing compounds is favored under alkaline conditions
3. Characteristics of the adsorbent, which is a result of the process used to generate and activate the carbon
4. Properties of the compound; for example, aliphatic compounds are less well adsorbed than aromatics and halogenated compounds

Effluent levels of between 1 and 10 ug/l can be achieved for many organics (Ehrenfeld and Bass, 1984). Partial adsorption of several heavy metals also occurs. Granular or powdered activated carbon has been used. Granular activated carbon is typically employed in reactors: the powdered carbon is added to the wastewater and then either settled or filtered for removal with the sludge.

Carbon adsorption removes low concentrations of organic contaminants or residual organics from other treatment systems. It is the best system for emergency response. Activated carbon systems can be batch, column, or fluidized-bed reactors (Lee and Ward, 1986).

5.4.1.7 Air and Steam Stripping

Under some circumstances, soil containing volatile hydrocarbons or solvents can be decontaminated by air stripping (Niaki, Pollock, Medlin,

Shealy, and Broscious, Draft). In this process, air is injected into the soil through a series of extraction wells. It is blown into the injection wells and pulled out of the extraction wells. As it flows through the soil, volatile materials are stripped off into the airstream. The organics are removed from the air in a vapor phase carbon adsorption system or by fume incineration. The success of in situ air stripping depends upon relatively unrestricted and uniform flow of air through the soil. Clay soils, packed soils, or soils with a high water table are not good candidates for in situ air stripping.

Volatile compounds can also be removed from withdrawn groundwater by air or steam stripping (Nielsen, 1983). Various configurations of equipment can be used in air stripping, including diffused aeration, countercurrent packed columns, cross-flow towers and coke tray aerators with countercurrent packed columns. The latter is probably the most useful for decontamination of groundwater, since the countercurrent packed columns provide the most interfacial liquid area, high air-to-water volumes, and can be easily connected to vapor recovery equipment (Knox, Canter, Kincannon, Stover, and Ward, 1984). In many applications of air strippers, the volatile compounds are transferred from the water to the air where the vapors can then be collected and treated by incineration or adsorption to carbon (Nielsen, 1983). Increased temperatures can improve the removal efficiency of the stripping process for some compounds, such as aldehydes and alcohols (Law Engineering Testing Company, 1982).

Steam stripping has the same effects as elevated temperature air stripping (Knox, Canter, Kincannon, Stover, and Ward, 1984). Costs for air stripping have been estimated to be between 9 and 90 cents/1000 gal of water treated for removal of 90 percent of trichloroethylene.

A technique similar to air stripping for the removal of some compounds from contaminated groundwater is that of dissolved air flotation, in which suspended fine particles or globules of oil and grease are floated to the surface by the action of pressurized air and then removed by skimming (Ehrenfeld and Bass, 1984). This technique has been used to remove up to 90 percent of the total suspended solids or oil and grease in wastewater containing 900 ppm of these substances.

5.4.1.8 Reverse Osmosis

This process may be used to concentrate inorganics and some high molecular weight organics from waste streams (Lee and Ward, 1986). It passes the contaminated water through a semipermiable membrane at high pressure. The clean water leaves behind the concentrated wastes and any particulates. Pretreatment of the waste stream is likely to be required to achieve a constant influent composition--pH is particularly important--to kill any organisms that might form a biological film that would reduce permeability, to remove suspended solids, and to remove chlorine, which might affect the membrane.

5.4.1.9 Mobilization/Immobilization

Surfactants may be added during soil washing to mobilize the contaminants (Ellis, Payne, Tafuri, and Freestone, 1984). A 4 percent solution of two nonionic surfactants removed greater than 90 percent of polychlorinated biphenyls and a high boiling distillation fraction of crude oil from test soil columns with 10 pore volume washes.

Contaminants can be immobilized by precipitation or encapsulation in an insoluble matrix (Pye, Patrick, and Quarles, 1983). A spill of acrylate monomer was treated with catalyst and activator in order to produce a solidified polymer, thereby, immobilizing an estimated 85 to 90 percent of the liquid monomer (Knox, Canter, Kincannon, Stover, and Ward, 1984).

5.4.2 Examples of the Use of Treatment Trains

- In a study of a site polluted by hydrocarbons, chlorinated hydrocarbons, and organo chloride pesticides, it was found that no single technology could remove or destroy all of the contaminants (Rickabaugh, Clement, Martin, and Sunderhaus, 1986). However, when microbial degradation, surfactant scrubbing, photolysis, and reverse osmosis were combined, nearly total destruction of these compounds could be attained on-site.
- This approach was also employed at a site where 130,000 gal of several organics had been spilled (Lee and Ward, 1985). Treatment of the site was by clarification, adsorption onto granular activated carbon, air stripping, and then reinjection. After levels of the contaminants had fallen below 1000 mg/l, a biodegradation program employing facultative hydrocarbon-degrading bacteria, nutrients, and oxygen was begun. Biodegradation by both the indigenous microbes and the added organisms reduced the levels of the contaminants in soil cores from 25,000 to 2,000 mg/l within two months. The monitoring wells showed no levels above 1 mg/l at the end of the program.
- Another example involving the use of multiple treatment processes was a bench-scale study on a site in Muskegon, MI, contaminated by several priority pollutants and at least 70 other organics at levels in the hundreds of

ppm (Lee and Ward, 1985). Acclimation of an activated sludge culture to the contaminated groundwater was unsuccessful, and a commercial microbial culture was ineffective at degrading the contaminants. However, coupling an activated sludge process to granular activated carbon treatment proved beneficial, as the organisms were able to degrade the organics that passed through the carbon system. This treatment train was able to remove up to 95 percent of the total organic carbon in the wastewater, as long as the activated carbon continued to function properly.

- The release of phenol and chlorinated derivates in the soil in the Midwest was treated by installing a recovery system and using activated carbon filters on the groundwater (Walton and Dobbs, 1980). Surface waters were contained in a pond. Mutant bacteria were injected into the pond and into the contaminated soil. After an incubation and adaptation period, the phenol was completely degraded in 40 days, while the orthochlorophenol was reduced from 120 to 30 ppm.

5.5 BIODEGRADATION IMPLEMENTATION PLAN

The basic steps in an in situ bioreclamation process are (Lee and Ward, 1985; Buckingham, 1981):

Free product recovery
Site investigation
Well design
Well installation
Microbial degradation optimization study
System design
Operation
Groundwater sampling
Sampling techniques
Groundwater testing
Monitoring
Oxygen management of groundwater
Nutrient management of groundwater

5.5.1 Free Product Recovery

Free product or the source of the contamination should be removed before a treatment program is initiated (Nielsen, 1983). Physical recovery methods are used to recover as much of the free product as possible (Lee and Ward, 1985).

The design of a free product recovery system is greatly affected by the geohydrologic characteristics of the site (Niaki, Pollock, Medlin, Shealy, and Broscious, Draft).   Groundwater flow, lithologic permeability, and gravity have the greatest effect on hydrocarbon migration in subsurface environments (Van Dam, 1967).   Use of equations can allow estimation of the actual thickness of the hydrocarbon saturation zone, based upon measured hydrocarbon thickness in monitor wells and lithology.   The hydrocarbon layer in a monitor well may be one to ten times as thick as the actual hydrocarbon saturated zone in the surrounding soil.   Geologic bedding dip, as well as structural features, may also control hydrocarbon movement (Osgood, 1974).

Free product can be recovered by several techniques (Canter and Knox, 1984).   Pumping operations to collect the floating lens include multiple-level pumps with hydrocarbon detectors and single-level skimming pumps (Jeter, 1985).   A single-pump system utilizing one or more wells requires minimal equipment and drilling costs, but produces a mixture of product and water that must be separated (Canter and Knox, 1984).   A two-pump, two-well system utilizes one well to produce a water table gradient that allows the second well to recover the floating product.   Another type of system utilizes a single well with two pumps--a lower pump to produce a gradient and an upper one to collect the free product.   Well systems represent a proven method for control of many hazardous waste contamination problems.   However, they do have high maintenance and operation costs; limited application to fine soils; removal of clean water along with contaminated water, which increases the costs of treatment; and long operation times, especially to remove sorbed contaminants.

There are also several other technologies for bulk gasoline recovery (Jeter, 1985).   Interceptor trenches dug below the water table can be used to collect floating hydrocarbon layers.   Dissolved product has been treated by air stripping and granular activated carbon (GAC).   However, air stripping may discharge these contaminants into the atmosphere.   A significant proportion of the product from a spill or leak will remain in the unsaturated soil zone and be unavailable for free product recovery under existing conditions.   This material can then be subjected to biological treatment (Hoag and Marley, 1986).

Local and federal agencies do not accept bulk fuel removal as a complete restoration.   If practical, it may be wise to continue to pump contaminated wells to contain the contaminants (Raymond, Jamison, and Hudson, 1976).

5.5.2 Site Investigation

Implementation of the most effective and cost-beneficial procedure for treating a hazardous waste requires adequate site assessment of both the saturated zone and the vadose zone (Buckingham, 1981). A thorough understanding of the hydrogeology and extent of contamination of the site must be obtained and used to design the treatment system (Lee and Ward, 1986; Knox, Canter, Kincannon, Stover, and Ward, 1968). This is usually accomplished by installation of a network of monitoring wells and characterization of the aquifer through pump tests, analysis of soil properties, and other methods (Roberts, Koff, and Karr, 1988). The stratigraphy beneath the site and a description of the soils and bedrock should be detailed, including the presence and extent of inhomogeneities in the soil (Lee and Ward, 1986; Quince and Gardner, 1982). The hydrogeologic data needed are formation porosity, hydraulic gradient, depths of both the aquifer and the contaminated zone, permeability, groundwater velocity and direction, specific yield of the aquifer, and recharge/discharge information. Knowledge of other environmental parameters, such as precipitation, temperature (Buckingham, 1981), soil moisture, pH, oxidation-reduction potential, and oxygen, nutrient, and organic matter content are necessary. The source, quantity, and nature of the spilled material must also be assessed (Knox, Canter, Kincannon, Stover, and Ward, 1968).

5.5.3 Well Design

A well is an excavated shaft in the ground that may be lined or cased with some material to keep it from caving in, often with plastic or metal pipe (Bitton and Gerba, 1985). When determining specific well structure, all inflows of water, except through the zone of interest, should be sealed out. Three well types are diagrammed in Figure A.5-4 (Absalon and Starr, 1980). Type I well is for use in preliminary studies, prior to planning and implementation of additional studies. A grout seal prevents groundwater contamination by surface waters, but the well has obvious limitation in that soil water can percolate down the borehole to the groundwater level. This well does not meet the Resource Conservation and Recovery Act's requirements for groundwater quality monitoring well construction. Type II well has a combination bentonite and grout seal around that section of the well above the water table. It meets EPA guidelines for monitoring well construction. Type III well is grouted and sealed from the ground surface to the top of the zone of interest. It is intended for difficult drilling applications and long-term monitoring.

Materials used in well construction typically depend upon the mechanical strength required. Shallow wells (less than 50 ft) are normally constructed of schedule 40 polyvinylchloride (PVC) plastic, and deeper wells are constructed of schedule 80 PVC. PVC is relatively inert, although certain contamination problems may be associated with its use. Organic constituents of groundwater are adsorbed on PVC casing. The casing may also contribute low levels of organic contaminants to groundwater samples, such as phthalic acid esters used as plasticizers in PVC manufacture, and solvents from cements used to join lengths of PVC tubing (Dunlap, McNabb, Scalf, and Cosby, 1977). Teflon is currently being investigated as an alternate well construction material. However, it has two major drawbacks--high cost and relative softness (Absalon and Starr, 1980). Type III well is constructed of PVC or stainless steel above the water table and PVC or Teflon below the water table where the casing is directly in contact with groundwater. If the level of contamination is high, it is probable that the effects of PVC on the water sample will be insignificant.

5.5.4 Well Installation

Installation of a network of monitoring wells and characterization of the aquifer through pump tests and analysis of soil properties will provide understanding of the hydrogeology of the site to plan the installation of an interceptor system. (Nielsen, 1983).

Consideration should be given to the use of extensive monitoring well networks to characterize more adequately the spatial and temporal variability that has been observed (Plumb, 1985). In addition, a substantial number of substances (including several organic priority pollutants) have been detected at a higher frequency or higher concentration at upgradient wells, which suggests that the direction of chemical migration may be influenced by factors other than groundwater flow. Therefore, it should not be assumed that background conditions are pristine and that upgradient monitoring can be minimized. There should also be a number of wells placed at upgradient locations.

The composition of well water will generally be influenced by well installation methods (Environmental Protection Agency, 1976). For example, care should be taken in the selection of a water supply, if the well is installed by the wash boring method. This technique introduces

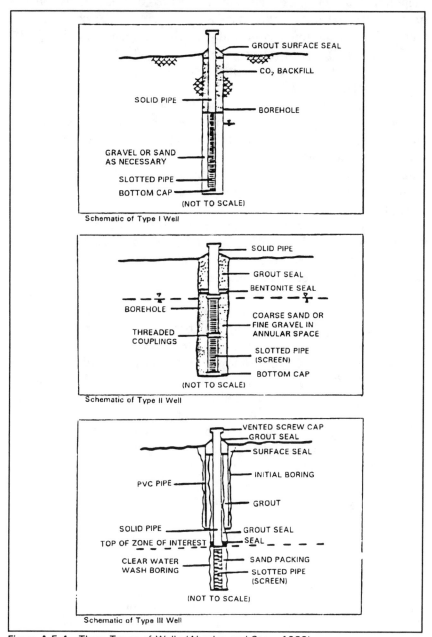

Figure A.5-4. Three Types of Wells (Absalon and Starr, 1980)

large quantities of surface waters into the groundwater, which could yield samples that are unrepresentative of subsurface conditions (Absalon and Starr, 1980). Drilling fluids may alter the subsurface chemical and microbiological environment in well vicinities (Dunlap, McNabb, Scalf, and Cosby, 1977). Drilling mud can inhibit groundwater flow by closing interstitial granular spaces in aquifers and reducing aquifer permeability (Absalon and Starr, 1980).

Several effective well construction methods, such as the air rotary method, appear to have little or no effect on groundwater quality. Augers are an efficient technique for drilling shallow wells (40 to 50 ft), in reasonably compact alluvial materials (Dunlap, McNabb, Scalf, and Cosby, 1977). These techniques should be used where they are physically and economically feasible, in order to minimize groundwater contamination.

With any well construction method, it is extremely important to clean all drilling equipment before the monitoring well is installed (Absalon and Starr, 1980). This prevents cross contamination of wells by drilling tools. It has been recommended that all casing materials be prepared before installation in a borehole (Dunlap, McNabb, Scalf, and Cosby, 1977). This involves a wash with soap and water, a water rinsing, a solvent rinsing, and a final rinsing with organic-free water.

Cation exchange capacity (CEC) is the sum total of exchangeable cations that a soil can adsorb (Brady, 1974). Finer textured soils tend to have higher CEC values than sandy soils. Cations of various elements are present in soil solution and are transported to groundwater. Bentonite, a sealer used in well construction, is composed predominantly of clay. Therefore, it has a high CEC, which may change the concentration of cations in the groundwater, thus, altering its chemical composition.

5.5.5 Microbial Degradation Optimization Study

The site must be investigated for the presence or absence of microorganisms (e.g., aerobic bacteria, anaerobic bacteria, actinomycetes, fungi, and algae) (Sims and Bass, 1984). A laboratory study must be conducted to determine whether the native microbial population can degrade the components of the spill (Raymond, 1978). If it cannot, the biodegradation approach should either be abandoned or additional studies conducted to determine what microorganisms or combination of organisms should be added for the particular contaminants involved.

Laboratory studies must also characterize the waste. The solubility of the contaminant may have to be increased by adding emulsifiers to allow microbial action on the compound (Knox, Canter, Kincannon, Stover, and Ward, 1968). The C:N:P ratio must be determined. If this ratio is not optimal, bench-scale experiments should be performed to ascertain whether adjusting the ratio will increase the numbers of organisms and the rate of biodegradation. The optimum oxygen, nutrient, and temperature requirements of the indigenous microbial population (or added organisms) must be found for degradation of the specific site contaminants (Lee and Ward, 1986; Knox, Canter, Kincannon, Stover, and Ward, 1968). The study determines what combination of nutrients will give the maximal cell growth on gasoline, for instance, in 96 hr under aerobic conditions at the ambient temperature of the groundwater (Lee and Ward, 1985). Nutrient formulations may differ among sites. The form of the nutrient added also varies, but will usually contain sources of nitrogen, phosphorus, magnesium, carbonate, calcium, manganese, sulfate, sodium, and iron. A laboratory investigation of the kinetics of biodegradation for the organisms should also be performed (Raymond, 1974).

5.5.6 System Design

The system for nutrient and oxygen addition and circulation through the aquifer will have to be designed and built, and should be done under the direction of an experienced groundwater geologist (Lee and Ward, 1986). This system usually consists of injection and production wells to control flow, and equipment to add and mix the nutrient solution (Raymond, 1978). Reinjecting pumped groundwater back into the system will recirculate any unused nutrients, thus, avoiding the problem of disposal of this water, if this approach is permissible in the state in which the contaminant incident has occurred (Lee and Ward, 1985). Creation of a closed loop of recovery, treatment, and recharge flushes the contaminants out of the soil rapidly and establishes hydrodynamic control separating the contaminated zone from the rest of the aquifer (Knox, Canter, Kincannon, Stover, and Ward, 1968). Another benefit of the closed loop is that acclimated bacteria can be added to the aquifer via the extracted groundwater and can act in situ to degrade the contaminants. The recharge water can be adjusted to provide optimal conditions for the growth of the added bacteria and the indigenous population, which may also act on the contaminants.

5.5.7 Operation

Once the system is constructed, nutrient and oxygen addition can begin (Raymond, Jamison, Hudson, Mitchell, and Farmer, 1978). Nutrients can be added by batch or by continuous feed; continuous feed supplies a more constant nutrient source, but requires more labor and equipment to implement. Large quantities of nutrients may be required. At one site, 16.65 tons of chemicals were added, and at another, a total of 87 tons of food-grade quality chemicals were purchased. Oxygen can be supplied to the aquifer by sparging air into wells.

5.5.8 Groundwater Sampling

Groundwater is obtained from a well by some sort of device to bring water to the surface (Davis, 1967).

There are problems associated with collecting subsurface samples. Subsurface organisms are essentially restricted to the spaces between soil particles (Davis, 1967). These organisms are either free floating in the interstitial water or are attached to the solid phase (Marxen, 1981). Samples of subsurface solids are often taken at only one point and cannot be repeated (Davis, 1967). The microorganisms in water samples from wells may be transient and may not be representative, quantitatively or qualitatively, of the indigenous microbial populations existing in the formation being sampled. However, properly collected well samples allow the investigator to determine the concentration of microorganisms and chemical constituents in the water at a particular point at a given time and over an extended period of time. This will provide useful information about processes occurring in the subsurface. However, the proper collection of subsurface water for microbiological analyses is extremely difficult.

It is hard to collect a representative sample, and it is extremely difficult to collect an uncontaminated one. The best way to study microbial activity in deep formations is to take core samples during drilling. An uncontaminated sample can be best obtained by subsampling a core sample. The outer, 1 to 2 cm of the contaminated portion of the samples are removed, leaving the inner, untouched core for bacterial analysis (Gilmore, 1959). Most coring procedures may prove to be too expensive and time consuming in a field investigation (Wilson, McNabb, Wilson, and Noonan, 1983; Wilson, Noonan, and McNabb, 1983).

It is assumed that free-floating organisms travel with the groundwater (Davis, 1967). Therefore, wells must be sited and screened to a depth that will intercept a particular flow path. Wells upsite can supply data on water quality for comparison. The depth of screening is important in interpreting data. Drilling fluids often contain various additives and may be a source of bacterial contamination (McNabb and Mallard, 1984) and should not be used when taking core samples (Scalf, McNabb, Dunlap, Cosby, and Fryberger, 1981). To minimize loss of a sample composed of loose material, pouring liquid nitrogen down the pipe will freeze the material (Danielpol, 1982).

Because water in the unsaturated zone is under pressures less than atmospheric, the water will not move into wells that are exposed to atmospheric pressure (Davis, 1967). Specially designed gravity lysimeters (Prill, Oaksford, and Potorti, 1979) have been used successfully to recover bacteria from the unsaturated zone (Vaughn, Landry, Beckwith, and Thomas, 1981z).

The rate of groundwater movement is generally slow, usually measured in feet/day or feet/year. Environmental conditions are relatively constant (Dunlap and McNabb, 1973). Temperatures in the earth's crust fluctuate only in the topmost layer of about 10 m (33 ft), which may be affected by seasonal temperature/variations (Kuznetsov, Ivanov, and Lyalikova, 1963). Therefore, frequent sampling would not be necessary. It may also take months for groundwater to move from a source of contamination to the nearest downgradient well (Davis, 1967).

5.5.8.1 Sampling Techniques

Wells should be pumped or bailed before sampling is initiated, in order to ensure that water samples are representative of the groundwater near the well (Buckingham, 19818). There are no universal rules for the quantity of well water to be extracted. Wells installed through use of drilling fluids generally require extraction of large quantities of water, since they alter groundwater quality more substantially than wells installed with augers or air rotary devices. As a general rule, sampled water should not be turbid (Absalon and Starr, 1980).

Groundwater samples can be retrieved by pumping or bailing techniques (Environmental Protection Agency, 1976). Pumping methods, specifically the use of portable submersible pumps, can be cost efficient, but they often release compressed air into the well. Reduced pressure pumping systems can strip samples of volatile constituents. In addition, pressure changes in pumped samples can cause changes in oxidation state, pH, and temperature of the sample (Absalon and Starr, 1980). These changes affect the sample's chemical constituents.

A barcad is a porous, hollow alumina cylinder that extracts water and conducts it to the surface through a small diameter rise tube (Gardner and Ayres, 1980). The sampler is activated from the surface by a valve inside the cylinder. Although the barcad uses compressed, inert gas, this technique minimizes agitation of the water sample and is believed to preserve volatile organic compounds better than most conventional methods.

Manual sampling techniques are useful under a variety of circumstances; i.e., when sampling is infrequent, biological or sediment samples are required, hazardous material spills (or other special incidents) are being investigated, or concentrations of substances are relatively constant (Shelley, 1977).

There are several manual sampling techniques. One method uses a small weight attached on the end of a sash cord (Raymond, Jamison, and Hudson, 1976). Properly cleaned and sterilized 6-oz glass bottles are fastened near the weight with half-hitch knots around the body and neck of the bottles. A cork is secured just above the bottle opening. To sample the groundwater, the cork is forced into the bottle, and the bottle is lowered to the desired depth. A sharp jerk of the cord removes the cork in order to permit filling of the bottle. Organic substance tends to cling to sample containers. Therefore, glass bottles are appropriate samplers because they are inert, relative to organic material, and they can withstand a rigorous cleaning procedure (Environmental Protection Agency, 1976).

The Teflon bailer, diagrammed in Figure A.5-5, is another manual sampler (Dunlap, McNabb, Scalf, and Cosby, 1977). It is 18 to 36 in. long, constructed of heavy wall Teflon, and is plugged at the bottom end with a short length of Teflon rod. Water enters the bailer through a 5/16-in. hole drilled through the end plug, as the bailer is lowered into the well. It is prevented from draining out by a 3/4-in. diameter glass marble, which fits into a conical seat machined into the top of the plug. The plug fits snugly inside the tube, making up the body of the bailer. Thus, no adhesives, which could contaminate sampled water, are necessary to hold it in place. Bailers are sterilized by autoclaving before they are used, to minimize well contamination. The bailer is raised and lowered with the use of a cable. Samples are poured from the bailer into clean serum bottles of appropriate size (125 ml). Caution is taken to avoid turbulence, which might result in loss of volatile organics or excessive oxygenation of samples. Serum bottles are topped off to avoid including gas spaces.

Preservation techniques retard chemical and biological changes after a sample is taken (Environmental Protection Agency, 1976). Bacteria are inhibited by refrigeration and acidic conditions (created by the addition of sulfuric acid), and chemical reaction rates are retarded by refrigeration (Environmental Protection Agency, 1974).

5.5.9 Monitoring

Contaminated groundwater must be analyzed not only for the extent of contamination, but also for its oxygen and nutrient levels. A system of monitoring wells is usually established for such purposes. Representative groundwater samples can be properly acquired through well design and construction techniques.

Regular monitoring of the levels of organics and inorganics in the effluent and groundwater is necessary to judge the effectiveness of the treatment and to ensure proper distribution of the nutrients. The response of the bacterial population to nutrient addition should also be monitored. The highest counts of up to 10^7 organisms/ml (Lee and Ward, 1985) will generally be in the areas of greatest contamination, with a 10- to 1000-fold increase in the numbers of gasoline-utilizing organisms possible after nutrient addition.

Monitoring a number of parameters is necessary to determine process performance (Environmental Protection Agency, 1985b). Monitoring of groundwater can be performed at the injection and extraction wells, as well as at monitoring wells. On-site wells will monitor process performance, and off-site wells will monitor for pollutant migration, as well as provide background information on changes in subsurface conditions due to seasonal fluctuation. Table A.5-13 lists parameters that should be monitored and suggests methods for monitoring them.

If a monitoring well is not screened properly, its intake might miss the contaminant plume altogether, since groundwater flow and contaminant transport can be extremely stratified (Josephson, 1983a). The plume's thickness might not be as great as that of the aquifer itself. Plume identification involves four dimensions. The monitoring scheme must consider plume spread in two horizontal and one vertical direction over time. This knowledge will help ensure that less uncontaminated water is incidentally pumped with the contaminated for treatment.

5.5.10 Groundwater Testing

The groundwater samples collected from the monitoring can be analyzed on-site or shipped to water testing laboratories for analysis after they have been properly preserved (Buckingham, 1981). Laboratories are available through federal, state, and municipal governments, as well as through universities and private companies.

This section outlines three specific groundwater tests for oxygen, nitrogen, and phosphorus. The U.S. EPA sampling requirements, preservation techniques, and allowable holding times are summarized in Table A.5-14 (Environmental Protection Agency, 1974).

a. Dissolved Oxygen (DO)

Dissolved oxygen can be determined by two methods (Rand, 1975). The Winkler or iodometric method is the most

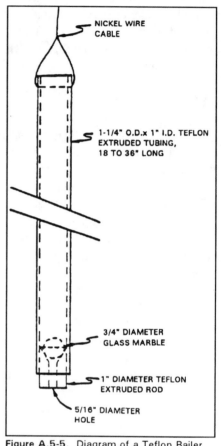

Figure A.5-5. Diagram of a Teflon Bailer (Dunlap, McNabb, Scalf, and Cosby, 1977)

NICKEL WIRE CABLE

1-1/4" O.D. x 1" I.D. TEFLON EXTRUDED TUBING, 18 TO 36" LONG

3/4" DIAMETER GLASS MARBLE

1" DIAMETER TEFLON EXTRUDED ROD

5/16" DIAMETER HOLE

precise and reliable titrimetric procedure for dissolved oxygen analysis. However, organic compounds and oxidizing or reducing agents can interfere with this technique. Therefore, certain modified methods have been developed to minimize these interferences.

The iodometric method is not ideally suited for field testing, continuous monitoring, or for DO determination in situ. However, electrodes provide an alternate testing method for such conditions.

Sensing elements of membrane-covered electrode systems are protected by an oxygen-permeable plastic membrane. This serves as a diffusion barrier against impurities that can interfere with results. They are, therefore, especially suited for use in polluted waters where iodometric methods, and even their modifications, are subject to serious errors caused by interferences. Electrode meters are available commercially and yield an accuracy of +-0.1 mg/l DO and a precision of +-0.05 mg/l DO.

b. Biochemical Oxygen Demand (BOD)

A test for biochemical oxygen demand measures the dissolved oxygen consumed by microbial life while assimilating and oxidizing the organic matter present (Environmental Protection Agency, 1976). Analysis involves a 5-day incubation period. The reduction of dissolved oxygen during that period yields a measure of BOD.

Samples for BOD analysis may undergo significant decay during handling and storage (Thibault and Elliot, 1980). A portion of the demand may be satisfied, if the sample is stored for several days before the test is initiated. This results in a lower BOD value than is actually present in groundwater.

c. Chemical Oxygen Demand (COD)

A test for chemical oxygen demand measures the oxygen equivalent of that fraction of organic matter in a sample that may undergo oxidation by a strong organic or inorganic chemical oxidant (Rand, et al., 1975). COD is determined by the dichromate reflux method. The oxidant in this test has advantages over other oxidants in applicability to a variety of samples, oxidizability, and ease of manipulation.

d. Nitrogen

Nitrate, nitrite, ammonia, and organic nitrogen are forms of nitrogen that are of interest in waters and wastewaters. They are interconvertible and are components of the nitrogen cycle.

Ammonia is naturally present both in surface and groundwater. It may be produced by the reduction of nitrate under anaerobic conditions. Therefore, a high ammonia concentration can indicate anaerobic conditions in groundwater.

Three methods are available to test for ammonia. The Nessler method is sensitive up to 20 ug/l ammonia nitrogen under prime conditions, and it is suitable for testing ammonia nitrogen concentrations

Table A.5-13. Recommended Parameters to Monitor (Environmental Protection Agency, 1985b)

Parameter	Location of Analysis	Media	Analytical Method
Total organic carbon (TOC)	Laboratory	Groundwater	TOC analyzer
Priority pollutant analysis or analysis of specific organics	Laboratory	Soil and groundwater	GC/MS[a]
Microbiology-cell counts	Laboratory	Soil and groundwater	Direct counts. Plate counts on groundwater media or enriched media
	Field	Groundwater	Plate counts with portable water test kits (e.g., Soil Test Inc., Evanston, IL)
Temperature conductivity dissolved oxygen (DO) pH	Field	Groundwater	In situ water quality monitoring instrument or prepackaged chemicals, field test kits
Alkalinity acidity, M&P chloride hardness (total) NH_3-N, NO_3-N PO_4, all forms SO_4, TDS (total dissolved solids)	Field	Groundwater	Prepackaged chemicals/ field test kits; water analyzer photometer (Soil Test, Inc.; Lamotte Chemical, Chestertown, MD)
Heavy metals (if present)	Field Laboratory	Groundwater Soil and groundwater	Prepackaged chemicals/ test kits; GC/MS; AAS[b]
Hydrogen peroxide	Field	Groundwater	Prepackaged chemicals for H_2O_2, test strips available, Titanium sulfate titration and spectrophotometer analysis for greater accuracy

[a] GS/MS = gas chromatography/mass spectrometry
[b] AAS = atomic absorption spectrometry

Table A.5-14. U.S.   EPA Sampling Requirements, Preservation Techniques, and Allowable Holding Times (Environmental Protection Agency, 1974)

Measurement	Volume Required (ml)	Container	Preservative	Holding Time
Dissolved oxygen (DO)				
Probe (_in situ_) DO concentration determined on-site				no holding
Winkler (in lab)	300	glass	[a]	4-8 hr
Biochemical oxygen demand (BOD)	1000	glass or plastic	cool, 4°C	6 hr[b]
Chemical oxygen demand (COD)	50	glass or plastic	H_2SO_4 to pH 2	7 d
Nitrogen				
Ammonia	400	glass or plastic	cool, 4°C H_2SO_4 to pH 2	24 hr
Nitrate	100	glass or plastic	cool, 4°C H_2SO_4	24 hr
Nitrite	50	glass or plastic	cool, 4°C	24 hr
Kjeldahl total	500	glass or plastic	cool, 4°C H_2SO_4 to pH 2	7 d
[c]Phosphorus				
Orthophosphate, dissolved	50	glass or plastic	filter on-site cool, 4°C	24 hr
hydrolyzable	50	glass or plastic	cool, 4°C H_2SO_4	24 hr
total	50	glass or plastic	cool, 4°C	7 d
Total, dissolved	50	glass or plastic	filter on-site cool, 4 °C	24 hr

a = To preserve the sample for Winkler determination, add 0.7 ml concentrated H_2SO_4 and 1 ml sodium azide solution to a 300 ml DO sample (Rand, et al., 1975).

b = If samples cannot be returned to the laboratory in less than 6 hr, and holding time exceeds this limit, the final reported data should indicate the actual holding time.

c = Do not store samples containing low concentrations of phosphorus in plastic bottles because phosphate may be adsorbed onto the walls of the bottles (Rand, et al., 1975).

of up to 5 mg/l. Interfering substances may be removed primarily by distillation. The phenate method has a sensitivity of 10 ug/l ammonia nitrogen, and it can test for ammonia nitrogen concentrations of up to 500 mg/l. Preliminary distillation is required, if the alkalinity exceeds 500 mg/l or if color or turbidity is present, if the sample has been preserved with acid, or if the ammonia nitrogen concentration is greater than 5 mg/l.

Nitrate is generally present in low concentration in surface waters, but it may attain high levels in some groundwater. It is a required nutrient for many photosynthetic autotrophs and has been recognized as the growth-limiting factor in some cases. In high concentrations, however, it contributes to the illness known as infant methemoglobinemia. Therefore, a limit of 10 mg/l nitrate has been imposed on drinking water, in attempts to prevent this disorder. A number of colorimetric procedures are available for determining nitrate levels.

Nitrate is difficult to determine because of the relatively complicated procedures required, the high probability that interfering substances will be present, and the limited concentration ranges of the various methods. Two screening techniques are available to determine initially the approximate range of nitrate in the sample. A method suitable for the concentration range of the sample is subsequently selected.

There are six methods available to test for nitrate nitrogen concentration. The ultraviolet spectrophotometric method is used for screening only those samples with low organic matter content. The cadmium reduction method reduces nitrate almost quantitatively to nitrite, when a sample is run through a column containing amalgamated cadmium filings. The brucine method involves a reaction between nitrate and brucine that produces a yellow color. Final nitrate concentration is subsequently determined by colorimetry. The chromotropic acid method is recommended for the concentration range 0.1 to 5 mg NO_3-N/1. In this method, nitrate reacts with chromotropic acid to form a yellow reaction product exhibiting maximum absorbance at 410 nm. The absorbance is read on a spectrophotometer to determine nitrate concentration. Devarda's alloy reduction technique is recommended for those samples in which the nitrate-nitrogen concentration is greater than 2 mg/l. In this method, nitrate and nitrite are reduced to ammonia under hot alkaline conditions in the presence of the reducing agent, Devarda's alloy. For field use, the nitrate electrode method is the simplest and most practical technique. It is used for screening water and wastewater samples.

Nitrite is an intermediate in the nitrification process. Its concentration is determined by the diazotization method in the nitrite-nitrogen range of 1 to 25 ug/l nitrogen. Photometric measurements can be made in the range of 5 to 50 ug/l.

Organic nitrogen and ammonia can be determined together and have been referred to as "total nitrogen" or "kjeldahl nitrogen," a term that reflects the method used in their determination. The "organic nitrogen" can be obtained by the difference of the individual kjeldahl nitrogen and ammonia nitrogen determinations.

e. Phosphorus

Phosphorus occurs principally as phosphate compounds in natural waters and wastewaters (Rand, et al., 1975). These compounds are usually categorized into orthophosphates, condensed phosphates, and organically bound phosphates. These may occur in the soluble form, or they may be incorporated into particles of detritus, or into the cells of organisms. Phosphorus fertilizer is generally applied as orthophosphate. Phosphorus is necessary for the growth of organisms, and it can be the nutrient that limits the productivity of a body of water.

Phosphate analyses comprise two general steps: 1) conversion of the phosphorus form of interest to soluble orthophosphate and 2) colorimetric determination of soluble orthophosphate. For analytical purposes, phosphate has been classified into four chemical types: total, ortho, acid-hydrolyzable, and organic. Each of these types can be tested in three different physical states: total, filterable (dissolved), and nonfilterable (particulate).

Phosphorus may exist in suspension and may combine with organic matter. As a result, a digestion method to determine total phosphorus must be effective both in oxidizing organic matter effectively, rupturing both C-P and C-O-P bonds, and in solubilizing suspended material to release the phosphorus as soluble orthophosphate. Three digestion methods are available. The perchloric acid technique is a time-consuming method and is suggested for only particularly difficult samples, such as sediments. The nitric acid-sulfuric acid technique is recommended for most samples. However, the persulfate oxidation technique appears to be the simplest digestion method.

Three methods are described to determine orthophosphate concentration. Selection of a method depends largely upon the concentration range of orthophosphate. The vanadomolybdic acid method is most useful for orthophosphate analyses in the range of 1 to

20 mg P/l. The stannous chloride and ascorbic acid methods are more suited for the range of 0.01 to 6 mg P/l.

Detailed procedures for the above tests are described in Standard Methods for the Examination of Water and Wastewater, 1975 (Standard Methods for the Examination of Water and Wastewater, 1975).

5.5.11 Oxygen Management of Groundwater

Oxygen management of the groundwater system is dependent upon existing levels of DO, BOD, and COD (Buckingham, 1981). If dissolved oxygen is low and chemical and biochemical oxygen demands are high, the water requires the addition of oxygen.

It is difficult to quantify the oxygen requirements of the groundwater system. One method is described, which was specifically evaluated for an oil spill (Jamison, Raymond, and Hudson, 1975). It is based upon the composition of bacterial cells in terms of oxygen and on the conversion rate of hydrocarbons to cells (see Appendix A, Section 5.1.2.4).

There are a number of existing devices that can be used to meet the oxygen requirements of bacteria in contaminated groundwater. Fluid and semisolid systems can be aerated with pumps, propellers, stirrers, spargers, sprayers, and cascades. A system has been described, which introduces air into wells by a diffuser tube connected to a small air compressor by a rubber hose. The compressor theoretically delivers around 0.07 m^3 of air/min. Diffusers are positioned 5 ft from the well bottom. Compressors are fitted with pressure gauges and relief valves to aid in determining that each diffuser is operating properly.

5.5.12 Nutrient Management of Groundwater

Phosphorus and nitrogen are major nutrients essential to a groundwater fertilization program. They are added to the groundwater in compounds available as commercial fertilizers.

Expense plays a major consideration in the selection of the type of fertilizer to add to a groundwater system (Brady, 1974). Production costs and application methods affect the overall economic feasibility of a groundwater fertilizer program. For example, low labor costs are associated with the application of liquid fertilizers. Materials are handled in tanks and pumped out for transfer or for application. Generally, cost per unit nutrient element is higher due to the demand for more sophisticated equipment for fertilizer storage and handling.

Fertilizers must be water soluble in order to be effective in releasing nutrients to the groundwater. Some fertilizers have a coating that times them for slow release into the environment. This feature is undesirable in the case of groundwater contamination because it is important to stimulate bacteria as quickly and efficiently as possible. Monitoring the nutrient levels, or other indicator of nutrient status (such as BOD) is required so as to be able to adjust nutrient input accordingly.

Safety is another major consideration in the selection of fertilizers. Some, such as ammonium nitrate, must be handled with caution because of their explosive hazard. In addition, for long-term treatment, it is extremely important that fertilizers be of food-grade quality, since treated groundwater may ultimately reach a public drinking system. This will aid in preventing unnecessary contamination.

It is probable that subsurface injection is economically feasible when compared with methods of surface application (Smith, McWhorter, and Ward, 1977). Greater capital costs are assumed, but labor costs are reduced because of the limited surface area involved. Pumping liquid fertilizer through a rubber hose and into the well is a possible method of nutrient injection. This technique may cause agitation or aeration of the groundwater, which needs to be avoided only during the acquisition of water samples for oxygen analysis.

Nutrient requirements can be determined by two methods. The first is based upon the phosphorus and nitrogen contents of bacterial cells in relation to their carbon content, i.e., their C:N:P ratio (see Appendix A, Section 5.1.2.5). The second method bases the nutrient requirements on the BOD of the groundwater system; e.g., BOD:N:P = 100:5:1 (Sherrard and Schroeder, 1975). This ratio is generally applied to industrial wastewater treatment in order to remove a significant fraction of the organic matter present in industrial effluent. It requires BOD analysis of the contaminated groundwater followed by subsequent adjustments to nutrient levels.

Metabolic abilities and nutrient requirements of groundwater microorganisms vary substantially within an aquifer (Swindoll, Aelion, and Pfaender, Draft). In some cases, mineralization may be limited by available inorganic nutrients. Other samples from the same aquifer mineralization may not be affected. Degradation can be more greatly enhanced by addition of multiple inorganic nutrients than by addition of single substances. Alternative carbon sources, such as glucose or amino acids, inhibit mineralization of the xenobiotic substrates, possibly because of preferential utilization of the more easily degradable carbon amendments.

Each in situ application is a research effort that must be customized to the particular site and contaminant characteristics (Amdurer, Fellman, and Abdelhamid, 1985). There are no generalized cases, and decisions must rely on engineering judgment. Nonuniform geological formations that impede the flow of waterborne compounds are not realistic candidates for in situ treatment. The greatest in situ success will be with a plume or a spill situation.

Index

Abiotic processes, 18, A36. <u>See</u> <u>also</u> specific types

Abiotic stress, A135

Above-ground biological treatment, 125

Acclimation, 108, 124, A37, A128-A129, A191, 200, A200
  biological enhancement and, 101-102
  in soil biodegradation, 74-76

Acetaldehyde, A7

Acetate, 52, 118, A39, A53, A98, A104, A139

Acetic acid, A101, A209

Acetic anhydride, A7

Acetobacterium woodii, A105

Acetogenic bacteria, 39, A101

Acetone, A7, A27, A47

Acetyl coenzyme A, A88

Achromobacter sp., 38, 52, 55, 56, A92, A140

Acid extractables, 8. <u>See also</u> specific types

Acinetobacter
  anitratum, A108
  calcoaceticus, A92 sp., 20, A83
    in groundwater biodegradation, 51
    PAHs and, A98
    in soil biodegradation, 38, 40, A56, A62, A132
    straight-chain alkanes and, A86

in water biodegradation, 55, 56

Acridine orange, A65

Acrylic acid, A7

Acrylonitrile, A47

Actinomyces
  candidus, A87 sp., 40

Actinomycetes, 19, 122, 124. <u>See</u> <u>also</u> specific types
  aerobic degradation and, 60
  in soil biodegradation, 48, 50, 89, A56-A57

Activated carbon, A205, A214

Activated sludge, xiii, 23, A2, A3-A4, A137

Activating tanks, 102

Activation, 18

Acylation, 60

Additives, 31, 134, A151. <u>See also</u> specific types

Adsorption, 17

Advection, 67, A26

Aerated lagoons, 23

Aeration, 102, 110, A172, A193, A195

Aerobacter sp., A52

Aerobic attached growth biological processes, A2

Aerobic biological systems, A2-A3

Aerobic bioreclamation, A85